Cardiac and Vascular Biology

Volume 4

The book series gives an overview on all aspects of state-of-the-art research on the cardiovascular system in health and disease. Basic research aspects of medically relevant topics are covered and the latest advances and methods covering diverse disciplines as epigenetics, genetics, mechanobiology, platelet research or stem cell biology are featured. The book series is intended for researchers, experts and graduates, both basic and clinically oriented, that look for a carefully selected collection of high quality review articles on their respective field of expertise.

More information about this series at http://www.springer.com/series/13128

Masaki Ieda
Wolfram-Hubertus Zimmermann
Editors

Cardiac Regeneration

Springer

Editors
Masaki Ieda
Keio University, School of Medicine
Department of Cardiology
Tokyo
Japan

Wolfram-Hubertus Zimmermann
Institute of Pharmacology and Toxicology
University Medical Center Göttingen
Georg-August University
Göttingen
Germany

ISSN 2509-7830 ISSN 2509-7849 (electronic)
Cardiac and Vascular Biology
ISBN 978-3-319-56104-2 ISBN 978-3-319-56106-6 (eBook)
DOI 10.1007/978-3-319-56106-6

Library of Congress Control Number: 2017948432

Printed on acid-free paper

This Springer imprint is published by Springer Nature
The registered company is Springer International Publishing AG
The registered company address is: Gewerbestrasse 11, 6330 Cham, Switzerland

Preface

Cardiovascular diseases are a leading cause of death worldwide. The failure to regenerate the diseased heart with current therapies creates also in light of the increasing numbers of affected patients a serious unmet medical need.

Cell-based therapies using skeletal myoblasts, bone marrow-derived cells, mesenchymal stem cells, and cardiac stem cells have been developed as potential regenerative therapies and are being tested in clinical trials. Results from these studies suggest relevant protective mechanisms but provide limited evidence for bona fide remuscularization of the failing heart.

Pluripotent stem cells, including embryonic stem cells and induced pluripotent stem cells (iPSCs), can be directed to differentiate into cardiomyocytes with excellent efficiency, making them an attractive source for the remuscularization of the failing heart. Recently, direct programming of somatic cells into cardiomyocytes has emerged, which may ultimately allow for in vivo conversion of scar into myocardium.

Common challenges to cell-based or cell-targeted therapeutics with an anticipated profound effect on therapeutic efficacy include the route of administration, cell/tissue targeting, dosing, and sustainability of regenerative therapies.

With this book we intended to address a broad spectrum of challenges and opportunities in the field of heart regeneration. We are thankful to the many contributing investigators as to their willingness to share their unique experiences and express their candid opinions on how myocardial regeneration may be advanced and ultimately translated to address one of the most concerning clinical issues to date, namely heart failure.

Tokyo, Japan
Göttingen, Germany

Masaki Ieda
Wolfram-Hubertus Zimmermann

The original version of the book was revised. Thc correction to the book is availale at DOI 10.1007/978-3-319-56106-6_13

Contents

Past and Future of Cell-Based Heart Repair

1

Ahmed I. Mahmoud and Richard T. Lee

Abstract
The field of heart regeneration has witnessed significant advancements toward developing new therapeutics in the past decade. Strategies to regenerate the adult human heart are in constant development in both the experimental and clinical arenas. Although stem cell therapies remain controversial, cell-based heart repair is a promising approach toward regenerating the adult human heart. Experience with cell therapy has resulted in several important milestones in clinical studies. There are still important roadblocks ahead before cell therapy can achieve the regeneration potential for broad numbers of patients. In this chapter, we focus on the history of cardiac cell repair and therapeutic strategies and discuss the lessons learned in cell-based heart regeneration.

1.1 Introduction

Cardiovascular diseases remain to be one of the leading causes of mortality worldwide and represent an enormous health and economic burden (Whelan et al. 2010). Identifying strategies to regenerate the adult human heart after injury has spurred a furiously paced experimental race toward this goal.

Historically, the mammalian heart has been considered to be a postmitotic organ, without any capacity for cell turnover and regeneration post-injury (Laflamme and Murry 2011). Instead of regenerating muscle, a scar is formed to maintain the integrity of the mammalian heart following injury; hypertrophy of the remaining myocardium takes place, but the loss of myocardium can eventually lead to the development

A.I. Mahmoud • R.T. Lee (✉)
Department of Stem Cell and Regenerative Biology, The Harvard Stem Cell Institute, Harvard University, Cambridge, MA 02138, USA
e-mail: Richard_Lee@harvard.edu

M. Ieda, W.-H. Zimmermann (eds.), *Cardiac Regeneration*, Cardiac and Vascular Biology 4, DOI 10.1007/978-3-319-56106-6_1

of heart failure (Jessup and Brozena 2003). Mechanical approaches for treatment of heart failure aimed to counteract the weakening of the heart muscle following injury include left ventricular assist devices (Terracciano et al. 2010). Other approaches include neurohormonal inhibition, which is widely used in clinical practice (Sharpe et al. 1991). These approaches are beneficial for patients with heart failure, but the potential to completely regenerate lost myocardium remains an important goal.

Early discoveries showed that endogenous cardiac regeneration can occur in some vertebrate organisms such as the newt and zebrafish (Oberpriller and Oberpriller 1974; Poss et al. 2002). Recently, the neonatal mouse heart was reported to regenerate in response to injury in a manner similar to lower vertebrates (Porrello et al. 2011, 2013). The regenerative response has been attributed to the ability of cardiomyocytes to proliferate with restoration of functional myocardium (Jopling et al. 2010; Kikuchi et al. 2010; Porrello et al. 2011). Cardiomyocyte cell cycle activity is maintained throughout the adult life of vertebrates, but rapidly declines with age in mammals (Li et al. 1996; Poss et al. 2002; Walsh et al. 2010). The lessons learned from lower vertebrates as well as the neonatal mouse suggest that endogenous heart regeneration can occur, and understanding this process could allow new therapeutic approaches to regenerate the human heart.

Early mouse studies showed that cardiomyocyte turnover in the adult murine heart occurs at low levels, around 1% annually (Soonpaa and Field 1997). Cardiomyocyte turnover in the adult mouse heart during aging and following injury was demonstrated at high resolution using mouse genetic lineage tracing and multi-imaging mass spectrometry showing similar levels of myocyte turnover (Senyo et al. 2013; Hsieh et al. 2007). To measure the levels of cell turnover in the adult human heart, a landmark study exploited the rise of ^{14}C levels during the cold war testing of nuclear weapons, which created an opportunity to trace the levels of ^{14}C from human heart samples and thus enabled the researchers to determine the rate of cardiomyocyte turnover in the adult human heart (Bergmann et al. 2009). Similar to the murine heart, the adult human heart showed cardiomyocyte turnover at extremely low levels, around 1% annually (Bergmann et al. 2009). Although the cardiomyocyte refreshment is insufficient for a substantial regenerative response following injury, this indicates that the heart is much more resilient than previously considered. Surprisingly, a recent study reported that the human neonatal heart can regenerate after a myocardial infarction (Haubner et al. 2015). The similarities between the neonatal mouse and neonatal human heart, as well as the adult mouse and adult human heart, suggest that regenerating the adult heart will be feasible.

1.2 Cell Therapy for Cardiac Repair

The development of many tools in regenerative medicine has inspired cardiovascular investigators to utilize these methods to regenerate the human heart to restore contractile function following injury. Stem cells have generated particular excitement for their potential for cell-based cardiac repair (Garbern and Lee 2013). The plasticity of stem cells and their ability to differentiate into multiple cell types has generated hope for the future of regenerative medicine. The past decade has witnessed numerous studies that used different cell types with varying abilities for cardiac repair, which led to many clinical trials. The results from these trials continue to generate controversy regarding the impact of cellular therapy, but it is clear that cell therapy may have an important future for human heart regeneration.

The collective knowledge of cellular plasticity in the mammalian heart as well as the explosion of the stem cell field fueled the hope to either harness the endogenous potential of the mammalian heart or utilize the potential of exogenous stem cells that can differentiate into functional myocardium. In the following paragraphs, we will discuss the utility of different types of exogenous stem cells, as well as the potential of different endogenous cardiac progenitors for cellular transplantation, in addition to cellular reprogramming, to regenerate the adult human heart.

1.2.1 Skeletal Myoblasts

Skeletal myoblasts were among the initial cell types to be introduced for clinical cardiac cell therapy. Skeletal myoblasts were reasonable candidates due to their resistance to ischemia, as well as their differentiation potential (Durrani et al. 2010). In addition, early results showed the promise of skeletal myoblasts in heart repair following injury in multiple experimental animal models (Durrani et al. 2010). However, it was shown that myoblasts fail to integrate with the host myocardium and thus fail to beat in sync with the heart (Leobon et al. 2003). Furthermore, the first multicenter, randomized, placebo-controlled human clinical trial for myoblast autologous grafting in ischemic cardiomyopathy (MAGIC) did not enhance cardiac contractile function (Menasche et al. 2008). These results led to a reduced enthusiasm toward the use of skeletal myoblasts, and regenerative approaches moved onward toward more promising cell types.

1.2.2 Bone Marrow-Derived Stem Cells

Bone marrow-derived stem cells have the capacity to differentiate into multiple cell types including vascular and cardiac cell fates both in vitro and in vivo (Hirschi and Goodell 2002). The detection of Y-chromosome-positive cardiomyocytes in female hearts that were transplanted into male patients suggested that bone marrow-derived stem cells can differentiate into cardiomyocytes (Quaini et al. 2002). Over a decade ago, bone marrow-derived stem cells that express the surface marker c-kit emerged as candidates for regenerating the heart following injury through transdifferentiation into cardiomyocytes (Orlic et al. 2001). The differentiation potential of c-kit + cells in vivo led to controversy. Although the initial report suggested transdifferentiation of these cells into cardiomyocytes, subsequent studies by other groups found no evidence of transdifferentiation into cardiomyocytes, but rather showed the formation of more mature hematopoietic cell lineages following transplantation (Murry et al. 2004; Balsam et al. 2004). Improved ventricular function was detected following bone marrow-derived stem cell injections in multiple studies; several reports suggested that this effect is due to a paracrine effect through enhancing proliferation and differentiation of endogenous cardiac progenitors, thus promoting cardiac repair indirectly (Loffredo et al. 2011; Hatzistergos et al. 2010; Urbich et al. 2005; Mathieu et al. 2009; Kinnaird et al. 2004; Gnecchi et al. 2006; Hong et al. 2014).

REPAIR-AMI was the first randomized, blinded clinical trial to use autologous bone marrow cells through intracoronary infusion for acute myocardial infarction (MI) patients (Schachinger et al. 2006). There was a significant improvement in the left ventricular function following bone marrow transplantation, an effect that persisted up to 2–5 years after transplantation but with no impact on survival, though the trial had insufficient power to study survival (Assmus et al. 2010, 2014). However, a subsequent trial (TIME) using autologous bone marrow cells in ST-segment elevation MI (STEMI) showed no effect on improving cardiac function (Traverse et al. 2012). These mixed results generated debate on the impact of bone marrow-derived cells on improving function and survival of MI patients (Marban and Malliaras 2012). Retrospective evaluation of these clinical trials revealed some of the discrepancies and potential pitfalls to be avoided for proper assessment of the value of these cells as a clinical treatment (Simari et al. 2014; Nowbar et al. 2014). A large phase 3 clinical trial to assess the value of bone marrow cells in myocardial infarction patients is currently underway (BAMI trial).

1.2.3 Endothelial Progenitors

Endothelial progenitor cells (EPCs) are a small population of adult hematopoietic CD34+ progenitors that were identified in 1997 and which have the capacity to differentiate into endothelial cells (Asahara et al. 1997). Preclinical studies of EPCs showed promising results in enhancing recovery following ischemia in different tissues, mainly by enhancing neovascularization (Kawamoto and Losordo 2008). Following myocardial infarction, EPC transplantation may enhance functional recovery and myocardial integrity in vivo (Iwasaki et al. 2006; Kawamoto et al. 2001). This requires the homing of the EPCs to the site of ischemia followed by proliferation and differentiation into functional endothelial cells (Hristov et al. 2007). In addition to neovascularization, endothelial cells enhance cardiomyocyte survival and organization and contraction of surrounding cardiomyocytes through paracrine signaling (Narmoneva et al. 2004). These data suggest that endothelial cells can promote cardiac repair through different mechanisms.

Early phase clinical trials using cell transplantation of EPCs showed promising results for functional recovery following a cardiac insult (Vrtovec et al. 2013; Stamm et al. 2007). Similarly, pharmacological mobilization of EPCs using granulocyte colony-stimulating factor (G-CSF) showed enhanced cardiac function post-injury (Achilli et al. 2010). A major impediment to the understanding the full potential of EPCs is the different isolation methods for EPCs between different groups; thus the identity and purity of the EPC populations have not been consistent. Owing to the limited size of previous trials, large studies would be required to establish the efficacy of EPCs for heart repair.

1.2.4 Mesenchymal Stem Cells

Another subset of progenitors within the bone marrow is mesenchymal stem cells (MSCs). MSCs are multipotent and can differentiate into adipocytes, chondrocytes, and osteoblasts (Pittenger et al. 1999). MSCs are found in multiple tissues, and they can be expanded to the large numbers necessary for transplantation. Furthermore, MSCs appear less immunogenic due to the absence of MHC-II complex and may have lower probability of rejection (Kuraitis et al. 2011). Allogeneic MSCs showed therapeutic benefits following transplantation in the injured rodent and swine heart (Williams and Hare 2011). Initially there was evidence that allogeneic MSCs can differentiate into cardiomyocytes in vivo following engraftment in the adult murine and swine heart (Toma et al. 2002; Quevedo et al. 2009). Subsequent studies showed that MSCs probably accomplish beneficial effects via paracrine mechanisms (Mirotsou et al. 2007; Gnecchi et al. 2005, 2008). MSC transplantation in large animals showed activation and differentiation of cardiac stem cells (Hatzistergos et al. 2010). The POSEIDON trial showed improved patient outcome and ventricular remodeling, but no significant improvement in ventricular function (Hare et al. 2012). A randomized phase 3 clinical trial using MSCs for ischemic heart failure is currently ongoing (CHART-1), and the results from this trial will shed light on the future of MSCs in the clinic.

1.2.5 Endogenous Cardiac Stem Cells

1.2.5.1 C-kit + Cardiac Progenitors

Following the developments in the hematopoietic stem cell (HSC) field, c-kit, the receptor for stem cell factor, was described as a surface marker of HSC stemness. It was reported that the heart has an endogenous cardiac progenitor cell (CPC) population that is c-kit + without any hematopoietic lineage marker expression (Lin-) (Beltrami et al. 2003). These cells were described as clonogenic and multipotent due to their ability to differentiate into cardiomyocytes, endothelial cells, and smooth muscle cells in vitro and in vivo. Expansion of the c-kit + CPCs ex vivo and injection of the cells in vivo following MI showed a dramatic regenerative effect on the heart (Beltrami et al. 2003). Furthermore, a similar population of c-kit + CPCs was reported in the adult human heart and could repopulate the infarcted murine myocardium (Bearzi et al. 2007). The ability of c-kit + CPCs to enhance myocardial regeneration was demonstrated by different groups in small and large animal studies (Linke et al. 2005; Fischer et al. 2009; Angert et al. 2011). In contrast, other groups showed that c-kit + CPCs from adult hearts do not differentiate into cardiomyocytes ex vivo (Zaruba et al. 2010; Jesty et al. 2012). One study reported that c-kit + CPCs are not only necessary but also sufficient for myocardial regeneration following cardiac injury (Ellison et al. 2013). These conflicting results regarding the differentiation potential and functional impact of c-kit + CPCs on myocardial regeneration were addressed by a very well-designed lineage tracing study aimed to label the putative c-kit + CPCs in the heart to trace their lineage during aging and following injury (Van Berlo et al. 2014). Although this study showed that c-kit + CPCs could differentiate into cardiomyocytes, this occurred at negligibly low rates, which suggested that c-kit + CPCs would have no impact on myocyte replenishment following injury (Van Berlo et al. 2014). In contrast, c-kit + CPCs produced a high percentage of cardiac endothelial cells, suggesting that the endogenous c-kit + CPCs are more likely to be endothelial progenitor cells rather than true cardiomyocyte progenitors.

Although it is still unclear how c-kit + CPCs can enhance myocardial repair, their impact over the past decade has led to a number of clinical trials to assess their therapeutic potential. An early clinical trial to test the safety of c-kit + CPCs was a phase 1, randomized clinical trial for CPC intracoronary infusion in patients with ischemic cardiomyopathy (SCIPIO) (Bolli et al. 2011). This trial showed that CPC injection is safe with no adverse effects up to 1 year, with an improvement in LV function. The phase 1 trial was very small, however.

1.2.5.2 Cardiosphere and Cardiosphere-Derived Cells

Cardiosphere-derived cells represent another subset of cardiac progenitor cells in both murine and human hearts that are multipotent and can differentiate into different cardiac cell types (Messina et al. 2004). Cardiosphere-derived cells (CDCs) can be isolated from cells cultured from endomyocardial biopsies, and injection of CDCs in a large animal model of infarct has led to enhancement of cardiac function (Smith et al. 2007). Cardiosphere-derived cells were isolated from human hearts as well, and intracoronary injection of human CDCs in a pig infarct model improved cardiac function and reduced scar formation (Johnston et al. 2009). A phase 1 clinical trial using cardiosphere-derived autologous stem cells to reverse ventricular dysfunction (CADUCEUS) for intracoronary injection in myocardial infarction patients suggested the safety of these cells for clinical use, with potential benefits on cardiac function (Makkar et al. 2012). Further mechanistic studies of cardiosphere-derived cells showed that intracoronary injection of CDCs post MI stimulate endogenous cardiomyocyte proliferation, as well as recruitment of endogenous progenitors (Malliaras et al. 2013). These dual mechanisms may explain the beneficial outcomes following CDC cell therapy. Larger studies will be necessary to reveal the impact of cardiospheres and CDCs on this cell type as a candidate for myocardial regeneration in humans.

1.2.5.3 Side Population Cells

Side population (SP) cells are a population of cells characterized by their ability to exclude the Hoechst dye, since they express the ATP-binding cassette transporter proteins. These cells were first characterized as a hematopoietic stem cell population in the bone marrow (Goodell et al. 1996). To determine whether SP cells from the bone marrow can enhance cardiac repair, SP cells were transplanted in mice following an ischemia reperfusion injury and shown to have some therapeutic benefit and enhance myocardial repair (Jackson et al. 2001). Cardiac SP cells were further isolated from the developing and adult mouse heart, with the capacity to differentiate into cardiomyocytes, endothelial cells, and smooth muscle cells (Martin et al. 2004). Intravenous infusion of cardiac SP cells into rats that underwent myocardial infarction demonstrated that cardiac SP cells were able to migrate and home to the injured myocardium, differentiate into different cardiac cell types, and enhance heart regeneration (Oyama et al. 2007). These preclinical studies suggest a potential benefit for SP cells for cell therapy, although human clinical trials have not been performed with SP cells injected into the heart.

1.2.5.4 Sca1+ Cardiac Progenitors

Stem cell antigen1 (Sca1) is a cell surface marker expressed on the surface of multiple tissue-specific resident stem cells (Holmes and Stanford 2007). A Sca1+ cardiac stem cell population has been identified in the adult mouse heart and can differentiate into cardiomyocytes in vitro following treatment with 5-azacytidine as well as oxytocin (Oh et al. 2003; Matsuura et al. 2004). Intravenously injected Sca1+ cells were able to home to injured myocardium, differentiate into cardiomyocytes or fuse with host cells, and enhance repair following ischemia reperfusion injury (Oh et al. 2003). Similarly, intramyocardial transplantation of Sca1+ cardiac stem cells improved LV function post MI, although this effect was probably mediated through a paracrine effect by increased neovascularization and enhanced cardiomyocyte function (Wang et al. 2006). Lineage tracing of Sca1+ cells in the heart suggested that Sca1+ cells contribute to cardiomyocyte renewal in the adult heart, as well as in response to injury (Uchida et al. 2013). A major impediment to the clinical potential of Sca1+ cardiac stem cells is the lack of the Sca1 antigen in humans, which limits the use of these cells for human therapy.

1.2.5.5 Islet1+ Cardiac Progenitors

Islet1 (Isl1) is a transcription factor that marks a cardiac progenitor population from the second heart field that can differentiate into multiple cardiac lineages during heart embryonic development (Moretti et al. 2006; Cai et al. 2003). Interestingly, an Isl1+ cardiac progenitor population was identified in the early postnatal heart that can differentiate into mature cardiac lineages (Laugwitz et al. 2005). However, the Isl1+ progenitor population only persists in the sinoatrial node in the adult murine heart and is nearly absent from the left ventricle either at baseline or following MI (Weinberger et al. 2012). Isl1+ cardiac progenitors have not been studied for human cell therapy.

1.2.5.6 Epicardial Progenitors

Epicardial progenitor cells that express the transcription factor Wt1 have an important role during murine heart development, as they contribute to the formation of functional cardiomyocytes (Zhou et al. 2008). This role of epicardial progenitors led to the search for a similar progenitor population in the adult mammalian heart. Lineage tracing of Wt1 in the adult heart showed that Wt1 epicardial progenitors are present in the adult mouse heart and can give rise to bona fide cardiomyocytes following MI (Smart et al. 2011). Priming the Wt1 epicardial progenitors with thymosin β4 before MI may be an important step for these progenitors to give rise to cardiomyocytes, as priming them with thymosin β4 after MI does not seem to contribute to the differentiation of epicardial progenitors into cardiomyocytes (Zhou et al. 2012). However, embryonic and adult epicardial progenitors seem to be different subpopulations, as adult epicardial progenitors are more heterogeneous and have a different expression profile at the molecular level than embryonic progenitors (Bollini et al. 2014). Interestingly, a recent protocol described the derivation of primary human epicardial-derived cells from right atrial appendage biopsies, which can serve as a platform to further identify the therapeutic potential of epicardial progenitors for adult cardiac cell repair (Clunie-O'Connor et al. 2015).

1.2.6 ES-Derived Cardiomyocytes

Generation of differentiated, mature, and functional cardiomyocytes from pluripotent ES cells is a promising approach to replenish lost myocardium following injury (Xu et al. 2002). The development of directed differentiation protocols of pluripotent embryonic stem cells (ES) into cardiomyocytes has witnessed significant advances (Mummery et al. 2012). Purified human ES cell-derived cardiomyocytes (hESC-CM) can be derived when cultured with activin A and bone morphogenetic protein 4, which can improve cardiac function of the infarcted rat heart (Laflamme et al. 2007). Transplantation of cardiovascular progenitors derived from hESC led to engraftment in the infarcted hearts of nonhuman primates (Blin et al. 2010).

To establish the electrophysiological properties of hESC-CM, purified hESC-CM were transplanted in guinea pigs following injury (Shiba et al. 2012). These grafts led to reduced arrhythmias and were able to electrically couple with the host myocardium and thus efficiently enhance myocardial function following cryoinjury (Shiba et al. 2012). Recently, a large animal study in macaques showed that hESC-CM was able to remuscularize and regenerate the infarcted monkey heart (Chong et al. 2014). These recent promising results show the significant potential of hESC-CM. However, the ability of the transplanted hESC-CM to integrate efficiently in syncytium and prevent arrhythmias is still a concern (Chong et al. 2014). These issues will need to be addressed before hESC-CMs can be used for clinical trials. While ethical concerns might hamper ES use in the clinic, induced pluripotent stem cells (iPSCs) could replace hESC as the source of cardiomyocytes for cell therapy.

1.2.7 Induced Pluripotent Stem Cells and Reprogramming

1.2.7.1 iPSCs

Reprogramming adult mouse and human fibroblasts into a pluripotent state by transduction of four transcription factors, OCT4, SOX2, KLF4, and c-MYC (OSKM), was a revolutionary moment in biomedicine (Takahashi and Yamanaka 2006; Takahashi et al. 2007). iPSCs resemble ES cells morphologically and molecularly, and thus they provide an alternative to ES cell use, in addition to the advantage of generating patient-specific cell lines for autologous regenerative therapies. Although there was an initial concern toward using iPSCs clinically due to the use of oncogenes and viral vectors which can lead to teratoma formation, new methods and protocols are emerging that utilize small molecules, episomes, or proteins for reprogramming, which will increase the safety of the generated iPSCs (Zhou et al. 2009; Lin et al. 2009; Okita et al. 2008). Human iPSCs (hiPSCs) have been successfully used to generate numerous cell types including cardiomyocytes (Karakikes et al. 2015; Zhang et al. 2009). More importantly, hiPSCs provide a novel platform to dissect the underlying mechanisms of disease in patients (Bellin et al. 2012; Wang et al. 2014; Davis et al. 2012). Furthermore, intramyocardial transplantation of hiPSC-derived cardiomyocytes (hiPSC-CM) in a large animal model following MI led to a significant improvement of ventricular function and reduction of scar size while abrogating ventricular arrhythmias (Ye et al. 2014). Although further studies are required to truly understand the optimal way to use hiPSC-CMs, the recent developments indicate that this approach holds significant promise for future cell therapy (Okano et al. 2013).

1.2.7.2 Direct Reprogramming into Cardiomyocytes

Reprogramming fibroblasts into pluripotent cells led to a race toward reprogramming one cell type to another differentiated cell type. Transdifferentiation of fibroblasts to cardiomyocytes is an appealing approach as it could use the fibroblasts in the scar region to generate new myocardium. Using multiple combinations of cardiac transcription factors, a combination of three transcription factors, GATA4, MEF2C, and TBX5 (known as GMT), was able to reprogram mouse fibroblasts into induced cardiac-like myocytes in vitro (Ieda et al. 2010). To further examine whether direct reprogramming can occur in vivo, GMT and GHMT (H for HAND2) retroviral injections successfully reprogrammed cardiac fibroblasts into cardiomyocytes in vivo that resulted in an improved cardiac function and reduced scar following MI (Qian et al. 2012; Song et al. 2012). Furthermore, reprogramming fibroblasts into cardiomyocytes was achieved using microRNAs both in vitro and in vivo with improved cardiac regeneration (Jayawardena et al. 2012, 2015). Similar to iPSCs, reprogramming fibroblasts into cardiomyocytes occurs at low efficiency and may lead to the formation of immature cardiomyocyte-like cells rather than bona fide mature cardiomyocytes. Interestingly, recent studies showed that reprogramming efficiency could be enhanced significantly via upregulation of Akt1, as well as through inhibition of pro-fibrotic signaling (Zhou et al. 2015; Zhao et al. 2015). Further studies are necessary in order to generate mature and functional cardiomyocytes, as well as to fully understand the molecular mechanisms of reprogramming before moving forward to the clinic.

1.3 Future of Cell Therapy

Cardiac cell therapy has witnessed enormous achievements over the past decade. Cardiac cell therapy appears to be safe, with minimal adverse effects, while showing potential therapeutic benefits. However, cardiac cell therapy is not yet a clear success, as some analyses revealed no therapeutic benefit in acute myocardial infarction patients (Gyongyosi et al. 2015; Fisher et al. 2015). There is no consensus on which cell type will prove to be most effective. A recent study aimed at comparing hESC-CMs, cardiovascular progenitors (CVPs), and bone marrow mononuclear cells in a nude rat model of myocardial infarction (Fernandes et al. 2015). Interestingly, hESC-CMs and CVPs showed comparable improvement of cardiac repair, while bone marrow cells were less efficient (Fernandes et al. 2015). Comparing different cells is an important step in order to understand the optimal cell therapy for humans.

The results from large outcome trials are highly anticipated in order to determine the impact and the usefulness of cells in cardiac therapy. Although the mechanism of action of different cell types may vary, whether through direct differentiation into new myocardium, neovascularization, or paracrine effects, we need to expand our understanding at the molecular level. Clinical trials are the gold standard for assessing any treatment, but owing to the controversies within the cell therapy field, it is important to take a step back and progress at both the bench and the bedside. The lessons learned from heart regeneration in lower vertebrates and neonatal mice should improve our understanding of the promising approaches to regenerating the adult heart. Cardiac cellular therapy does lead to the complete regenerative response seen in animal models of endogenous heart regeneration, and thus there are many lessons to be learned from nature.

Acknowledgments This work was supported by an American Heart Association Postdoctoral Fellowship (15POST21870000) to AIM. This work was supported by NIH grants (AG040019) and (HL117986) to RTL and by the Leducq Foundation.

Compliance with Ethical Standards

Conflict of Interest The authors declare that they have no conflict of interest.

Ethical approval This article does not contain any studies with human participants performed by any of the authors.

References

Achilli F, Malafronte C, Lenatti L, Gentile F, Dadone V, Gibelli G, Maggiolini S, Squadroni L, Di Leo C, Burba I, Pesce M, Mircoli L, Capogrossi MC, Di Lelio A, Camisasca P, Morabito A, Colombo G, Pompilio G, Investigators S-A (2010) Granulocyte colony-stimulating factor attenuates left ventricular remodelling after acute anterior STEMI: results of the single-blind, randomized, placebo-controlled multicentre stem cell Mobilization in acute myocardial infarction (STEM-AMI) trial. Eur J Heart Fail 12:1111–1121

Angert D, Berretta RM, Kubo H, Zhang H, Chen X, Wang W, Ogorek B, Barbe M, Houser SR (2011) Repair of the injured adult heart involves new myocytes potentially derived from resident cardiac stem cells. Circ Res 108:1226–1237

Asahara T, Murohara T, Sullivan A, Silver M, van der Zee R, Li T, Witzenbichler B, Schatteman G, Isner JM (1997) Isolation of putative progenitor endothelial cells for angiogenesis. Science 275:964–967

Assmus B, Leistner DM, Schachinger V, Erbs S, Elsasser A, Haberbosch W, Hambrecht R, Sedding D, Yu J, Corti R, Mathey DG, Barth C, Mayer-Wehrstein C, Burck I, Sueselbeck T,

Dill T, Hamm CW, Tonn T, Dimmeler S, Zeiher AM, Group R-AS (2014) Long-term clinical outcome after intracoronary application of bone marrow-derived mononuclear cells for acute myocardial infarction: migratory capacity of administered cells determines event-free survival. Eur Heart J 35:1275–1283

Assmus B, Rolf A, Erbs S, Elsasser A, Haberbosch W, Hambrecht R, Tillmanns H, Yu J, Corti R, Mathey DG, Hamm CW, Suselbeck T, Tonn T, Dimmeler S, Dill T, Zeiher AM, Schachinger V, Investigators R-A (2010) Clinical outcome 2 years after intracoronary administration of bone marrow-derived progenitor cells in acute myocardial infarction. Circ Heart Fail 3:89–96

Balsam LB, Wagers AJ, Christensen JL, Kofidis T, Weissman IL, Robbins RC (2004) Haematopoietic stem cells adopt mature haematopoietic fates in ischaemic myocardium. Nature 428:668–673

Bearzi C, Rota M, Hosoda T, Tillmanns J, Nascimbene A, De Angelis A, Yasuzawa-Amano S, Trofimova I, Siggins RW, Lecapitaine N, Cascapera S, Beltrami AP, D'Alessandro DA, Zias E, Quaini F, Urbanek K, Michler RE, Bolli R, Kajstura J, Leri A, Anversa P (2007) Human cardiac stem cells. Proc Natl Acad Sci U S A 104:14068–14073

Bellin M, Marchetto MC, Gage FH, Mummery CL (2012) Induced pluripotent stem cells: the new patient? Nat Rev Mol Cell Biol 13:713–726

Beltrami AP, Barlucchi L, Torella D, Baker M, Limana F, Chimenti S, Kasahara H, Rota M, Musso E, Urbanek K, Leri A, Kajstura J, Nadal-Ginard B, Anversa P (2003) Adult cardiac stem cells are multipotent and support myocardial regeneration. Cell 114:763–776

Bergmann O, Bhardwaj RD, Bernard S, Zdunek S, Barnabe-Heider F, Walsh S, Zupicich J, Alkass K, Buchholz BA, Druid H, Jovinge S, Frisen J (2009) Evidence for cardiomyocyte renewal in humans. Science 324:98–102

Blin G, Nury D, Stefanovic S, Neri T, Guillevic O, Brinon B, Bellamy V, Rucker-Martin C, Barbry P, Bel A, Bruneval P, Cowan C, Pouly J, Mitalipov S, Gouadon E, Binder P, Hagege A, Desnos M, Renaud JF, Menasche P, Puceat M (2010) A purified population of multipotent cardiovascular progenitors derived from primate pluripotent stem cells engrafts in postmyocardial infarcted nonhuman primates. J Clin Invest 120:1125–1139

Bolli R, Chugh AR, D'Amario D, Loughran JH, Stoddard MF, Ikram S, Beache GM, Wagner SG, Leri A, Hosoda T, Sanada F, Elmore JB, Goichberg P, Cappetta D, Solankhi NK, Fahsah I, Rokosh DG, Slaughter MS, Kajstura J, Anversa P (2011) Cardiac stem cells in patients with ischaemic cardiomyopathy (SCIPIO): initial results of a randomised phase 1 trial. Lancet 378:1847–1857

Bollini S, Vieira JM, Howard S, Dube KN, Balmer GM, Smart N, Riley PR (2014) Re-activated adult epicardial progenitor cells are a heterogeneous population molecularly distinct from their embryonic counterparts. Stem Cells Dev 23:1719–1730

Cai CL, Liang X, Shi Y, Chu PH, Pfaff SL, Chen J, Evans S (2003) Isl1 identifies a cardiac progenitor population that proliferates prior to differentiation and contributes a majority of cells to the heart. Dev Cell 5:877–889

Chong JJ, Yang X, Don CW, Minami E, Liu YW, Weyers JJ, Mahoney WM, van Biber B, Cook SM, Palpant NJ, Gantz JA, Fugate JA, Muskheli V, Gough GM, Vogel KW, Astley CA, Hotchkiss CE, Baldessari A, Pabon L, Reinecke H, Gill EA, Nelson V, Kiem HP, Laflamme MA, Murry CE (2014) Human embryonic-stem-cell-derived cardiomyocytes regenerate non-human primate hearts. Nature 510:273–277

Clunie-O'Connor C, Smits AM, Antoniades C, Russell AJ, Yellon DM, Goumans MJ, Riley PR (2015) The derivation of primary human epicardium-derived cells. Curr Protoc Stem Cell Biol 35:2C.5.1–2C.5.12

Davis RP, Casini S, van den Berg CW, Hoekstra M, Remme CA, Dambrot C, Salvatori D, Oostwaard DW, Wilde AA, Bezzina CR, Verkerk AO, Freund C, Mummery CL (2012) Cardiomyocytes derived from pluripotent stem cells recapitulate electrophysiological characteristics of an overlap syndrome of cardiac sodium channel disease. Circulation 125:3079–3091

Durrani S, Konoplyannikov M, Ashraf M, Haider KH (2010) Skeletal myoblasts for cardiac repair. Regen Med 5:919–932

Ellison GM, Vicinanza C, Smith AJ, Aquila I, Leone A, Waring CD, Henning BJ, Stirparo GG, Papait R, Scarfo M, Agosti V, Viglietto G, Condorelli G, Indolfi C, Ottolenghi S, Torella D, Nadal-Ginard B (2013) Adult c-kit(pos) cardiac stem cells are necessary and sufficient for functional cardiac regeneration and repair. Cell 154:827–842

Fernandes S, Chong JJ, Paige SL, Iwata M, Torok-Storb B, Keller G, Reinecke H, Murry CE (2015) Comparison of human embryonic stem cell-derived cardiomyocytes, cardiovascular progenitors, and bone marrow mononuclear cells for cardiac repair. Stem Cell Reports 5(5):P753–P762

Fischer KM, Cottage CT, Wu W, Din S, Gude NA, Avitabile D, Quijada P, Collins BL, Fransioli J, Sussman MA (2009) Enhancement of myocardial regeneration through genetic engineering of cardiac progenitor cells expressing Pim-1 kinase. Circulation 120:2077–2087

Fisher SA, Doree C, Mathur A, Martin-Rendon E (2015) Meta-analysis of cell therapy trials for patients with heart failure. Circ Res 116:1361–1377

Garbern JC, Lee RT (2013) Cardiac stem cell therapy and the promise of heart regeneration. Cell Stem Cell 12:689–698

Gnecchi M, He H, Liang OD, Melo LG, Morello F, Mu H, Noiseux N, Zhang L, Pratt RE, Ingwall JS, Dzau VJ (2005) Paracrine action accounts for marked protection of ischemic heart by Akt-modified mesenchymal stem cells. Nat Med 11:367–368

Gnecchi M, He H, Noiseux N, Liang OD, Zhang L, Morello F, Mu H, Melo LG, Pratt RE, Ingwall JS, Dzau VJ (2006) Evidence supporting paracrine hypothesis for Akt-modified mesenchymal stem cell-mediated cardiac protection and functional improvement. FASEB J 20:661–669

Gnecchi M, Zhang Z, Ni A, Dzau VJ (2008) Paracrine mechanisms in adult stem cell signaling and therapy. Circ Res 103:1204–1219

Goodell MA, Brose K, Paradis G, Conner AS, Mulligan RC (1996) Isolation and functional properties of murine hematopoietic stem cells that are replicating in vivo. J Exp Med 183:1797–1806

Gyongyosi M, Wojakowski W, Lemarchand P, Lunde K, Tendera M, Bartunek J, Marban E, Assmus B, Henry TD, Traverse JH, Moye LA, Surder D, Corti R, Huikuri H, Miettinen J, Wohrle J, Obradovic S, Roncalli J, Malliaras K, Pokushalov E, Romanov A, Kastrup J, Bergmann MW, Atsma DE, Diederichsen A, Edes I, Benedek I, Benedek T, Pejkov H, Nyolczas N, Pavo N, Bergler-Klein J, Pavo IJ, Sylven C, Berti S, Navarese EP, Maurer G, Investigators A (2015) Meta-Analysis of Cell-based CaRdiac stUdiEs (ACCRUE) in patients with acute myocardial infarction based on individual patient data. Circ Res 116:1346–1360

Hare JM, Fishman JE, Gerstenblith G, Difede Velazquez DL, Zambrano JP, Suncion VY, Tracy M, Ghersin E, Johnston PV, Brinker JA, Breton E, Davis-Sproul J, Schulman IH, Byrnes J, Mendizabal AM, Lowery MH, Rouy D, Altman P, Wong Po Foo C, Ruiz P, Amador A, da Silva J, Mcniece IK, Heldman AW, George R, Lardo A (2012) Comparison of allogeneic vs autologous bone marrow-derived mesenchymal stem cells delivered by transendocardial injection in patients with ischemic cardiomyopathy: the POSEIDON randomized trial. JAMA 308:2369–2379

Hatzistergos KE, Quevedo H, Oskouei BN, Hu Q, Feigenbaum GS, Margitich IS, Mazhari R, Boyle AJ, Zambrano JP, Rodriguez JE, Dulce R, Pattany PM, Valdes D, Revilla C, Heldman AW, Mcniece I, Hare JM (2010) Bone marrow mesenchymal stem cells stimulate cardiac stem cell proliferation and differentiation. Circ Res 107:913–922

Haubner BJ, Schneider J, Schweigmann UU, Schuetz T, Dichtl W, Velik-Salchner C, Stein JI, Penninger JM (2015) Functional recovery of a human neonatal heart after severe myocardial infarction. Circ Res 118(2):216–221

Hirschi KK, Goodell MA (2002) Hematopoietic, vascular and cardiac fates of bone marrow-derived stem cells. Gene Ther 9:648–652

Holmes C, Stanford WL (2007) Concise review: stem cell antigen-1: expression, function, and enigma. Stem Cells 25:1339–1347

Hong KU, Guo Y, Li QH, Cao P, Al-Maqtari T, Vajravelu BN, Du J, Book MJ, Zhu X, Nong Y, Bhatnagar A, Bolli R (2014) c-kit+ Cardiac stem cells alleviate post-myocardial infarction left ventricular dysfunction despite poor engraftment and negligible retention in the recipient heart. PLoS One 9:e96725

Hristov M, Zernecke A, Liehn EA, Weber C (2007) Regulation of endothelial progenitor cell homing after arterial injury. Thromb Haemost 98:274–277

Hsieh PC, Segers VF, Davis ME, Macgillivray C, Gannon J, Molkentin JD, Robbins J, Lee RT (2007) Evidence from a genetic fate-mapping study that stem cells refresh adult mammalian cardiomyocytes after injury. Nat Med 13:970–974

Ieda M, Fu JD, Delgado-Olguin P, Vedantham V, Hayashi Y, Bruneau BG, Srivastava D (2010) Direct reprogramming of fibroblasts into functional cardiomyocytes by defined factors. Cell 142:375–386

Iwasaki H, Kawamoto A, Ishikawa M, Oyamada A, Nakamori S, Nishimura H, Sadamoto K, Horii M, Matsumoto T, Murasawa S, Shibata T, Suehiro S, Asahara T (2006) Dose-dependent contribution of CD34-positive cell transplantation to concurrent vasculogenesis and cardiomyogenesis for functional regenerative recovery after myocardial infarction. Circulation 113:1311–1325

Jackson KA, Majka SM, Wang H, Pocius J, Hartley CJ, Majesky MW, Entman ML, Michael LH, Hirschi KK, Goodell MA (2001) Regeneration of ischemic cardiac muscle and vascular endothelium by adult stem cells. J Clin Invest 107:1395–1402

Jayawardena TM, Egemnazarov B, Finch EA, Zhang L, Payne JA, Pandya K, Zhang Z, Rosenberg P, Mirotsou M, Dzau VJ (2012) MicroRNA-mediated in vitro and in vivo direct reprogramming of cardiac fibroblasts to cardiomyocytes. Circ Res 110:1465–1473

Jayawardena TM, Finch EA, Zhang L, Zhang H, Hodgkinson CP, Pratt RE, Rosenberg PB, Mirotsou M, Dzau VJ (2015) MicroRNA induced cardiac reprogramming in vivo: evidence for mature cardiac myocytes and improved cardiac function. Circ Res 116:418–424

Jessup M, Brozena S (2003) Heart failure. N Engl J Med 348:2007–2018

Jesty SA, Steffey MA, Lee FK, Breitbach M, Hesse M, Reining S, Lee JC, Doran RM, Nikitin AY, Fleischmann BK, Kotlikoff MI (2012) c-kit+ precursors support postinfarction myogenesis in the neonatal, but not adult, heart. Proc Natl Acad Sci U S A 109:13380–13385

Johnston PV, Sasano T, Mills K, Evers R, Lee ST, Smith RR, Lardo AC, Lai S, Steenbergen C, Gerstenblith G, Lange R, Marban E (2009) Engraftment, differentiation, and functional benefits of autologous cardiosphere-derived cells in porcine ischemic cardiomyopathy. Circulation 120:1075–1083

Jopling C, Sleep E, Raya M, Marti M, Raya A, Izpisua Belmonte JC (2010) Zebrafish heart regeneration occurs by cardiomyocyte dedifferentiation and proliferation. Nature 464:606–609

Karakikes I, Ameen M, Termglinchan V, Wu JC (2015) Human induced pluripotent stem cell-derived cardiomyocytes: insights into molecular, cellular, and functional phenotypes. Circ Res 117:80–88

Kawamoto A, Gwon HC, Iwaguro H, Yamaguchi JI, Uchida S, Masuda H, Silver M, Ma H, Kearney M, Isner JM, Asahara T (2001) Therapeutic potential of ex vivo expanded endothelial progenitor cells for myocardial ischemia. Circulation 103:634–637

Kawamoto A, Losordo DW (2008) Endothelial progenitor cells for cardiovascular regeneration. Trends Cardiovasc Med 18:33–37

Kikuchi K, Holdway JE, Werdich AA, Anderson RM, Fang Y, Egnaczyk GF, Evans T, Macrae CA, Stainier DY, Poss KD (2010) Primary contribution to zebrafish heart regeneration by gata4(+) cardiomyocytes. Nature 464:601–605

Kinnaird T, Stabile E, Burnett MS, Lee CW, Barr S, Fuchs S, Epstein SE (2004) Marrow-derived stromal cells express genes encoding a broad spectrum of arteriogenic cytokines and promote in vitro and in vivo arteriogenesis through paracrine mechanisms. Circ Res 94:678–685

Kuraitis D, Ruel M, Suuronen EJ (2011) Mesenchymal stem cells for cardiovascular regeneration. Cardiovasc Drugs Ther 25:349–362

Laflamme MA, Chen KY, Naumova AV, Muskheli V, Fugate JA, Dupras SK, Reinecke H, Xu C, Hassanipour M, Police S, O'Sullivan C, Collins L, Chen Y, Minami E, Gill EA, Ueno S, Yuan C, Gold J, Murry CE (2007) Cardiomyocytes derived from human embryonic stem cells in pro-survival factors enhance function of infarcted rat hearts. Nat Biotechnol 25:1015–1024

Laflamme MA, Murry CE (2011) Heart regeneration. Nature 473:326–335

Laugwitz KL, Moretti A, Lam J, Gruber P, Chen Y, Woodard S, Lin LZ, Cai CL, Lu MM, Reth M, Platoshyn O, Yuan JX, Evans S, Chien KR (2005) Postnatal isl1+ cardioblasts enter fully differentiated cardiomyocyte lineages. Nature 433:647–653

Leobon B, Garcin I, Menasche P, Vilquin JT, Audinat E, Charpak S (2003) Myoblasts transplanted into rat infarcted myocardium are functionally isolated from their host. Proc Natl Acad Sci U S A 100:7808–7811

Li F, Wang X, Capasso JM, Gerdes AM (1996) Rapid transition of cardiac myocytes from hyperplasia to hypertrophy during postnatal development. J Mol Cell Cardiol 28:1737–1746

Lin T, Ambasudhan R, Yuan X, Li W, Hilcove S, Abujarour R, Lin X, Hahm HS, Hao E, Hayek A, Ding S (2009) A chemical platform for improved induction of human iPSCs. Nat Methods 6:805–808

Linke A, Muller P, Nurzynska D, Casarsa C, Torella D, Nascimbene A, Castaldo C, Cascapera S, Bohm M, Quaini F, Urbanek K, Leri A, Hintze TH, Kajstura J, Anversa P (2005) Stem cells in the dog heart are self-renewing, clonogenic, and multipotent and regenerate infarcted myocardium, improving cardiac function. Proc Natl Acad Sci U S A 102:8966–8971

Loffredo FS, Steinhauser ML, Gannon J, Lee RT (2011) Bone marrow-derived cell therapy stimulates endogenous cardiomyocyte progenitors and promotes cardiac repair. Cell Stem Cell 8:389–398

Makkar RR, Smith RR, Cheng K, Malliaras K, Thomson LE, Berman D, Czer LS, Marban L, Mendizabal A, Johnston PV, Russell SD, Schuleri KH, Lardo AC, Gerstenblith G, Marban E (2012) Intracoronary cardiosphere-derived cells for heart regeneration after myocardial infarction (Caduceus): a prospective, randomised phase 1 trial. Lancet 379:895–904

Malliaras K, Zhang Y, Seinfeld J, Galang G, Tseliou E, Cheng K, Sun B, Aminzadeh M, Marban E (2013) Cardiomyocyte proliferation and progenitor cell recruitment underlie therapeutic regeneration after myocardial infarction in the adult mouse heart. EMBO Mol Med 5:191–209

Marban E, Malliaras K (2012) Mixed results for bone marrow-derived cell therapy for ischemic heart disease. JAMA 308:2405–2406

Martin CM, Meeson AP, Robertson SM, Hawke TJ, Richardson JA, Bates S, Goetsch SC, Gallardo TD, Garry DJ (2004) Persistent expression of the Atp-binding cassette transporter, Abcg2, identifies cardiac Sp cells in the developing and adult heart. Dev Biol 265:262–275

Mathieu M, Bartunek J, El Oumeiri B, Touihri K, Hadad I, Thoma P, Metens T, Da Costa AM, Mahmoudabady M, Egrise D, Blocklet D, Mazouz N, Naeije R, Heyndrickx G, Mcentee K (2009) Cell therapy with autologous bone marrow mononuclear stem cells is associated with superior cardiac recovery compared with use of nonmodified mesenchymal stem cells in a canine model of chronic myocardial infarction. J Thorac Cardiovasc Surg 138:646–653

Matsuura K, Nagai T, Nishigaki N, Oyama T, Nishi J, Wada H, Sano M, Toko H, Akazawa H, Sato T, Nakaya H, Kasanuki H, Komuro I (2004) Adult cardiac Sca-1-positive cells differentiate into beating cardiomyocytes. J Biol Chem 279:11384–11391

Menasche P, Alfieri O, Janssens S, Mckenna W, Reichenspurner H, Trinquart L, Vilquin JT, Marolleau JP, Seymour B, Larghero J, Lake S, Chatellier G, Solomon S, Desnos M, Hagege AA (2008) The myoblast autologous grafting in ischemic cardiomyopathy (magic) trial: first randomized placebo-controlled study of myoblast transplantation. Circulation 117:1189–1200

Messina E, De Angelis L, Frati G, Morrone S, Chimenti S, Fiordaliso F, Salio M, Battaglia M, Latronico MV, Coletta M, Vivarelli E, Frati L, Cossu G, Giacomello A (2004) Isolation and expansion of adult cardiac stem cells from human and murine heart. Circ Res 95:911–921

Mirotsou M, Zhang Z, Deb A, Zhang L, Gnecchi M, Noiseux N, Mu H, Pachori A, Dzau V (2007) Secreted frizzled related protein 2 (Sfrp2) is the key Akt-mesenchymal stem cell-released paracrine factor mediating myocardial survival and repair. Proc Natl Acad Sci U S A 104:1643–1648

Moretti A, Caron L, Nakano A, Lam JT, Bernshausen A, Chen Y, Qyang Y, Bu L, Sasaki M, Martin-Puig S, Sun Y, Evans SM, Laugwitz KL, Chien KR (2006) Multipotent embryonic isl1+ progenitor cells lead to cardiac, smooth muscle, and endothelial cell diversification. Cell 127:1151–1165

Mummery CL, Zhang J, Ng ES, Elliott DA, Elefanty AG, Kamp TJ (2012) Differentiation of human embryonic stem cells and induced pluripotent stem cells to cardiomyocytes: a methods overview. Circ Res 111:344–358

Murry CE, Soonpaa MH, Reinecke H, Nakajima H, Nakajima HO, Rubart M, Pasumarthi KB, Virag JI, Bartelmez SH, Poppa V, Bradford G, Dowell JD, Williams DA, Field LJ (2004) Haematopoietic stem cells do not transdifferentiate into cardiac myocytes in myocardial infarcts. Nature 428:664–668

Narmoneva DA, Vukmirovic R, Davis ME, Kamm RD, Lee RT (2004) Endothelial cells promote cardiac myocyte survival and spatial reorganization: implications for cardiac regeneration. Circulation 110:962–968

Nowbar AN, Mielewczik M, Karavassilis M, Dehbi HM, Shun-Shin MJ, Jones S, Howard JP, Cole GD, Francis DP, Group DW (2014) Discrepancies in autologous bone marrow stem cell trials and enhancement of ejection fraction (Damascene): weighted regression and meta-analysis. BMJ 348:g2688

Oberpriller JO, Oberpriller JC (1974) Response of the adult newt ventricle to injury. J Exp Zool 187:249–253

Oh H, Bradfute SB, Gallardo TD, Nakamura T, Gaussin V, Mishina Y, Pocius J, Michael LH, Behringer RR, Garry DJ, Entman ML, Schneider MD (2003) Cardiac progenitor cells from adult myocardium: homing, differentiation, and fusion after infarction. Proc Natl Acad Sci U S A 100:12313–12318

Okano H, Nakamura M, Yoshida K, Okada Y, Tsuji O, Nori S, Ikeda E, Yamanaka S, Miura K (2013) Steps toward safe cell therapy using induced pluripotent stem cells. Circ Res 112:523–533

Okita K, Nakagawa M, Hyenjong H, Ichisaka T, Yamanaka S (2008) Generation of mouse induced pluripotent stem cells without viral vectors. Science 322:949–953

Orlic D, Kajstura J, Chimenti S, Jakoniuk I, Anderson SM, Li B, Pickel J, Mckay R, Nadal-Ginard B, Bodine DM, Leri A, Anversa P (2001) Bone marrow cells regenerate infarcted myocardium. Nature 410:701–705

Oyama T, Nagai T, Wada H, Naito AT, Matsuura K, Iwanaga K, Takahashi T, Goto M, Mikami Y, Yasuda N, Akazawa H, Uezumi A, Takeda S, Komuro I (2007) Cardiac side population cells have a potential to migrate and differentiate into cardiomyocytes in vitro and in vivo. J Cell Biol 176:329–341

Pittenger MF, Mackay AM, Beck SC, Jaiswal RK, Douglas R, Mosca JD, Moorman MA, Simonetti DW, Craig S, Marshak DR (1999) Multilineage potential of adult human mesenchymal stem cells. Science 284:143–147

Porrello ER, Mahmoud AI, Simpson E, Hill JA, Richardson JA, Olson EN, Sadek HA (2011) Transient regenerative potential of the neonatal mouse heart. Science 331:1078–1080

Porrello ER, Mahmoud AI, Simpson E, Johnson BA, Grinsfelder D, Canseco D, Mammen PP, Rothermel BA, Olson EN, Sadek HA (2013) Regulation of neonatal and adult mammalian heart regeneration by the miR-15 family. Proc Natl Acad Sci U S A 110:187–192

Poss KD, Wilson LG, Keating MT (2002) Heart regeneration in zebrafish. Science 298:2188–2190

Qian L, Huang Y, Spencer CI, Foley A, Vedantham V, Liu L, Conway SJ, Fu JD, Srivastava D (2012) In vivo reprogramming of murine cardiac fibroblasts into induced cardiomyocytes. Nature 485:593–598

Quaini F, Urbanek K, Beltrami AP, Finato N, Beltrami CA, Nadal-Ginard B, Kajstura J, Leri A, Anversa P (2002) Chimerism of the transplanted heart. N Engl J Med 346:5–15

Quevedo HC, Hatzistergos KE, Oskouei BN, Feigenbaum GS, Rodriguez JE, Valdes D, Pattany PM, Zambrano JP, Hu Q, Mcniece I, Heldman AW, Hare JM (2009) Allogeneic mesenchymal stem cells restore cardiac function in chronic ischemic cardiomyopathy via trilineage differentiating capacity. Proc Natl Acad Sci U S A 106:14022–14027

Schachinger V, Erbs S, Elsasser A, Haberbosch W, Hambrecht R, Holschermann H, Yu J, Corti R, Mathey DG, Hamm CW, Suselbeck T, Assmus B, Tonn T, Dimmeler S, Zeiher AM, Investigators R-A (2006) Intracoronary bone marrow-derived progenitor cells in acute myocardial infarction. N Engl J Med 355:1210–1221

Senyo SE, Steinhauser ML, Pizzimenti CL, Yang VK, Cai L, Wang M, Wu TD, Guerquin-Kern JL, Lechene CP, Lee RT (2013) Mammalian heart renewal by pre-existing cardiomyocytes. Nature 493:433–436

Sharpe N, Smith H, Murphy J, Greaves S, Hart H, Gamble G (1991) Early prevention of left ventricular dysfunction after myocardial infarction with angiotensin-converting-enzyme inhibition. Lancet 337:872–876

Shiba Y, Fernandes S, Zhu WZ, Filice D, Muskheli V, Kim J, Palpant NJ, Gantz J, Moyes KW, Reinecke H, Van Biber B, Dardas T, Mignone JL, Izawa A, Hanna R, Viswanathan M, Gold JD, Kotlikoff MI, Sarvazyan N, Kay MW, Murry CE, Laflamme MA (2012) Human ES-cell-derived cardiomyocytes electrically couple and suppress arrhythmias in injured hearts. Nature 489:322–325

Simari RD, Pepine CJ, Traverse JH, Henry TD, Bolli R, Spoon DB, Yeh E, Hare JM, Schulman IH, Anderson RD, Lambert C, Sayre SL, Taylor DA, Ebert RF, Moye LA (2014) Bone marrow mononuclear cell therapy for acute myocardial infarction: a perspective from the cardiovascular cell therapy research network. Circ Res 114:1564–1568

Smart N, Bollini S, Dube KN, Vieira JM, Zhou B, Davidson S, Yellon D, Riegler J, Price AN, Lythgoe MF, Pu WT, Riley PR (2011) De novo cardiomyocytes from within the activated adult heart after injury. Nature 474:640–644

Smith RR, Barile L, Cho HC, Leppo MK, Hare JM, Messina E, Giacomello A, Abraham MR, Marban E (2007) Regenerative potential of cardiosphere-derived cells expanded from percutaneous endomyocardial biopsy specimens. Circulation 115:896–908

Song K, Nam YJ, Luo X, Qi X, Tan W, Huang GN, Acharya A, Smith CL, Tallquist MD, Neilson EG, Hill JA, Bassel-Duby R, Olson EN (2012) Heart repair by reprogramming non-myocytes with cardiac transcription factors. Nature 485:599–604

Soonpaa MH, Field LJ (1997) Assessment of cardiomyocyte DNA synthesis in normal and injured adult mouse hearts. Am J Phys 272:H220–H226

Stamm C, Kleine HD, Choi YH, Dunkelmann S, Lauffs JA, Lorenzen B, David A, Liebold A, Nienaber C, Zurakowski D, Freund M, Steinhoff G (2007) Intramyocardial delivery of CD133+ bone marrow cells and coronary artery bypass grafting for chronic ischemic heart disease: safety and efficacy studies. J Thorac Cardiovasc Surg 133:717–725

Takahashi K, Tanabe K, Ohnuki M, Narita M, Ichisaka T, Tomoda K, Yamanaka S (2007) Induction of pluripotent stem cells from adult human fibroblasts by defined factors. Cell 131:861–872

Takahashi K, Yamanaka S (2006) Induction of pluripotent stem cells from mouse embryonic and adult fibroblast cultures by defined factors. Cell 126:663–676

Terracciano CM, Miller LW, Yacoub MH (2010) Contemporary use of ventricular assist devices. Annu Rev Med 61:255–270

Toma C, Pittenger MF, Cahill KS, Byrne BJ, Kessler PD (2002) Human mesenchymal stem cells differentiate to a cardiomyocyte phenotype in the adult murine heart. Circulation 105:93–98

Traverse JH, Henry TD, Pepine CJ, Willerson JT, Zhao DX, Ellis SG, Forder JR, Anderson RD, Hatzopoulos AK, Penn MS, Perin EC, Chambers J, Baran KW, Raveendran G, Lambert C, Lerman A, Simon DI, Vaughan DE, Lai D, Gee AP, Taylor DA, Cogle CR, Thomas JD, Olson RE, Bowman S, Francescon J, Geither C, Handberg E, Kappenman C, Westbrook L, Piller LB, Simpson LM, Baraniuk S, Loghin C, Aguilar D, Richman S, Zierold C, Spoon DB, Bettencourt J, Sayre SL, Vojvodic RW, Skarlatos SI, Gordon DJ, Ebert RF, Kwak M, Moye LA, Simari RD, Cardiovascular Cell Therapy Research, N (2012) Effect of the use and timing of bone marrow mononuclear cell delivery on left ventricular function after acute myocardial infarction: the TIME randomized trial. JAMA 308:2380–2389

Uchida S, De Gaspari P, Kostin S, Jenniches K, Kilic A, Izumiya Y, Shiojima I, Grosse Kreymborg K, Renz H, Walsh K, Braun T (2013) Sca1-derived cells are a source of myocardial renewal in the murine adult heart. Stem Cell Reports 1:397–410

Urbich C, Aicher A, Heeschen C, Dernbach E, Hofmann WK, Zeiher AM, Dimmeler S (2005) Soluble factors released by endothelial progenitor cells promote migration of endothelial cells and cardiac resident progenitor cells. J Mol Cell Cardiol 39:733–742

Van Berlo JH, Kanisicak O, Maillet M, Vagnozzi RJ, Karch J, Lin SC, Middleton RC, Marban E, Molkentin JD (2014) c-kit+ cells minimally contribute cardiomyocytes to the heart. Nature 509:337–341

Vrtovec B, Poglajen G, Lezaic L, Sever M, Domanovic D, Cernelc P, Socan A, Schrepfer S, Torre-Amione G, Haddad F, Wu JC (2013) Effects of intracoronary CD34+ stem cell transplantation in nonischemic dilated cardiomyopathy patients: 5-year follow-up. Circ Res 112:165–173

Walsh S, Ponten A, Fleischmann BK, Jovinge S (2010) Cardiomyocyte cell cycle control and growth estimation in vivo--an analysis based on cardiomyocyte nuclei. Cardiovasc Res 86:365–373

Wang G, Mccain ML, Yang L, He A, Pasqualini FS, Agarwal A, Yuan H, Jiang D, Zhang D, Zangi L, Geva J, Roberts AE, Ma Q, Ding J, Chen J, Wang DZ, Li K, Wang J, Wanders RJ, Kulik W, Vaz FM, Laflamme MA, Murry CE, Chien KR, Kelley RI, Church GM, Parker KK, Pu WT (2014) Modeling the mitochondrial cardiomyopathy of Barth syndrome with induced pluripotent stem cell and heart-on-chip technologies. Nat Med 20:616–623

Wang X, Hu Q, Nakamura Y, Lee J, Zhang G, From AH, Zhang J (2006) The role of the sca-1+/CD31- cardiac progenitor cell population in postinfarction left ventricular remodeling. Stem Cells 24:1779–1788

Weinberger F, Mehrkens D, Friedrich FW, Stubbendorff M, Hua X, Muller JC, Schrepfer S, Evans SM, Carrier L, Eschenhagen T (2012) Localization of Islet-1-positive cells in the healthy and infarcted adult murine heart. Circ Res 110:1303–1310

Whelan RS, Kaplinskiy V, Kitsis RN (2010) Cell death in the pathogenesis of heart disease: mechanisms and significance. Annu Rev Physiol 72:19–44

Williams AR, Hare JM (2011) Mesenchymal stem cells: biology, pathophysiology, translational findings, and therapeutic implications for cardiac disease. Circ Res 109:923–940

Xu C, Police S, Rao N, Carpenter MK (2002) Characterization and enrichment of cardiomyocytes derived from human embryonic stem cells. Circ Res 91:501–508

Ye L, Chang YH, Xiong Q, Zhang P, Zhang L, Somasundaram P, Lepley M, Swingen C, Su L, Wendel JS, Guo J, Jang A, Rosenbush D, Greder L, Dutton JR, Zhang J, Kamp TJ, Kaufman DS, Ge Y, Zhang J (2014) Cardiac repair in a porcine model of acute myocardial infarction with human induced pluripotent stem cell-derived cardiovascular cells. Cell Stem Cell 15:750–761

Zaruba MM, Soonpaa M, Reuter S, Field LJ (2010) Cardiomyogenic potential of C-kit(+)-expressing cells derived from neonatal and adult mouse hearts. Circulation 121:1992–2000

Zhang J, Wilson GF, Soerens AG, Koonce CH, Yu J, Palecek SP, Thomson JA, Kamp TJ (2009) Functional cardiomyocytes derived from human induced pluripotent stem cells. Circ Res 104:e30–e41

Zhao Y, Londono P, Cao Y, Sharpe EJ, Proenza C, O'Rourke R, Jones KL, Jeong MY, Walker LA, Buttrick PM, Mckinsey TA, Song K (2015) High-efficiency reprogramming of fibroblasts into cardiomyocytes requires suppression of pro-fibrotic signalling. Nat Commun 6:8243

Zhou B, Honor LB, Ma Q, Oh JH, Lin RZ, Melero-Martin JM, Von Gise A, Zhou P, Hu T, He L, Wu KH, Zhang H, Zhang Y, Pu WT (2012) Thymosin beta 4 treatment after myocardial infarction does not reprogram epicardial cells into cardiomyocytes. J Mol Cell Cardiol 52:43–47

Zhou B, Ma Q, Rajagopal S, Wu SM, Domian I, Rivera-Feliciano J, Jiang D, Von Gise A, Ikeda S, Chien KR, Pu WT (2008) Epicardial progenitors contribute to the cardiomyocyte lineage in the developing heart. Nature 454:109–113

Zhou H, Dickson ME, Kim MS, Bassel-Duby R, Olson EN (2015) Akt1/protein kinase B enhances transcriptional reprogramming of fibroblasts to functional cardiomyocytes. Proc Natl Acad Sci U S A 112:11864–11869

Zhou H, Wu S, Joo JY, Zhu S, Han DW, Lin T, Trauger S, Bien G, Yao S, Zhu Y, Siuzdak G, Scholer HR, Duan L, Ding S (2009) Generation of induced pluripotent stem cells using recombinant proteins. Cell Stem Cell 4:381–384

Progenitor Cells from the Adult Heart

2

Georgina M. Ellison-Hughes and Fiona C. Lewis

Abstract

The adult myocardium harbours a population of resident (endogenous) multipotent cardiac stem and progenitor cells (eCSCs). Manipulation of these cells in situ and ex vivo has opened new therapeutic avenues for anatomical and functional myocardial regeneration. However, recently the ability of the c-kit^{pos} stem and progenitor cells to transdifferentiate into new cardiomyocytes has been disputed. Within an already highly controversial research field, these publications have caused significant confusion in their interpretation. Importantly, identifying, tracing and characterising stem and progenitor cells according to expression of a single surface receptor such as c-kit do not identify eCSCs. As discussed in this chapter, eCSCs isolated from the adult heart have a specific phenotype, being negative for blood lineage markers such as CD34, CD45 and CD31, and exhibit properties of stem and progenitor cells, being clonogenic, self-renewing and multipotent. Under the appropriate conditions, eCSCs differentiate into fully functional beating cardiomyocytes and regenerate cardiomyocytes lost from damage in vivo. Finally, eCSCs are susceptible to the effects of ageing, making regulation of this parameter highly impactful in the efficacy of myocardial regenerative therapies.

Despite the adult mammalian heart being composed of terminally differentiated cardiomyocytes that are permanently withdrawn from the cell cycle (Nadal-Ginard 1978; Chien and Olson 2002), it is now apparent that the adult heart has the capacity, albeit low, to self-renew cardiomyocytes over the human lifespan (Bergmann

G.M. Ellison-Hughes (✉) • F.C. Lewis
Centre of Human and Aerospace Physiological Sciences and Centre for Stem Cells and Regenerative Medicine, Faculty of Life Sciences and Medicine, King's College London, London SE1 1UL, UK
e-mail: georgina.ellison@kcl.ac.uk

M. Ieda, W.-H. Zimmermann (eds.), *Cardiac Regeneration*, Cardiac and Vascular Biology 4, DOI 10.1007/978-3-319-56106-6_2

et al. 2009, 2015). This is supported by the detection of small, newly formed, immature cardiomyocytes, which incorporate BrdU/EdU and/or stain positive for Ki67, Aurora B and embryonic/neonatal myosin heavy chain (Fig. 2.1), as well as cardiomyocytes undergoing mitosis, under normal conditions and in response to diverse pathological and physiological stimuli (Urbanek et al. 2003, 2005; Bergmann et al. 2009; Boström et al. 2010; Overy and Priest 1966; Kajstura et al. 1998; Waring et al. 2014). The source of these newly formed cardiomyocytes is still a matter of debate (Laflamme and Murry 2011). Three main sources of origin of the new cardiomyocytes have been claimed: (a) circulating progenitors, which through the bloodstream home to the myocardium and differentiate into cardiomyocytes (Quaini et al. 2002); (b) mitotic division of the pre-existing cardiomyocytes (Boström et al.

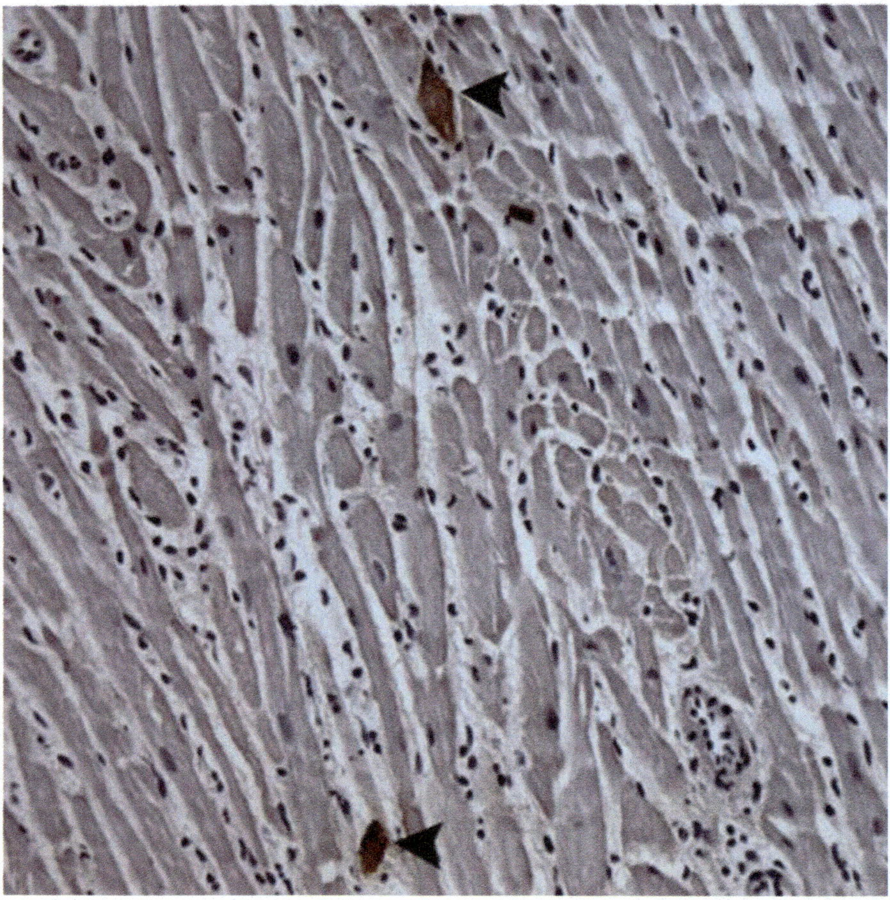

Fig. 2.1 Regenerating cardiomyocytes in the adult rat heart. Two small regenerating cardiomyocytes (*arrowheads*) detected using a mouse monoclonal myosin heavy chain (developmental) primary antibody (Novocastra, Leica Biosystems). This antibody recognises a MHC present during the embryonic and period in the development of skeletal muscle, and the same MHC is re-expressed during regeneration of new skeletal muscle fibres (Ecob-Prince et al. 1989; Williams et al. 2001)

2010; Bersell et al. 2009; Kühn et al. 2007; Senyo et al. 2013); and (c) a small population of resident multipotent stem cells able to differentiate into the main cell types of the heart (i.e. cardiomyocytes, smooth and endothelial vascular and connective tissue cells) (Torella et al. 2007; Rasmussen et al. 2011).

Blood-borne precursors are well documented for having a role in inflammation and healing. When adult mouse bone marrow cells were injected into the chick embryo, they converted to a myocardial phenotype (Eisenberg et al. 2006). Their cardiomyogenic potential in the damaged adult heart is however at best very much limited (Loffredo et al. 2011; Ellison et al. 2013). The evidence so far presented in support of re-entry of terminally differentiated cardiomyocytes into the cell cycle has been limited to show division of cells that express proteins of the contractile apparatus in their cytoplasm (Boström et al. 2010; Bersell et al. 2009; Kühn et al. 2007; Senyo et al. 2013). This evidence is equally compatible with new myocyte formation from the pool of multipotent cardiac stem/progenitor cells, which as precursor cells express contractile proteins, and because newly born myocytes are not yet terminally differentiated, they are capable of a few rounds of division before irreversibly withdrawing from the cell cycle (Nadal-Ginard et al. 2003).

The best documented source of the small, immature, newly formed cardiomyocytes in the adult mammalian heart, including the human, is a small population of endogenous cardiac stem and progenitor cells (eCSCs) distributed throughout the atria and ventricles, which can give rise to functional cardiomyocytes and vasculature in vitro and in vivo (Torella et al. 2007; Ellison et al. 2007a). Importantly, owing to genetic labelling and transitional tracking, it is now documented that the newly formed cardiomyocytes observed in the adult mammalian heart are the product of eCSC differentiation (Hsieh et al. 2007; Ellison et al. 2013; van Berlo et al. 2014).

2.1 Phenotype and Characteristics of eCSCs

The first report of endogenous cardiac stem and progenitor cells in the adult mammalian heart was in 2003 (Beltrami et al. 2003), and since then their existence has been confirmed by a number of independent groups. Although a variety of markers (c-kit, Sca-1, PDGFRα, Wt1) have been proposed to identify eCSCs in different species and throughout development (Oh et al. 2003; Matsuura et al. 2004; Messina et al. 2004; Martin et al. 2004; Laugwitz et al. 2005; Moretti et al. 2006; Kattman et al. 2006; Wu et al. 2006; Smart et al. 2011; Chong et al. 2011; Noseda et al. 2015), it still remains to be determined whether these markers identify different populations of eCSCs or, more likely, different developmental and/or physiological stages of the same cell type (Ellison et al. 2010; Keith and Bolli 2015).

The progeny of a single eCSC is able to differentiate into cardiomyocytes, smooth muscle and endothelial vascular cells and, when transplanted into the border zone of an infarct, regenerates functional contractile muscle and the microvasculature of the tissue (Beltrami et al. 2003; Ellison et al. 2013). In a normal adult myocardium, at any given time, most of the eCSCs are quiescent, and only a small fraction is active to replace the cardiomyocytes and vascular cells lost by wear and tear. In response to stress (hypoxia, exercise, work overload or diffuse damage), however, a proportion of the resident eCSCs are rapidly activated; they multiply and generate new muscle and vascular cells (Urbanek et al. 2003; Ellison et al. 2007a, 2013; Waring et al. 2014), contributing to cardiac remodelling. The activation of the eCSCs is able to regenerate the myocardial cells lost as a consequence of diffuse myocardial damage, which kills up to 10% of the myocardial mass (Ellison et al. 2013), and their transplantation can regenerate the contractile cells lost as a consequence of a major acute myocardial infarction (AMI) affecting up to 25% of the left ventricular mass (Beltrami et al. 2003; Ellison et al. 2013).

2.1.1 c-kit^{pos} Stem and Progenitor Cells

c-kit, also known as CD117, is a tyrosine kinase type III receptor, which is expressed in several cell types and plays a significant role in a variety of cell functions, including identifying haematopoietic stem cells while regulating their cell fate (Roskoski 2005). c-kit positive (c-kit^{pos}) stem, and progenitor cells have been identified in the myocardium that are also positive for Sca-1 and MDR-1 (ABCG2) yet are negative for markers of the blood cell lineage, CD31, CD34 and CD45 (described as lineage-negative). They are self-renewing, clonogenic and multipotent and exhibit significant regenerative potential when injected into the adult rat heart following a myocardial infarction (MI), forming new cardiomyocytes and vasculature and restoring cardiac function (Beltrami et al. 2003; Ellison et al. 2013). c-kit^{pos} eCSCs with similar properties to those originally identified in the rat have been identified and characterised in the mouse (Messina et al. 2004; Fransioli et al. 2008), dog (Linke et al. 2005), pig (Ellison et al. 2011) and human (Messina et al. 2004; Torella et al. 2006; Bearzi et al. 2007; Arsalan et al. 2012). These cells are present at a similar density in all species (~1 eCSC per 1000 cardiomyocytes or 45,000 human eCSCs per gram of tissue) (Torella et al. 2007). Similar to the rodent heart, the distribution of c-kit^{pos} eCSCs in the pig and human heart varies with cardiac chamber, and this will differ in the human depending on disease status (our unpublished findings). The adult-derived c-kit^{pos} eCSCs are very similar in their characteristics and potential to a population of cardiac-specific (c-kit^{pos}/Nkx2.5^{pos}) cells identified in the mouse embryo that differentiate into cardiomyocytes and also smooth muscle cells (Wu et al. 2006). Indeed, similar to adult-derived eCSCs, embryonic cardiac c-kit^{pos}/Nkx2.5^{pos} cells possessed the capacity for long-term expansion in vitro, clonogenicity and differentiation into both cardiomyocytes and smooth muscle cells from a single-cell-derived colony (Wu et al. 2006).

Recently considerable confusion has mounted because of the development of genetic lineage tracing mouse models according to the expression of c-kit. Stem cells, as defined by Potten and Loeffler, are "undifferentiated cells capable of (1) proliferation, (2) self-maintenance, (3) production of large number of differentiated progeny, (4) regeneration of the tissue after injury, and (5) flexibility in the use of these options" (Potten and Loeffler 1990). It is important to iterate that a cell must possess these characteristics to be defined a stem cell; identifying, tracing and characterising stem cells according to expression of a single surface receptor (Tallini et al. 2009; Jesty et al. 2012; van Berlo et al. 2014; Sultana et al. 2015), such as c-kit, do not identify eCSCs. Thus, relying on genetic labelling of c-kit^{pos} cells or quantifying c-kit^{pos} cells within any tissue, including the heart, to extrapolate their plasticity or regenerative potential is in our view a major biological and practical pitfall that brings data which require a careful interpretation.

In the adult heart, the total population of c-kit^{pos} cells (including the CD45^{pos} fraction representing cardiac mast cells and CD34/CD31^{pos} cells representing vascular progenitors and cells; Fig. 2.2) have little cardiomyogenic potential (van Berlo et al. 2014; Sultana et al. 2015) and following cryogenic injury (induced by touching a 1 mm diameter copper probe that is equilibrated in liquid nitrogen to the apex of the left ventricle) or myocardial infarction (induced by ligation of left anterior descending coronary artery) contribute predominantly through revascularisation of the damaged tissue (Tallini et al. 2009; Jesty et al. 2012; van Berlo et al. 2014; Sultana et al. 2015). These cells express Flk-1 and/or Pecam-1 (CD31) suggesting they are primarily vascular progenitors and bear more resemblance to the bone marrow-derived c-kit^{pos}/Sca-1^{pos}/Flk-1^{pos} cells identified by Fazel and colleagues, which following a myocardial infarction home to the heart and contribute to the revascularisation of the infarcted/damaged area by establishing a pro-angiogenic milieu (Fazel et al. 2006).

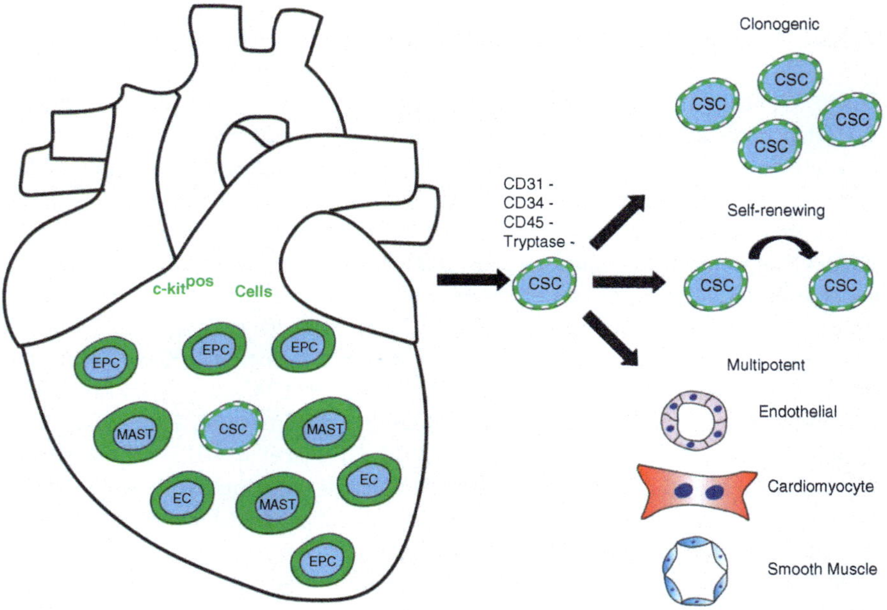

Fig. 2.2 c-kit^{pos} cells in the adult heart are not all stem/progenitor cells. c-kit-positive cells in the adult represent cardiac mast cells (MAST), endothelial cells (EC), endothelial progenitor cells (EPC) and cardiac stem cells (CSC). CSCs express c-kit at a lower level compared to mast cells, endothelial cells and endothelial progenitor cells. CSCs have a phenotype of c-kit$^{pos/low}$, CD45^{neg}, tryptase^{neg}, CD31^{neg} and CD34^{neg} and are clonogenic, self-renewing and multipotent, differentiating into the three cardiac lineages: cardiomyocyte, endothelial and smooth muscle cells

Importantly, the c-kit^{pos} eCSCs are CD34 and CD31 negative (Smith et al. 2014) making them distinguishable from these vascular progenitor c-kit^{pos} cells (Fang et al. 2012) (Fig. 2.2).

Despite the extensive characterisation of c-kit^{pos} eCSCs, where they meet all five properties of the 'stem cell' definition given above, their role and significance in the adult mammalian heart have been continually questioned (Passier et al. 2008; Pouly et al. 2008; Zaruba et al. 2010; van Berlo and Molkentin 2014). Pouly et al. investigated c-kit^{pos} cells in endomyocardial, right ventricular (RV) biopsies and right atrial appendages of heart transplant recipients 73.5 months post-transplantation. Using immunohistochemistry they found that c-kit^{pos} cells were rare (1/mm^2 atrial tissue and 2.7/mm^2 RV tissue). None of the c-kit^{pos} cells identified expressed Nxk2.5 or CD105; however, all of these cells expressed CD45 and tryptase, identifying them as cardiac mast cells. It is not surprising that the authors only identified mast cells, as cardiac mast cells account for ~80% of the total number of c-kit^{pos} cells in the atria (Ellison et al. 2011).

An important consideration when isolating c-kit^{pos} eCSCs using the enzymatic tissue digestion method (Smith et al. 2014) is to allow liberation of all eCSCs from deep within the myocardium, but also being aware that the c-kit receptor can be affected by over-enzymatic digestion becoming internalised (Lévesque et al. 2003).

2.1.2 Sca-1^{pos} and Side Population Progenitor Cells

Sca-1^{pos}, lineage-negative cardiac progenitor cells (CPCs) were first described in 2003 and are resident non-myocyte cells from the adult murine heart expressing stem cell antigen 1 (Sca-1). While the total Sca-1 CPCs express early cardiac-specific factors such as Gata-4 and MEF2C (Oh et al. 2003; Matsuura et al. 2004), only a fraction of them exhibit stem cell properties of self-renewal and clonogenicity (Ye et al. 2012; Chong et al. 2011; Matsuura et al. 2004; Noseda et al. 2015). Sca-1^{pos} CPCs are capable of cardiomyogenic differentiation in vitro (Oh et al. 2003; Ye et al. 2012; Matsuura et al. 2004; Takamiya et al. 2011; Chong et al. 2011; Wang et al. 2006) and exhibit in vivo cardiomyogenic regenerative potential (Oh et al. 2003; Noseda et al. 2015; Wang et al. 2006; Takamiya et al. 2011). Sca1^{pos} CPCs also show differentiation into both endothelial and smooth muscle lineages (Ye et al. 2012; Wang et al. 2006; Takamiya et al. 2011; Iwakura et al. 2011; Noseda et al. 2015). It is worth noting that there is also a population of Sca1^{pos} vascular progenitor cells which resides within the arterial adventitia (AdvSca1 cells) that have been shown to be regulated by sonic hedgehog signalling (Shh) (Passman et al. 2008).

Side population (SP) cells were first characterised as a primitive population of haematopoietic stem cells characterised by their unique ability to efflux the DNA-binding dye, Hoechst 33342 (Goodell et al. 1996). SP cells have since been isolated from extra-haematopoietic tissues, including bone marrow, skeletal muscle, liver, brain, heart and lung (Asakura and Rudnicki 2002), and the ATP-binding cassette transporter (ABCG2) has been identified as a molecular determinant of the SP phenotype (Zhou et al. 2001; Martin et al. 2004). Hierlihy et al. first reported that the adult myocardium contained an endogenous cardiac SP with stem cell-like activity and identified that this Hoechst dye-excluding population constituted ~1% of total cardiac cells in the mouse postnatal heart (Hierlihy et al. 2002). Transcriptional profiling revealed that the cardiac SP exhibits a Sca-1^{pos}, c-kit^{low}, $CD34^{neg}$ and $CD45^{neg}$ phenotype (Martin et al. 2004), and further interrogation of these cells revealed that 75% express the endothelial marker, CD31. However, the Sca-1^{pos} $CD31^{neg}$ population was subsequently identified as having the greatest cardiomyogenic potential and was found to represent ~10% of the total cardiac SP (Martin et al. 2004; Wang et al. 2006; Pfister et al. 2005; Oyama et al. 2007).

Although Sca-1 appears to be an ideal marker for isolating and identifying CPCs, its homology hasn't been confirmed in any species, other than mouse. This poses a significant problem when translating research to develop human regenerative therapies. As c-kit^{pos} eCSCs express Sca-1 (Smith et al. 2014) and the c-$kit^{pos}CD45^{neg}CD31^{neg}$ and Sca-$1^{pos}CD31^{neg}$ cell populations exhibit similar number, self-renewing, clonogenicity and differentiation potential in vitro and in vivo, it can be concluded that they are probably the same cell population and will only differ in their level of expression of c-kit and/or Sca-1 depending on their physiological/differentiation state.

2.2 Cardiac Differentiation Potential of eCSCs

Despite the extensive published data from different groups in support of the regenerative cardiomyogenic potential of the eCSCs in vivo (Ellison et al. 2013; Beltrami et al. 2003; Li et al. 2011; Noseda et al. 2015; Hsieh et al. 2007; Mohsin et al. 2012; Fischer et al. 2009; Angert et al. 2011), scepticism exists over an eCSC's potential to differentiate into a fully functional synchronised beating cardiomyocyte. The first demonstration that a cardiosphere-forming progenitor cell type isolated from the mouse heart could form spontaneous beating myocyte colonies in vitro was from Messina et al. (2004). Then it was shown that Sca-1^{pos}/$CD31^{neg}$/$CD34^{neg}$/$CD45^{neg}$ eCSCs isolated from adult mice hearts differentiated into active contracting cardiomyocytes in vitro (Pfister et al. 2005). We have also shown that clonal c-kit^{pos} eCSCs differentiate into functionally competent beating cardiomyocytes following supplementation with a stage-specific growth factor cocktail targeting TGFβ and Wnt signalling pathways, recapitulating the morphogens present during embryonic development (Smith et al. 2014). This stage-specific regime is not dissimilar to that used to induce differentiation of ESCs and iPSCs into the functional cardiomyogenic embryoid bodies in vitro (Yang et al. 2007). Therefore, like other stem cells, under the appropriate conditions eCSCs do have cardiomyogenic capability, differentiating into functionally competent, beating cardiomyocytes in vitro.

When c-kit^{pos} cells are transplanted intramyocardially in the border/infarct zone of myocardial infarcted hearts, reports have also shown lack of their ability to differentiate into cardiomyocytes. This lack of differentiation capability is most likely due to lack of characterisation of the transplanted cell type, poor cell survival and retention, hostile host environment and subsequent restriction of cell proliferation and integration and differentiation in this damage-regeneration infarct model. Similar findings have been shown for Sca-1^{pos} CPCs (Noseda et al. 2015) and other stem/progenitor cells, including ESCs (Don and Murry 2013). Furthermore, whether the cells are injected as freshly isolated or pre-cultured and expanded in vitro or clonogenic cells will influence their survival and subsequent proliferation, integration and differentiation post-transplantation. Stem cells are maintained in a quiescent state until activated by injury in vivo (Ellison et al. 2007b) or another stimulus ex vivo (i.e. cell culture). Therefore, a freshly isolated stem cell, as well as being highly stressed following isolation from its niche, is quiescent and, unless activated, will not exit from G0 and, upon transplantation, coupled with the hostile host environment, will be more prone to death and/or not likely to proliferate. A cycling-competent stem cell that has been propagated in vitro is more robust and shows increased survival and proliferation post-transplantation (our unpublished findings). Additionally, a clonogenic population, derived from a single cell, is multipotent and able to give rise to cells of all three cardiac lineages. We have shown that clonogenic eCSCs injected intramyocardially following myocardial infarction can replenish up to 20% of cardiomyocytes in the infarct zone, resulting in improved LV function (Ellison et al. 2013).

As stated above certain criteria need to be met to ensure that a cell can be defined as a 'stem/progenitor' cell. These include being self-renewing, clonogenic and multipotent. A cell that is injected in vivo to test its regenerative potential should at least show these characteristics in vitro and prior to transplantation. Unfortunately, only a few publications show that the cells they inject have the properties of stem and progenitor cells. Instead because they express stem cell markers such as c-kit or Sca-1 and have been isolated from myocardial tissue, they assume that they are eCSCs, when in fact they are very likely not, but rather CD34^{pos}/CD31^{pos} vascular progenitors and will give rise to new vasculature once transplanted.

It is currently disputed if adult tissue-specific stem cells possess true pluripotency. Indeed, Sca1^{pos} CPCs and c-kit^{pos} eCSCs have shown capability of differentiation into noncardiac lineages in vitro and in vivo (Takamiya et al. 2011; Chong et al. 2011; Miyamoto et al. 2010). Interestingly it has been reported that the level of Sca-1 expression may actually play a role in their differentiation potential with Sca-1 high CPCs having a broader differentiation potential, showing osteogenic, chondrogenic, smooth muscle, endothelial and cardiac differentiation in vitro than Sca-1 low CPCs (Takamiya et al. 2011). In vivo teratoma formation assays have also shown that while Sca-1^{pos} CPCs alone do not form tumours, when injected alongside ESCs, they differentiate into cells of the three germ layers (Chong et al. 2011), although this broad developmental plasticity is yet to be shown in tissue regeneration and repair in vivo.

2.3 The Controversy

As outlined above eCSCs are small primitive cells, positive for stem cell surface receptor markers (i.e. c-kit, Sca-1) and negative for markers of the haematopoietic and endothelial lineage (i.e. CD45 and CD31) and mast cells (i.e. tryptase). They exhibit properties of stem cells, being clonogenic and self-renewing, and differentiate into cardiomyocytes, smooth muscle and endothelial cells, both in vitro and in vivo. Despite these reputable published data, recently, by targeting the c-kit locus with multiple reporter genes in mice, the significance of c-kit^{pos} eCSCs to give rise to cardiomyocytes in vivo has been challenged (van Berlo et al. 2014; Sultana et al. 2015). Instead, these papers suggest a largely vasculogenic and advential lineage predisposition of c-kit^{pos} cells, which isn't surprising considering 90% of c-kit^{pos} cells from the adult heart are CD31-positive (our unpublished data).

In the following sections we like to point out specific limitations of previous studies questioning the existence of eCSCs:

1. *c-kitpos cells vs. c-kitpos eCSCs*: c-kit is expressed in numerous cell types in the bone marrow (haematopoietic stem and progenitor cells and mast cells), endothelial (and circulating progenitor) cells, prostate stem cells and interstitial cells of Cajal. Elimination of these cells, and in particular of CD45^{pos}/c-kit^{pos}/tryptase^{pos} mast cells and CD34^{pos}/CD31^{pos}/c-kit^{pos} endothelial progenitors which are several-fold higher in number in the heart than the eCSCs, from analysis is essential (Ellison et al. 2011, 2013; Smith et al. 2014). The CD45^{neg}/CD31^{neg}/CD34^{neg}/tryptase^{neg}/c-kit^{pos} eCSCs make up a small population (~2–8%) of the total c-kit^{pos} cells (Smith et al. 2014). Therefore, when using genetic lineage tracing to target the c-kit locus at large, definitive conclusions cannot be drawn on the cardiomyogenic potential of the eCSCs per se. Finally, the presented data is in agreement with our observation that a very small percentage of the tagged c-kit^{pos} cells can generate cardiomyocytes and can therefore be considered.
2. *Level of c-kit expression*: Our preliminary, unpublished data show that c-kit in eCSCs is expressed at a significantly lower level than in the mast cells and endothelial progenitor cells. Whether c-kit/cre lineage tracing models are able to tag and enable effective cre recombination to occur over the time periods tested in the lower c-kit-expressing eCSC cohort has not yet been determined (Nadal-Ginard et al. 2014).
3. *Injury model*: The adult heart has a low cardiomyocyte renewal rate, and although this rate may increase somewhat after injury, the heart itself is unable to effect large-scale cardiac regeneration, as would be expected from it following a myocardial infarction. Since the discovery of eCSCs, investigators have used the small animal myocardial infarction model to claim the lack of significance and cardiomyogenic regenerative potential of eCSCs. We question whether this specific model in light of the naturally low abundance of eCSCs is suited to support this claim. No solid organ, even with a large stem cell reserve and renewal capability, can regenerate itself from ligation of its main artery resulting in large segmental loss of tissue (Ellison et al. 2012). Therefore, when using the myocardial infarction model there will be very little spontaneous regeneration of cardiomyocytes (<0.01%), whether coming from resident, endogenous stem cells (Smart et al. 2011; van Berlo et al. 2014; Sultana et al. 2015; Hsieh et al. 2007) or proliferation of the survived cardiomyocytes (Senyo et al. 2013).

We have developed in our view a more physiologically relevant cardiac damage model that is in the presence of a patent coronary circulation and more recapitulates muscle wear and tear. When a single excessive (200 mg/kg for mouse; 5 mg/kg for rat) dose injection (s.c.) of the synthetic catecholamine, isoproterenol (ISO), is administered, there is significant diffuse sub-endocardial and apical cardiomyocyte necrotic death, resulting in a dropout of ~10% cardiomyocytes at 24 h post ISO (Goldspink et al. 2004; Ellison et al. 2007b, 2013). This leads to the development of acute cardiac failure; however, the myocardial damage and heart failure spontaneously reverse anatomically and functionally by 28 days (Ellison et al. 2013). Using the acute ISO model, we showed that the adult heart has intrinsic regenerative capacity, where the eCSCs restore cardiac function by regenerating the lost cardiomyocytes. When ISO injury was followed by a 4-week regime of the anti-proliferative agent 5-FU for ablation of eCSC expansion and consequent differentiation, no cardiac regeneration and functional recovery was apparent with animals ending in overt heart failure. However, the regenerative process is completely restored by replacing the ablated eCSCs with the progeny of one eCSC. After regeneration, selective suicide of these exogenous CSCs and their progeny abolishes regeneration, severely impairing ventricular performance. Thus, eCSCs are necessary and sufficient for the regeneration and repair of myocardial damage (Ellison et al. 2013). Incidentally, the acute ISO model should not be confused with chronic administration of ISO over a minimum of 7 days leading to heart failure and cardiac remodelling with significant fibrotic scar formation and as used by van Berlo et al. (2014).

2.4 Origin of eCSCs

An intriguing question concerning eCSCs resident in the heart is whether they are directly descended from lineages which have been present since early development or have possibly 'migrated' to the heart later in life. c-kit^{pos}/Nkx2.5^{pos} eCSCs have been identified in early cardiogenic mesoderm (Wu et al. 2006) and in murine embryonic hearts at E6.5 (Ferreira-Martins et al. 2012), a period of development currently thought to be confined solely to first heart field progenitors during primitive heart tube formation.

A study of Nkx2.5-positive, multipotent cardiac stem/progenitor cells early in development found expression of c-kit in ~28% of these cells, which were also negative for CD45, demonstrating that c-kit expression marks a major subset of cardiac progenitors during development (Wu et al. 2006). Furthermore, Nkx2.5^{pos}, c-kit^{pos} cells were more proliferative and less differentiated than Nkx2.5^{pos}, c-kit^{neg} cells; this correlation was not found with Sca-1 expression levels in Nkx2.5^{pos} cells (Wu et al. 2006). However, it has not been determined if the adult c-kit^{pos} eCSCs are directly descended from these cells. Analysis of GFP-positive cells in the embryo of a c-kit-GFP transgenic mouse during cardiac development showed a c-kit-expressing population of progenitor cells that was resident in the heart and did not migrate from extra-cardiac tissue (although a contribution to the c-kit-positive population from extra-cardiac sources could not be excluded) and were present in the postnatal period (Ferreira-Martins et al. 2012). These cells were also shown to have comparable properties to c-kit^{pos} eCSCs in adult life in terms of proliferation, multipotency and myocardial regenerative capacity (Ferreira-Martins et al. 2012).

Recently, using high-resolution genetic fate-mapping approaches with c-kit$^{CreERT2/+}$ and Wnt1::Flpe mouse lines, Hare and colleagues have shown that c-kit identifies a population of multipotent progenitors of cardiac neural crest origin (Hatzistergos et al. 2015). Recent evidence reviewed by Keith and Bolli (Keith and Bolli 2015) support the concept that c-kit-expressing cells in the heart are not limited to originating from one progenitor cell; rather c-kit expression is a property of cells that originate from multiple pools of progenitors in the developing and postnatal heart (e.g. FHF, proepicardium). Moreover, c-kit expression by itself does not define the embryonic origin, lineage commitment capabilities or differentiation potential of the various groups of progenitors (Keith and Bolli 2015).

2.5 Impact of Ageing and Senescence on eCSCs

Ageing poses the largest risk factor for cardiovascular disease (North and Sinclair 2012). Although long-term exposure to known cardiovascular risk factors strongly drives the development of cardiovascular pathologies, intrinsic cardiac ageing is considered to highly influence the pathogenesis of heart disease (Dutta et al. 2012). However, the fields of the biology of ageing and cardiovascular disease have been studied separately, and only recently their intersection has begun to receive the appropriate attention.

Over the course of ageing, the heart undergoes a number of anatomical, functional and cellular alterations. Early diastolic left ventricular (LV) filling, LV contractility and ejection fraction all decrease during ageing leading to a reduced cardiac output (Schulman et al. 1992; Fleg et al. 1995; Lakatta and Levy 2003a). In an attempt to compensate for the reduction in cardiac output, the myocardium is triggered to increase its muscle mass by undergoing hypertrophy, which in the long-term results in weakened cardiac function. Ageing of the arterial system is exemplified by increased arterial thickening and stiffness, luminal enlargement and dysfunctional endothelium with decreased responsiveness to stress and injury (Lakatta and Levy 2003b). Arterial stiffness contributes to LV pathological hypertrophy and stimulates fibroblast proliferation causing myocardial and arterial fibrosis. Impaired heart rate is another characteristic of the ageing heart. Loss of sinoatrial node cells together with fibrosis and hypertrophy, slow electric impulse propagation throughout the heart causes decreased maximum heart rate and arrhythmias (Antelmi et al. 2004). Thus, age-imposed anomalies of the cardiovasculature led to the onset of a variety of age-related pathologies, including ischemia, hypertension, atherosclerosis, age-related macular degeneration and stroke (North and Sinclair 2012).

Mammalian ageing has been defined as a gradual loss of the capacity to maintain tissue homeostasis or to repair tissues after injury or stress (Jeyapalan and Sedivy 2008). It is now well known that tissue regeneration and homeostasis are controlled by the tissue-specific stem-progenitor cell compartment present in every tissue (Weissman 2000; Li and Clevers 2010). Therefore, it is logical to postulate that pathological and pathophysiological conditions associated with distorted homeostasis and regenerative capacity, such as ageing, correlate with impairments in the corresponding stem cell pool (Sharpless and DePinho 2007; Rossi et al. 2008; Beltrami et al. 2011). Indeed, there is an already well-established overlap between ageing and stem cell impairment, observed in a number of organs and tissues (Martin et al. 1998; Flores et al. 2005; Liang et al. 2005; Nishimura et al. 2005; Janzen et al. 2006; Krishnamurthy et al. 2004; Molofsky et al. 2006; Beerman et al. 2010). Tissue-specific stem cells decline with age due to several factors including telomere shortening, DNA damage and external influences affecting stem cell niche homeostasis (Sharpless and DePinho 2007). In recent years, accumulated evidence signified that cardiac ageing and pathology affects eCSC activity and potency, and therefore this diminishes the capacity of the myocardium to maintain homeostasis (Chimenti et al. 2003; Torella et al. 2004; Urbanek et al. 2005; Sharpless and DePinho 2007; Rossi et al. 2008; Thijssen et al. 2009; Kajstura et al. 2010; Cesselli et al. 2011). As the majority of cardiovascular disease patients are of advanced age, we should focus on the biology of aged CSCs to reflect the aetiology of cardiovascular disease observed in the clinic.

In the heart, ageing and disease are shown to be associated with a significant accumulation of senescent and dysfunctional cardiomyocytes and eCSCs displaying attenuated telomerase activity, telomeric erosion, high incidence of telomere-induced dysfunction foci and elevated expression of the cyclin-dependent kinase inhibitors (CDKIs) p16INK4a and p21Cip1 (Chimenti et al. 2003; Torella et al. 2004; Gonzalez et al. 2008; Kajstura et al. 2010; Cesselli et al. 2011; Rota et al. 2006; Urbanek et al. 2005) (Fig. 2.3). Nevertheless, a population of functional eCSCs, which express telomerase, lack expression of senescent markers and express the cycling protein, Ki67, have been shown to persist in aged hearts (Urbanek et al. 2003; Dawn et al. 2005). Indeed, in the setting of pathophysiological ageing, telomerase-competent eCSCs with normal telomerases can still be found in various cardiac regions, which have the capacity to migrate to injured zones and generate a healthy progeny partly reversing the senescent phenotype and improving cardiac performance (Gonzalez et al. 2008). Unpublished data from our lab has found that the number of eCSCs that can be isolated from human myocardial samples is similar regardless of age, gender and pathology (~45,000/g of tissue). While eCSCs isolated from human hearts showed age-correlated increased expression of ageing/senescence markers and decreased expression of stemness/multipotency and proliferation markers. Moreover, 'aged-senescent' eCSCs show limited cloning and growth capacity and impaired cardiac differentiation capacity. Importantly, although the cloning efficiency was inversely age-related, single-cell-derived eCSC clones obtained from younger and older human hearts are indistinguishable by their gene expression and differentiation potential. These data suggest that while the loss of functionally competent eCSCs may underlie the progressive functional deterioration documented with age, eCSC ageing itself may be a stochastic process that does not affect all eCSCs in a cell autonomous manner.

Senescent cells are characterised by impaired proliferation, an altered gene expression profile, resistance to apoptosis and epigenetic modifications, as well as

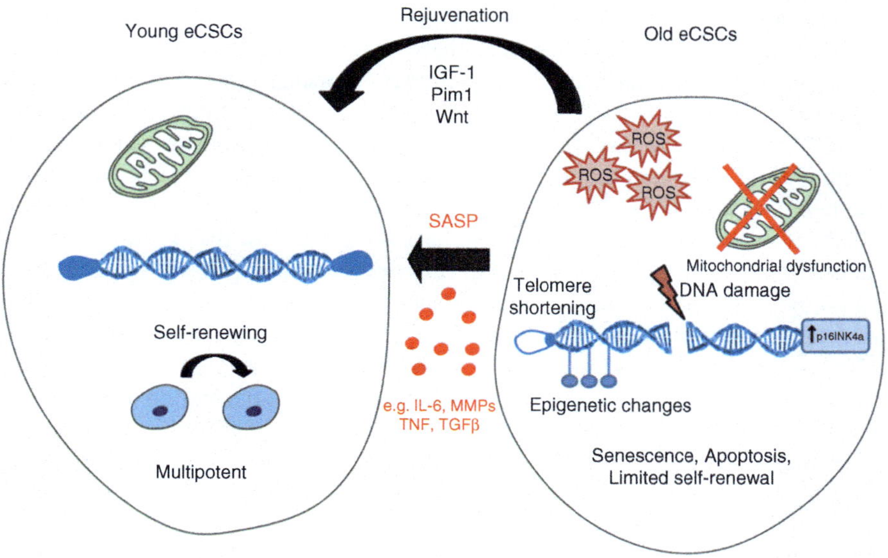

Fig. 2.3 Pathways contributing to eCSC dysfunction in the ageing process. eCSC ageing is regulated by a combination of intrinsic and extrinsic factors. Intrinsic changes include increased senescent marker expression, e.g. p16INK4a, DNA damage, telomere attrition, increased intracellular ROS, mitochondrial dysfunction and ageing-associated epigenetic changes, all of which are challenging to reverse in a clinically translatable manner. Extracellular changes include systemic circulating factors, local factors secreted by the niche and the SASP, which can negatively modulate cell function. These extrinsic pathways are potentially reversible and provide potential therapeutic targets to rejuvenate eCSCs and reverse the senescent, dysfunctional phenotype

producing an altered secretome, which acts on adjacent as well as distant cells, causing fibrosis, inflammation and a possible carcinogenic response (Kuilman and Peeper 2009; Kuilman et al. 2010; Rodier and Campisi 2011; Baker et al. 2011; Tchkonia et al. 2013). Although a universal marker exclusively expressed in senescent cells has not been identified, most senescent cells express p16INK4a, which is not commonly expressed by quiescent or terminally differentiated cells (Baker et al. 2011; Rodier and Campisi 2011). p16INK4a, which becomes progressively expressed with age, enforces cell-cycle arrest by activating retinoblastoma (RB) tumour-suppressor protein (Krishnamurthy et al. 2004; Kim and Sharpless 2006).

Interestingly, a recent study demonstrated silencing of p16INK4a in geriatric satellite cells restored their quiescence and regenerative potential (Sousa-Victor et al. 2014). Similarly, induction of p16INK4a has been shown to induce features of ageing and inhibit proliferation of intestinal stem cells; however, subsequent withdrawal of p16INK4a even after several weeks of induction is sufficient to allow rapid recovery of the affected cells (Boquoi et al. 2015). A recent study demonstrated that genetic reduction of p16INK4a reverses the pathology observed in dilated cardiomyopathy (Gonzalez-Valdes et al. 2015). Together these findings suggest that p16INK4a-expressing cells may exist in a pre-senescent state, which is potentially reversible.

Accumulation of p16INK4a-positive senescent cells within a tissue has been reported to exacerbate dysfunction as these impaired cells have an altered secretome consisting of matrix metalloproteinases, growth factors and inflammatory cytokines, known collectively as the senescent-associated secretory phenotype (SASP) (Coppé et al. 2010) (Fig. 2.3). The SASP can promote senescence of neighbouring cells, and this bystander effect has been shown to negatively affect the host tissue composition in a paracrine fashion (Acosta et al. 2008; Campisi 2005). Researchers at the Mayo clinic have shown that in the BubR1 progeroid mouse, removal of p16Ink4a senescent cells delayed the acquisition of age-related pathologies in adipose, skeletal muscle and eye, while late-life clearance attenuated progression of already established age-related disorders (Baker et al. 2011). Moreover, recently Kirkland and colleagues have also shown that the SASP can be supressed by targeting the JAK pathway and activin A, contributing to alleviating frailty (Xu et al. 2015a, b). Once the process of senescence is initiated in an organ of limited regenerative potential, such as the heart, this can lead to widespread cellular deterioration with the remaining unaffected cells unable to compensate for this cellular loss, ultimately leading to impaired cardiac function (Siddiqi and Sussman 2013). Therefore, therapeutic approaches inhibiting the SASP-mediated decline may improve eCSC, cardiomyocyte and vascular function and alleviate global cardiac deterioration.

2.6 Therapeutic Targets to Activate eCSCs and Reverse the Senescent, Dysfunctional eCSC Phenotype

To reverse the senescent eCSC phenotype, targeting extracellular signals appears to be a promising therapeutic avenue with early work showing that exposure of old skeletal muscle satellite cells to a youthful environment promotes restoration of their function (Conboy et al. 2005). Thus, manipulation of the cardiac microenvironment could alleviate eCSC dysfunction (Fig. 2.3). The IGF-1 signalling pathway has been implicated as a mediator of eCSC senescence, with increased IGF-1 signalling shown to attenuate ageing-associated markers (Torella et al. 2004). In a 22-month-old mice, c-kit^{pos} eCSCs show senescence, evidenced through impaired proliferation and differentiation potential, p16INK4a expression, reduced telomerase activity, telomere shortening, senescence and increased apoptosis (Torella et al. 2004). Senescent eCSCs become largely unable to generate new functionally competent myocytes, compromising cardiomyocyte turnover and favouring the accumulation of old poorly contracting cardiomyocytes (Torella et al. 2004). These findings show that cardiovascular ageing impairs eCSCs, leading to their decline and dysfunction, which leads to the development of cardiac dysfunction and failure. Interestingly, this progression is altered favourably in IGF-1 transgenic mice (Torella et al. 2004). Moreover, reduced phos-Akt expression associated with ageing is now thought to act as a main modulator of telomerase activity; thus, therapies aimed at counteracting this through stimulation of Akt have been shown to circumvent some of the effects of ageing (Torella et al. 2004; D'Amario et al. 2011). Another promising study focused on ex vivo modification with Pim-1, a serine/

threonine kinase to alleviate senescent characteristics. Mohsin et al. (2013) showed that Pim-1 rejuvenated the phenotypic and functional properties of eCSCs with restoration of youthful telomeric length, enhanced replicative capacity and decreased levels of p16Ink4a and p53 (Mohsin et al. 2013). More recently the cardioprotective effects of Pim-1 have been shown to be most effective when targeted to nuclear or mitochondrial compartments of eCSCs (Samse et al. 2015).

A number of molecular pathways involved in the reversal of eCSC senescence still remain unexplored; however, given the evidence available for other self-renewing tissues, potential future directions can be identified. One potential target is the Wnt signalling pathway, with a shift from canonical to non-canonical Wnt signalling reported in aged haematopoietic stem cells (HSCs) due to elevated expression of Wnt5a. Conversely, stem cell-intrinsic reduction of Wnt5a expression resulted in functionally rejuvenated aged HSCs (Florian et al. 2013). Mice which overexpressed the Wnt receptor, Frizzled, had reduced infarct size and improved cardiac function (Barandon et al. 2003), suggesting that this pathway may have a role to play in maintaining eCSC regenerative capacity. Bmi-1, necessary for self-renewal and regulator of p16Ink4a and p19, has also been shown to limit dilated cardiomyopathy by limiting heart senescence (Gonzalez-Valdes et al. 2015). The precise role played by reactive oxygen species, mitochondrial dysfunction and epigenetic changes associated with aged eCSCs also remains to be determined. Moreover, to date many of these pathways have only been studied either in cardiomyocytes in vitro or in rodent models; therefore, it is vital that we begin to uncover the mechanisms regulating senescence in human eCSCs and cardiomyocytes in order to move towards translation into clinical therapies.

In summary, the adult heart harbours a small population of cells, which exhibit all the necessary properties to be defined as bona fide stem and progenitor cells, being clonogenic, self-renewing and multipotent, in vitro and in vivo. In order to assess the role of the eCSCs in the adult myocardium, it is indispensable to first be able to identify and track the fate of these cells in contradistinction from other myocardial cells with which they share some specific marker(s), particularly expression of c-kit. Therefore, alternative markers should be sought and an in-depth sequencing analysis carried out. Finally, eCSCs are affected by ageing rendering a proportion of them senescent and dysfunctional. Regulation of eCSC ageing and senescence will impact the efficacy of regenerative therapies, considering the majority of patients in need of treatment are of advanced age. This should not be overlooked and should be considered at the forefront when designing and optimising protocols to repair and regenerate the injured and old myocardium.

Compliance with Ethical Standards

Conflict of Interest The authors declare that they have no conflict of interest.

Ethical Approval All experimental procedures were performed in accordance with the British Home Office Animals (Scientific Procedures) Act 1986 by appropriately qualified staff and approved by the institutional animal welfare and ethical review board.

References

Acosta JC, O'Loghlen A, Banito A et al (2008) Chemokine signaling via the CXCR2 receptor reinforces senescence. Cell 133(6):1006–1018

Angert D, Berretta RM, Kubo H et al (2011) Repair of the injured adult heart involves new myocytes potentially derived from resident cardiac stem cells. Circ Res 108(10):1226–1237

Antelmi I, de Paula RS, Shinzato AR et al (2004) Influence of age, gender, body mass index, and functional capacity on heart rate variability in a cohort of subjects without heart disease. Am J Cardiol 93(3):381–385

Arsalan M, Woitek F, Adams V et al (2012) Distribution of cardiac stem cells in the human heart. ISRN Cardiol 2012:483407

Asakura A, Rudnicki MA (2002) Side population cells from diverse adult tissues are capable of in vitro hematopoietic differentiation. Exp Hematol 30(11):1339–1345

Baker DJ, Wijshake T, Tchkonia T et al (2011) Clearance of p16Ink4a-positive senescent cells delays ageing-associated disorders. Nature 479:232–236

Barandon L, Couffinhal T, Ezan J et al (2003) Reduction of infarct size and prevention of cardiac rupture in transgenic mice overexpressing FrzA. Circulation 108(18):2282–2289

Bearzi C, Rota M, Hosoda T et al (2007) Human cardiac stem cells. PNAS 104(35):14068–14073

Beerman I, Maloney WJ, Weissmann IL et al (2010) Stem cells and the aging hematopoietic system. Curr Opin Immunol 22(4):500–506

Beltrami AP, Barlucchi L, Torella D et al (2003) Adult cardiac stem cells are multipotent and support myocardial regeneration. Cell 114:763–776

Beltrami AP, Cesselli D, Beltrami CA (2011) At the stem of youth and health. Pharmacol Ther 129(1):3–20

Bergmann O, Bhardwaj RD, Bernard S et al (2009) Evidence for cardiomyocyte renewal in humans. Science 324(5923):98–102

Bergmann O, Zdunek S, Felker A et al (2015) Dynamics of cell generation and turnover in the human heart. Cell 161(7):1566–1575

Bersell K, Arab S, Haring B et al (2009) Neuregulin1/ErbB4 signaling induces cardiomyocyte proliferation and repair of heart injury. Cell 138:257–270

Boquoi A, Arora S, Chen T et al (2015) Reversible cell cycle inhibition and premature aging features imposed by conditional expression of p16Ink4a. Aging Cell 14(1):139–147

Boström P, Mann N, Wu J et al (2010) C/EBPβ controls exercise-induced cardiac growth and protects against pathological cardiac remodeling. Cell 143:1072–1083

Campisi J (2005) Senescent cells, tumor suppression, and organismal aging: good citizens, bad neighbors. Cell 120:513–522

Cesselli D, Beltrami AP, D'Aurizio F et al (2011) Effects of age and heart failure on human cardiac stem cell function. Am J Pathol 179(1):349–366

Chien KR, Olson EN (2002) Converging pathways and principles in heart development and disease: CV@CSH. Cell 110:153–162

Chimenti C, Kajstura J, Torella D et al (2003) Senescence and death of primitive cells and myocytes lead to premature cardiac aging and heart failure. Circ Res 93(7):604–613

Chong JJ, Chandrakanthan V, Xaymardan M et al (2011) Adult cardiac-resident MSC-like stem cells with a proepicardial origin. Cell Stem Cell 9:527–540

Conboy IM, Conboy MJ, Wagers AJ et al (2005) Rejuvenation of aged progenitor cells by exposure to a young systemic environment. Nature 433:760–764

Coppé J, Desprez P, Krtolica A et al (2010) The Senescence-Associated Secretory Phenotype: the dark side of tumor suppression. Annu Rev Pathol 5:99–118

D'Amario D, Cabral-Da-Silva MC, Zheng H et al (2011) Insulin-like growth factor-1 receptor identifies a pool of human cardiac stem cells with superior therapeutic potential for myocardial regeneration. Circ Res 108(12):1467–1481

Dawn B, Stein AB, Urbanek K et al (2005) Cardiac stem cells delivered intravascularly traverse the vessel barrier, regenerate infarcted myocardium, and improve cardiac function. PNAS 102:3766–3771

Don CW, Murry CE (2013) Improving survival and efficacy of pluripotent stem cell-derived cardiac grafts. J Cell Mol Med 17(11):1355–1362
Dutta D, Calvani R, Bernabei R et al (2012) Contribution of impaired mitochondrial autophagy to cardiac aging: mechanisms and therapeutic opportunities. Circ Res 110(8):1125–1138
Ecob-Prince M, Hill M, Brown W (1989) Immunocytochemical demonstration of myosin heavy chain expression in human muscle. J Neurol Sci 91:71–78
Eisenberg CA, Burch JB, Eisenberg LM (2006) Bone marrow cells transdifferentiate to cardiomyocytes when introduced into the embryonic heart. Stem Cells 24:1236–1245
Ellison GM, Torella D, Karakikes I et al (2007a) Acute beta-adrenergic overload produces myocyte damage through calcium leakage from the ryanodine receptor 2 but spares cardiac stem cells. J Biol Chem 282:11397–11409
Ellison GM, Torella D, Karakikes I et al (2007b) Myocyte death and renewal: modern concepts of cardiac cellular homeostasis. Nat Clin Pract Cardiovasc Med 4(Suppl 1):S52–S59
Ellison GM, Galuppo V, Vicinanza C et al (2010) Cardiac stem and progenitor cell identification: different markers for the same cell? Front Biosci 2:641–652
Ellison GM, Torella D, Dellegrottaglie S et al (2011) Endogenous cardiac stem cell activation by insulin-like growth factor-1/hepatocyte growth factor intracoronary injection fosters survival and regeneration of the infarcted pig heart. J Am Coll Cardiol 58(9):977–986
Ellison GM, Nadal-Ginard B, Torella D (2012) Optimizing cardiac repair and regeneration through activation of the endogenous cardiac stem cell compartment. J Cardiovasc Transl Res 5(5):667–677
Ellison GM, Vicinanza C, Smith AJ et al (2013) Adult c-kit(pos) cardiac stem cells are necessary and sufficient for functional cardiac regeneration and repair. Cell 154(4):827–842
Fang S, Wei J, Pentinmikko N et al (2012) Generation of functional blood vessels from a single c-kit+ adult vascular endothelial stem cell. PLoS Biol 10(10):e1001407
Fazel S, Cimini M, Chen L et al (2006) Cardioprotective c-kit+ cells are from the bone marrow and regulate the myocardial balance of angiogenic cytokines. J Clin Invest 116(7):1865–1877
Ferreira-Martins J, Ogórek B, Cappetta D et al (2012) Cardiomyogenesis in the developing heart is regulated by c-kit-positive cardiac stem cells. Circ Res 110(5):701–715
Fischer KM, Cottage CT, Wu W et al (2009) Enhancement of myocardial regeneration through genetic engineering of cardiac progenitor cells expressing Pim-1 kinase. Circulation 120(21):2077–2087
Fleg JL, O'Connor F, Gerstenblith G et al (1995) Impact of age on the cardiovascular response to dynamic upright exercise in healthy men and women. J Appl Physiol 78(3):890–900
Flores I, Cayuela ML, Blasco MA (2005) Effects of telomerase and telomere length on epidermal stem cell behavior. Science 309(5738):1253–1256
Florian MC, Nattamai KJ, Dörr K (2013) A canonical to non-canonical Wnt signalling switch in haematopoietic stem-cell ageing. Nature 503:392–396
Fransioli J, Bailey B, Gude NA et al (2008) Evolution of the c-kit-positive cell response to pathological challenge in the myocardium. Stem Cells 26(5):1315–1324
Goldspink DF, Burniston JG, Ellison GM et al (2004) Catecholamine-induced apoptosis and necrosis in cardiac and skeletal myocytes of the rat in vivo: the same or separate death pathways? Exp Physiol 89(4):407–416
Gonzalez A, Rota M, Nurzynska D et al (2008) Activation of cardiac progenitor cells reverses the failing heart senescent phenotype and prolongs lifespan. Circ Res 102(5):597–606
Gonzalez-Valdes I, Hidalgo I, Bujarrabal A et al (2015) Bmi1 limits dilated cardiomyopathy and heart failure by inhibiting cardiac senescence. Nat Commun 6:6473
Goodell MA, Brose K, Paradis G et al (1996) Isolation and functional properties of murine hematopoietic stem cells that are replicating in vivo. J Exp Med 183(4):1797–1806
Hatzistergos KE, Takeuchi LM, Saur D et al (2015) cKit+ cardiac progenitors of neural crest origin. PNAS 112(42):13051–13056
Hierlihy AM, Seale P, Lobe CG et al (2002) The post-natal heart contains a myocardial stem cell population. FEBS Lett 530(1–3):239–243

Hsieh PC, Segers VF, Davis ME et al (2007) Evidence from a genetic fate-mapping study that stem cells refresh adult mammalian cardiomyocytes after injury. Nat Med 13(8):970–974
Iwakura T, Mohri T, Hamatani T et al (2011) STAT3/Pim-1 signaling pathway plays a crucial role in endothelial differentiation of cardiac resident Sca-1+ cells both in vitro and in vivo. J Mol Cell Cardiol 51(2):207–214
Janzen V, Forkert R, Fleming HE et al (2006) Stem-cell ageing modified by the cyclin-dependent kinase inhibitor p16INK4a. Nature 443(7110):421–426
Jesty SA, Steffey MA, Lee FK et al (2012) c-kit+ precursors support postinfarction myogenesis in the neonatal, but not adult, heart. PNAS 109(33):13380–13385
Jeyapalan JC, Sedivy JM (2008) Cellular senescence and organismal aging. Mech Ageing Dev 129(7–8):467–474
Kajstura J, Leri A, Finato N et al (1998) Myocyte proliferation in end-stage cardiac failure in humans. PNAS 95(15):8801–8805
Kajstura J, Gurusamy N, Ogórek B et al (2010) Myocyte turnover in the aging human heart. Circ Res 107:1374–1386
Kattman SJ, Huber TL, Keller GM (2006) Multipotent flk-1+ cardiovascular progenitor cells give rise to the cardiomyocyte, endothelial, and vascular smooth muscle lineages. Dev Cell 11:723–732
Keith MC, Bolli R (2015) "String theory" of c-kit(pos) cardiac cells: a new paradigm regarding the nature of these cells that may reconcile apparently discrepant results. Circ Res 116(7):1216–1230
Kim WY, Sharpless NE (2006) The regulation of INK4/ARF in cancer and aging. Cell 127(2):265–275
Krishnamurthy J, Torrice C, Ramsey MR et al (2004) Ink4a/Arf expression is a biomarker of aging. J Clin Invest 114:1299–1307
Kühn B, del Monte F, Hajjar RJ et al (2007) Periostin induces proliferation of differentiated cardiomyocytes and promotes cardiac repair. Nat Med 13:962–969
Kuilman T, Peeper DS (2009) Senescence-messaging secretome: SMS-ing cellular stress. Nat Rev Cancer 9(2):81–94
Kuilman T, Michaloglou C, Mooi WJ et al (2010) The essence of senescence. Genes Dev 24(22):2463–2479
Laflamme MA, Murry CE (2011) Heart regeneration. Nature 473:326–335
Lakatta EG, Levy D (2003a) Arterial and cardiac aging: major shareholders in cardiovascular disease enterprises: Part I: aging arteries: a "set up" for vascular disease. Circulation 107(1):139–146
Lakatta EG, Levy D (2003b) Arterial and cardiac aging: major shareholders in cardiovascular disease enterprises, part II: the aging heart in health: links to heart disease. Circulation 107:346–354
Laugwitz KL, Moretti A, Lam J et al (2005) Postnatal isl1+ cardioblasts enter fully differentiated cardiomyocyte lineages. Nature 433:647–653
Lévesque JP, Hendy J, Winkler IG et al (2003) Granulocyte colony-stimulating factor induces the release in the bone marrow of proteases that cleave c-KIT receptor (CD117) from the surface of hematopoietic progenitor cells. Exp Hematol 31(2):109–117
Li L, Clevers H (2010) Coexistence of quiescent and active adult stem cells in mammals. Science 327(5965):542–545
Li TS, Cheng K, Malliaras K et al (2011) Expansion of human cardiac stem cells in physiological oxygen improves cell production efficiency and potency for myocardial repair. Cardiovasc Res 89(1):157–165
Liang Y, Van Zant G, Szilvassy SJ (2005) Effects of aging on the homing and engraftment of murine hematopoietic stem and progenitor cells. Blood 106(4):1479–1487
Linke A, Müller P, Nurzynska D et al (2005) Stem cells in the dog heart are self-renewing, clonogenic, and multipotent and regenerate infarcted myocardium, improving cardiac function. PNAS 102(25):8966–8971
Loffredo FS, Steinhauser ML, Gannon J et al (2011) Bone marrow-derived cell therapy stimulates endogenous cardiomyocyte progenitors and promotes cardiac repair. Cell Stem Cell 8:389–398

Martin K, Kirkwood TB, Potten CS (1998) Age changes in stem cells of murine small intestinal crypts. Exp Cell Res 241(2):316–323
Martin CM, Meeson AP, Robertson SM et al (2004) Persistent expression of the ATP-binding cassette transporter, Abcg2, identifies cardiac SP cells in the developing and adult heart. Dev Biol 265:262–275
Matsuura K, Nagai T, Nishigaki N et al (2004) Adult cardiac Sca-1-positive cells differentiate into beating cardiomyocytes. J Biol Chem 279:11384–11391
Messina E, De Angelis L, Frati G et al (2004) Isolation and expansion of adult cardiac stem cells from human and murine heart. Circ Res 95:911–921
Miyamoto S, Kawaguchi N, Ellison GM et al (2010) Characterization of long-term cultured c-kit+ cardiac stem cells derived from adult rat hearts. Stem Cells Dev 19(1):105–116
Mohsin S, Khan M, Toko H et al (2012) Human cardiac progenitor cells engineered with Pim-I kinase enhance myocardial repair. J Am Coll Cardiol 60(14):1278–1287
Mohsin S, Khan M, Nguyen J et al (2013) Rejuvenation of human cardiac progenitor cells with pim-1 kinase. Circ Res 113(10):1169–1179
Molofsky AV, Slutsky SG, Joseph NM et al (2006) Increasing p16INK4a expression decreases forebrain progenitors and neurogenesis during ageing. Nature 443:448–452
Moretti A, Caron L, Nakano A et al (2006) Multipotent embryonic isl1+ progenitor cells lead to cardiac, smooth muscle, and endothelial cell diversification. Cell 127:1151–1165
Nadal-Ginard B (1978) Commitment, fusion and biochemical differentiation of a myogenic cell line in the absence of DNA synthesis. Cell 15:855–864
Nadal-Ginard B, Kajstura J, Leri A et al (2003) Myocyte death, growth, and regeneration in cardiac hypertrophy and failure. Circ Res 92:139–150
Nadal-Ginard B, Ellison GM, Torella D (2014) Absence of evidence is not evidence of absence: pitfalls of cre knock-ins in the c-Kit locus. Circ Res 115(4):415–418
Nishimura EK, Granter SR, Fisher DE (2005) Mechanisms of hair graying: incomplete melanocyte stem cell maintenance in the niche. Science 307(5710):720–724
North BJ, Sinclair DA (2012) The intersection between aging and cardiovascular disease. Circ Res 110(8):1097–1108
Noseda M, Harada M, McSweeney S et al (2015) PDGFRα demarcates the cardiogenic clonogenic Sca1+ stem/progenitor cell in adult murine myocardium. Nat Commun 6:6930
Oh H, Bradfute SB, Gallardo TD et al (2003) Cardiac progenitor cells from adult myocardium: homing, differentiation, and fusion after infarction. PNAS 100:12313–12318
Overy HR, Priest RE (1966) Mitotic cell division in postnatal cardiac growth. Lab Investig 15(6):1100–1103
Oyama T, Nagai T, Wada H et al (2007) Cardiac side population cells have a potential to migrate and differentiate into cardiomyocytes in vitro and in vivo. J Cell Biol 176(3):329–341
Passier R, van Laake LW, Mummery CL (2008) Stem-cell-based therapy and lessons from the heart. Nature 453(7193):322–329
Passman JN, Dong XR, Wu SP et al (2008) A sonic hedgehog signaling domain in the arterial adventitia supports resident Sca1+ smooth muscle progenitor cells. PNAS 105(27):9349–9354
Pfister O, Mouquet F, Jain M et al (2005) CD31- but Not CD31+ cardiac side population cells exhibit functional cardiomyogenic differentiation. Circ Res 97(1):52–61
Potten CS, Loeffler M (1990) Stem cells: attributes, cycles, spirals, pitfalls and uncertainties. Lessons for and from the crypt. Development 110(4):1001–1020
Pouly J, Bruneval P, Mandet C et al (2008) Cardiac stem cells in the real world. J Thorac Cardiovasc Surg 135(3):673–678
Quaini F, Urbanek K, Beltrami AP et al (2002) Chimerism of the transplanted heart. N Engl J Med 346:5–15
Rasmussen TL, Raveendran G, Zhang J et al (2011) Getting to the heart of myocardial stem cells and cell therapy. Circulation 123:1771–1779
Rodier F, Campisi J (2011) Four faces of cellular senescence. J Cell Biol 192(4):547–556
Roskoski R Jr (2005) Signaling by Kit protein-tyrosine kinase--the stem cell factor receptor. Biochem Biophys Res Commun 337(1):1–13

Rossi DJ, Jamieson CH, Weissman IL (2008) Stems cells and the pathways to aging and cancer. Cell 132(4):681–696

Rota M, LeCapitaine N, Hosoda T et al (2006) Diabetes promotes cardiac stem cell aging and heart failure, which are prevented by deletion of the p66shc gene. Circ Res 99(1):42–52

Samse K, Emathinger J, Hariharan N (2015) Functional effect of Pim1 depends upon intracellular localization in human cardiac progenitor cells. J Biol Chem 290(22):13935–13947

Schulman SP, Lakatta EG, Fleg JL (1992) Age-related decline in left ventricular filling at rest and exercise. Am J Phys 263(6 Pt 2):H1932–H1938

Senyo SE, Steinhauser ML, Pizzimenti CL et al (2013) Mammalian heart renewal by pre-existing cardiomyocytes. Nature 493(7432):433–436

Sharpless NE, DePinho RA (2007) How stem cells age and why this makes us grow old. Nat Rev Mol Cell Biol 8(9):703–713

Siddiqi S, Sussman MA (2013) Cardiac Hegemony of Senescence. Curr Transl Geriatr Exp Gerontol Rep 2(4):247–254

Smart N, Bollini S, Dubé KN et al (2011) De novo cardiomyocytes from within the activated adult heart after injury. Nature 474:640–644

Smith AJ, Lewis FC, Aquila I (2014) Isolation and characterization of resident endogenous c-Kit+ cardiac stem cells from the adult mouse and rat heart. Nat Protoc 9(7):1662–1681

Sousa-Victor P, Gutarra S, García-Prat L et al (2014) Geriatric muscle stem cells switch reversible quiescence into senescence. Nature 506:316–321

Sultana N, Zhang L, Yan J et al (2015) Resident c-kit(+) cells in the heart are not cardiac stem cells. Nat Commun 6:8701

Takamiya M, Haider KH, Ashraf M (2011) Identification and characterization of a novel multipotent sub-population of Sca-1^{+} cardiac progenitor cells for myocardial regeneration. PLoS One 6(9):e25265

Tallini YN, Greene KS, Craven M et al (2009) c-kit expression identifies cardiovascular precursors in the neonatal heart. PNAS 106(6):1808–1813

Tchkonia T, Zhu Y, van Deursen J et al (2013) Cellular senescence and the senescent secretory phenotype: therapeutic opportunities. J Clin Invest 123(3):966–972

Thijssen DH, Bullens LM, van Bemmel MM et al (2009) Does arterial shear explain the magnitude of flow-mediated dilation?: a comparison between young and older humans. Am J Physiol Heart Circ Physiol 296(1):H57–H64

Torella D, Rota M, Nurzynska D et al (2004) Cardiac stem cell and myocyte aging, heart failure, and insulin-like growth factor-1 overexpression. Circ Res 94:514–524

Torella D, Ellison GM, Méndez-Ferrer S et al (2006) Resident human cardiac stem cells: role in cardiac cellular homeostasis and potential for myocardial regeneration. Nat Clin Pract Cardiovasc Med 3(Suppl 1):S8–13

Torella D, Ellison GM, Karakikes I et al (2007) Resident cardiac stem cells. Cell Mol Life Sci 64:661–673

Urbanek K, Quaini F, Tasca G et al (2003) Intense myocyte formation from cardiac stem cells in human cardiac hypertrophy. PNAS 100:10440–10445

Urbanek K, Torella D, Sheikh F et al (2005) Myocardial regeneration by activation of multipotent cardiac stem cells in ischemic heart failure. PNAS 102(24):8692–8697

van Berlo JH, Molkentin JD (2014) An emerging consensus on cardiac regeneration. Nat Med 20(12):1386–1393

van Berlo JH, Kanisicak O, Maillet M et al (2014) c-kit+ cells minimally contribute cardiomyocytes to the heart. Nature 509(7500):337–341

Wang X, Hu Q, Nakamura Y et al (2006) The role of the sca-1+/CD31–cardiac progenitor cell population in postinfarction left ventricular remodeling. Stem Cells 24(7):1779–1788

Waring CD, Vicinanza C, Papalamprou A et al (2014) The adult heart responds to increased workload with physiologic hypertrophy, cardiac stem cell activation, and new myocyte formation. Eur Heart J 35:2722–2731

Weissman IL (2000) Stem cells: units of development, units of regeneration, and units in evolution. Cell 100(1):157–168

Williams P, Simpson H, Kenwright J, Goldspink G (2001) Muscle fibre damage and regeneration resulting from surgical limb distraction. Cells Tissues Organs 169:395–400
Wu SM, Fujiwara Y, Cibulsky SM et al (2006) Developmental origin of a bipotential myocardial and smooth muscle cell precursor in the mammalian heart. Cell 127:1137–1150
Xu M, Tchkonia T, Ding H, Ogrodnik M, Lubbers ER, Pirtskhalava T, White TA, Johnson KO, Stout MB, Mezera V, Giorgadze N, Jensen MD, LeBrasseur NK, Kirkland JL (2015a) JAK inhibition alleviates the cellular senescence-associated secretory phenotype and frailty in old age. Proc Natl Acad Sci U S A 112:E6301–E6310
Xu M, Palmer AK, Ding H, Weivoda MM, Pirtskhalava T, White TA, Sepe A, Johnson KO, Stout MB, Giorgadze N, Jensen MD, LeBrasseur NK, Tchkonia T, Kirkland JL (2015b) Targeting senescent cells enhances adipogenesis and metabolic function in old age. elife 4:pii: e12997
Yang MJ, Chen CH, Lin PJ et al (2007) Novel method of forming human embryoid bodies in a polystyrene dish surface-coated with a temperature-responsive methylcellulose hydrogel. Biomacromolecules 8(9):2746–2752
Ye J, Boyle A, Shih H et al (2012) Sca-1+ cardiosphere-derived cells are enriched for Isl1-expressing cardiac precursors and improve cardiac function after myocardial injury. PLoS One 7(1):e30329
Zaruba MM, Soonpaa M, Reuter S et al (2010) Cardiomyogenic potential of C-kit(+)-expressing cells derived from neonatal and adult mouse hearts. Circulation 121(18):1992–2000
Zhou S, Schuetz JD, Bunting KD et al (2001) The ABC transporter Bcrp1/ABCG2 is expressed in a wide variety of stem cells and is a molecular determinant of the side-population phenotype. Nat Med 7(9):1028–1034

Epicardial Progenitors in the Embryonic and Adult Heart

3

Cristina Villa del Campo, Joaquim Miguel Vieira, and Paul R. Riley

Abstract

Over the last decade, our knowledge of the function(s) of the epicardium in vertebrate heart development and repair has increased considerably. In development, the epicardium is required for proper heart formation by regulating myocardial compaction and contributing major cardiovascular cell types. In the adult heart, the idea of a dormant epicardium has been challenged by the observation that it can reacquire embryonic properties after heart injury and may contribute to tissue repair and regeneration. This has elevated the status of the adult epicardium to a resident source of regenerative cells with potential to restore cardiac structure and function after injury. Yet, many questions remain to be answered, in particular whether the observations arising from studies on model organisms are applicable to the human (diseased) heart. Here, we review the key established and emerging findings regarding epicardium formation, heterogeneity, and its therapeutic potential in heart repair. Moreover, we draw attention to studies focusing on the human epicardium, highlighting new tools that are being developed to promote further insight into the epicardium and its regenerative potential.

3.1 Epicardial Development Across Species

Different animal models including zebrafish, chick, and mouse have been used to study the development of the epicardium. Historically, the chicken embryo was the most widely used model to study the proepicardium and epicardium development.

C.V. del Campo • J.M. Vieira • P.R. Riley (✉)
Burdon Sanderson Cardiac Science Centre, Department of Physiology, Anatomy and Genetics, University of Oxford, Oxford OX1 3PT, UK
e-mail: paul.riley@dpag.ox.ac.uk

M. Ieda, W.-H. Zimmermann (eds.), *Cardiac Regeneration*, Cardiac and Vascular Biology 4, DOI 10.1007/978-3-319-56106-6_3

More recently, these findings have been validated in teleost and mammalian models, and much of the current knowledge regarding the formation of the epicardium in the human heart has been extrapolated from findings in the latter. The formation and role of the epicardium in general seems to be a conserved process in vertebrate heart development, but some species-specific features have been identified. The epicardium arises from the transient proepicardium (PE), which forms around embryonic day (E) 8.5 in the mouse (Komiyama et al. 1987; Schulte et al. 2007), Hamburger-Hamilton stage (HH) 14 in chick embryos, and 48 h postfertilization (hpf) in zebrafish (Serluca 2008). The proepicardium appears as a bilateral bulge forming close to the sinus venosus and fusing at the midline in both zebrafish (Serluca 2008) and mouse (Schulte et al. 2007), whereas in chick, the left region regresses (Schulte et al. 2007). Whether this difference has functional implications is not known, but in the chick embryo both proepicardium and epicardium formation seem to differ to that of teleosts and mammals, suggesting an evolutionary divergence in avians. The mechanism by which proepicardial cells migrate to cover the surface of the heart has also been reported to be different between species. In chick, cells are transferred to the surface of the heart via the formation of a cellular bridge, as was shown in early studies using electron microscopy (Manner 1992) (Fig. 3.1). In mice and fish, based on histological assays, this transfer was initially thought to happen via proepicardial vesicles floating in the pericardial cavity (Serluca 2008; Komiyama et al. 1987). Subsequently, it was documented, using combined optical and electron

Fig. 3.1 Epicardial development across species. (**a**) Schematic of heart development in the zebrafish at forty-eight hours postfertilization (hpf). Ventral to the yolk (Y) is the pericardial cavity containing the heart (H) consisting of a single atrium (A) and ventricle (V). The pericardial cavity is enlarged in the boxed image. The proepicardium releases vesicles that attach to the myocardial surface. The proepicardium arises from different regions in the dorsal pericardial wall: a cluster next to the atrioventricular canal (avcPE), another one close to the venous pole (vpEP), and a third source generated close to the arterial pole. The attachment of vesicles to the myocardial surface requires fluid forces generated by the heartbeat. (**b**) Schematic of heart development in the chick embryo. The heart (H) formed by the developing atria (A), ventricles (left ventricle, LV; right ventricle, RV), and outflow tract (OFT) is enlarged in the boxed image. At Hamburger-Hamilton stage (HH) 14, the proepicardium (PE) appears as a structure that forms close to the sinus venosus myocardium. The proepicardium is transferred to the heart by formation of a cellular bridge that adheres to the surface (PEb). (**c**) Schematic of murine heart development. The pericardial cavity is enlarged in the boxed image. Around embryonic day (E) 8.5, the proepicardium (PE), located close to the sinus venosus, migrates by emitting "fingerlike" protrusions through differential growth (PEb) and contacting the surface of the heart (H). It is also possible that the proepicardium migrates by shedding of vesicles and that both mechanisms play a role in epicardial formation in mammals. *Y* yolk, *H* heart, *vpPE* venous pole proepicardium, *avcPE* atrioventricular canal proepicardium, *apEP* arterial pole epicardium, *OFT* outflow tract, *PE* proepicardium, *PEb* proepicardium bridge, *A* atria, *BA* branchial arch, *LV* left ventricle, *RV* right ventricle

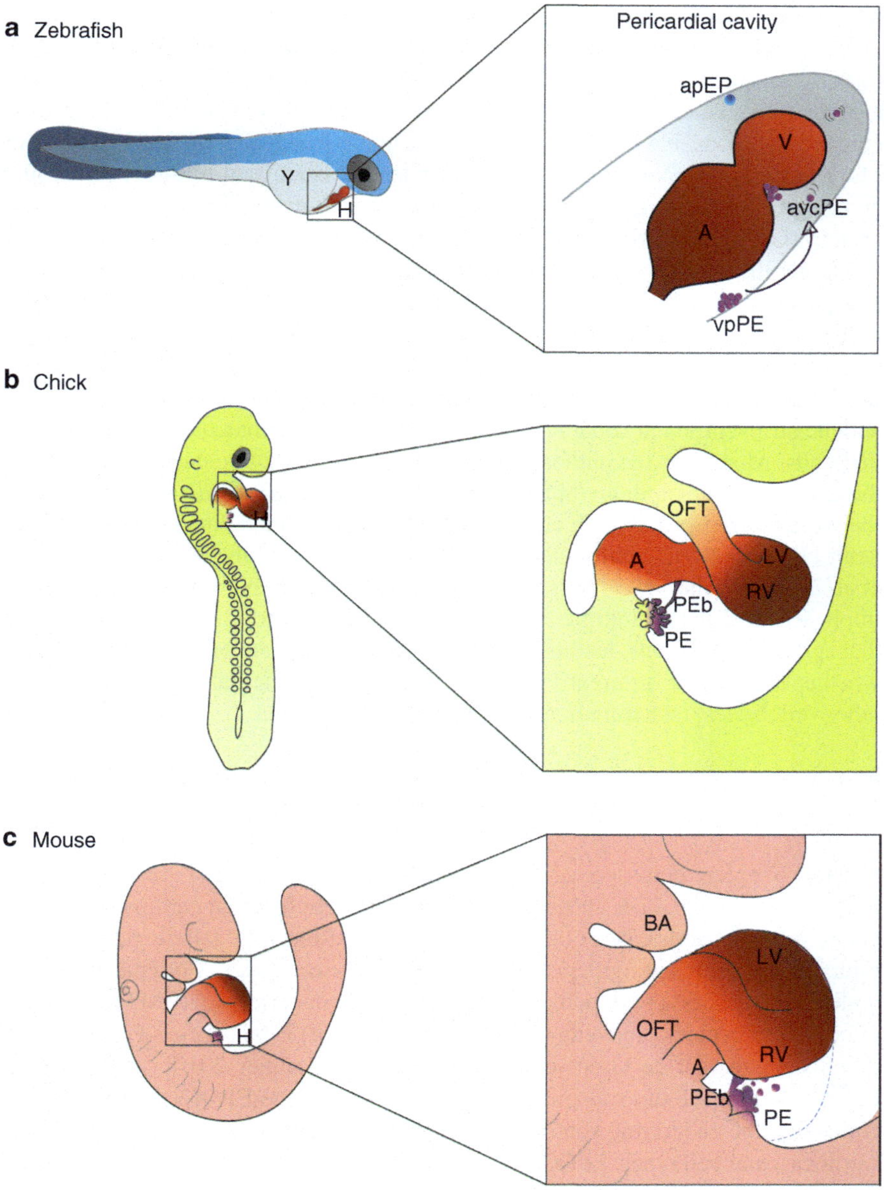
a Zebrafish
Pericardial cavity
apEP
V
Y
H
A
avcPE
vpPE
b Chick
OFT
A
LV
RV
H
PEb
PE
c Mouse
BA
LV
OFT
RV
H
A
PEb
PE

microscopy, that in the mouse embryo, proepicardial protrusions reach the surface of the developing heart by differential growth and then flatten out, expanding to form the epicardial layer (Rodgers et al. 2008). The authors of the latter study, however, did not rule out a contribution from free-floating proepicardial vesicles to the epicardium, alongside these fingerlike protrusions (Rodgers et al. 2008) (Fig. 3.1). More recently, in vivo fluorescent microscopy has been used to directly visualize epicardial formation in a zebrafish transgenic model expressing green fluorescent protein (GFP) driven by a regulatory element of *Wt1b* (*Wilms' tumor 1b*), which labels the PE (Peralta et al. 2013). This elegant approach confirmed the release of proepicardial vesicles into the pericardial cavity, through a process dependent on fluid forces generated by the heart beating (Fig. 3.1). Of significance, the same study also revealed that the PE appeared at two distinct sites in the zebrafish embryo, the venous pole of the heart (vpPE), which contributed 10% of the epicardial cells, and the atrioventricular canal region (avcPE), which contributed 80% of the epicardial cells. Moreover, the authors reported a third source of epicardial progenitors, the arterial pole epicardium (apEP) that derives from the pericardial wall adjacent to the ventricle, rather than a PE cluster, and is transferred upon direct contact with the ventricle. These cells contributed to the remaining 10% of the epicardium (Peralta et al. 2013). Further research is required to gain insight into the possible functional implications of these multiple PE origins and whether it results in different epicardial cell subpopulations, leading to heterogeneity in cell fate or function. Likewise, whether multiple or different PE-derived sources of epicardial cells are present in other vertebrates, including humans, is yet to be elucidated.

3.2 EPDC Derivatives and Epicardial Heterogeneity

The fate of epicardium-derived cells (EPDCs) has been a topic of great interest and debate in the field. EPDCs have been shown to be a multipotent cardiovascular precursor not only during heart development but also following pathological stress or injury in the adult setting. Upon epicardium formation, a subset of epicardial cells undergoes a process of epithelial-to-mesenchymal transformation (EMT) at E12.5 in mice, HH20 in chick, and 4 days postfertilization (dpf) in fish (Perez-Pomares et al. 1997; Gittenberger-de Groot et al. 1998; Liu and Stainier 2010), giving rise to EPDCs that invade the subepicardial compartment. Some EPDCs migrate further, colonizing the underlying ventricular layer and differentiating in situ into the major cardiovascular cell types. EPDCs have been shown to give rise to the smooth muscle and endothelium of the coronaries, fibroblasts, myocardium, and, more recently, adipose tissue. However, the extent of each of these contributions is still not clear, and the molecular signature of EPDCs is ill-defined. Whether all epicardial cells share a multipotent progenitor phenotype or whether they are a heterogeneous population of already pre-committed cells is yet to be answered.

The first studies investigating the contribution of EPDCs to the developing heart were performed in chick embryos using a diverse array of techniques such as cell labeling with virus injection, cellular transplants, and tissue engraftment and explants. These studies revealed EPDCs giving rise to coronary smooth muscle cells (Mikawa and Gourdie 1996; Perez-Pomares et al. 1997; Dettman et al. 1998; Manner 1999) and perivascular fibroblasts (Dettman et al. 1998; Manner 1999), mesenchymal cells of the valve cushions (Gittenberger-de Groot et al. 1998), interstitial fibroblasts, and a subset of endocardial cells (Dettman et al. 1998). In mice, most fate-mapping studies of EPDCs within the developing heart have been performed taking advantage of the Cre-Lox system (Sauer and Henderson 1988) to perform genetic lineage tracing. Cre-Lox allows tracing the progeny of a given cell and is based on the expression of the Cre recombinase under the control of a tissue-specific promoter/enhancer region of a gene expressed in the lineage of interest. The first epicardial marker used to study EPDC fate in mouse was the transcription factor Gata5 (Merki et al. 2005), followed by the transcription factors Tbx18 (Cai et al. 2008) and Wt1 (Wilm et al. 2005; Zhou et al. 2008; Wessels et al. 2012). Through the use of the aforementioned Cre lines, a contribution of EPDCs to coronary smooth muscle, interstitial and vascular fibroblasts, and valve mesenchyme has been established (Perez-Pomares et al. 2002a; Zhou et al. 2010), in agreement with fate-mapping studies previously performed in chick (Fig. 3.2). These findings were corroborated by the use of another epicardial transcription factor, Tcf21. Indeed, lineage tracing using *Tcf21*-mediated Cre expression showed a contribution to coronary smooth muscle cells and interstitial and perivascular fibroblasts (Acharya et al. 2012). Of significance, fate mapping in zebrafish using *Tcf21*-based lineage tracing supports these findings by reporting EPDC contribution as restricted to fibroblasts and smooth muscle cells (Kikuchi et al. 2011).

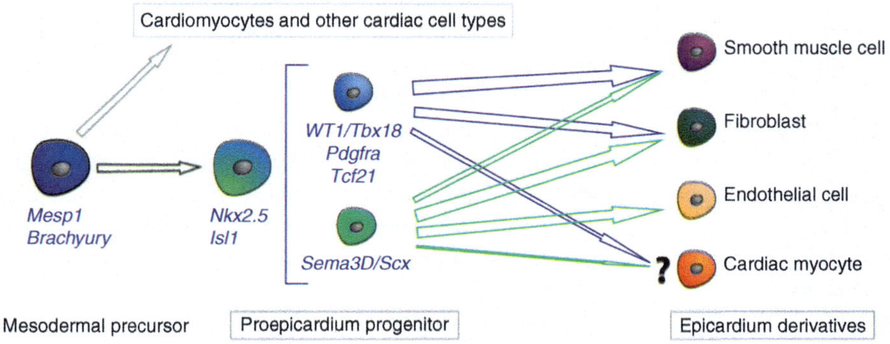

Fig. 3.2 (Pro)epicardial heterogeneity and EPDC potential. Proepicardial precursors arise from the lateral plate mesoderm and share a common origin with cardiomyocytes of the first heart field and other cell types. In the proepicardium, distinct proepicardial compartments characterized by heterogeneous marker expression give rise to coronary endothelial cells, vascular smooth muscle cells, fibroblasts, and, potentially, cardiomyocytes. Arrows indicate potential contribution; the thickness of arrows represents the relative extent of epicardial cell contribution

3.2.1 Epicardium Contribution to Coronary Endothelial Cells

Besides the aforementioned derivatives, EPDCs have been suggested to contribute to the endothelial lineage in the chick embryo based on retroviral labeling, dye labeling, and quail-chick chimeras (Dettman et al. 1998; Mikawa and Fischman 1992; Perez-Pomares et al. 1998, 2002a, b). However, this has since been disputed on the basis that endothelial cells in the transplanted grafts were liver-derived (Poelmann et al. 1993). In mice, an epicardial origin for endothelial cells has also been reported. Contribution to the coronary endothelium was detected using *Wt1* to drive Cre expression (Zhou et al. 2008); but subsequently, endogenous activity of *Wt1* was reported in endothelial cells during development (Rudat and Kispert 2012; Duim et al. 2015) and in the adult heart after injury in rats (Wagner et al. 2002), confounding data interpretation based on the use of the *Wt1*-Cre driver. An epicardial contribution to the coronary endothelium has been further suggested in a study that also highlighted the heterogeneity of the proepicardium (Katz et al. 2012) (also, Fig. 3.2). Katz and colleagues described distinct cell populations within the proepicardium that contribute differently to cardiovascular cell lineages. Using two different Cre lines, driven by the transcription factor Scx (scleraxis) and the secreted guidance factor Sema3D (semaphorin 3D), the authors reported a distinctive proepicardial population, which only partially overlapped with the Wt1-/Tbx18-expressing cells and contributed to the (coronary) endothelial lineage (Katz et al. 2012). Both Scx and Sema3D are expressed in the proepicardium, and their expression is maintained in the epicardium. Sema3D-based lineage tracing showed derivatives contributing to endothelial cells (6.9%). The derivatives of Scx contributing to endothelium were even in larger number, up to 24%. It is still unclear whether this contribution takes place via an initial differentiation of $Scx^+/Sema3D^+$ progenitors into endocardium and posterior contribution of the latter to the coronary endothelium, as described in the chick (Dettman et al. 1998), or a direct conversion of EPDCs into endothelial cells. More recently, Cano and colleagues showed a contribution from the septum transversum/proepicardium to endothelial cells, which is necessary for proper coronary formation (Cano et al. 2016). Using a PE-specific *Gata4* enhancer to perform lineage tracing, the authors revealed that 20% of embryonic endothelial cells in the coronary vasculature are derived from the proepicardium. Interestingly, deletion of *Wt1* expression in these PE-derived cells resulted in abnormal formation of the coronaries, with irregular vessels that fail to migrate and colonize the myocardial layer, further suggesting a role for the (pro)epicardium in coronary vasculogenesis (Cano et al. 2016).

In addition to a (pro)epicardial contribution, other origins for coronary endothelium have been proposed, fueling a debate on the precise spatiotemporal development of coronary vessels arising from multiple sources. Red-Horse and colleagues reported that endothelial cells, sprouting from the sinus venosus endothelium in the ventral region of the heart, colonized both subepicardial and deeper myocardial regions (Red-Horse et al. 2010). This angiogenic process has been shown to be mediated by epicardial-derived vascular endothelial growth factor C (VEGF-C) signaling (Red-Horse et al. 2010; Chen et al. 2014). A second source of endothelial cells was shown to colonize the dorsal region of the heart, arising from the endocardium (Red-Horse et al. 2010). Indeed, the endocardium has also been proposed to be a source of coronary endothelial cells through a differentiation process dependent on VEGF-A (Wu et al. 2012). The latter was shown by using an endocardium-specific transcription factor to drive Cre expression, *Nfatc1* (nuclear factor of activated T cells, cytoplasmic, calcineurin-dependent 1). In addition, a third wave of coronary vasculogenesis has been proposed to take place postnatally (between postnatal (P) days P3 and P7), with endothelial cells arising from both the subepicardial and endocardial layers (Tian et al. 2014).

Taken together, the emerging picture suggests heterogeneity in origin for the endothelial compartment of the coronary vasculature. The implications for different sources contributing to specific regional vascularization of the developing heart remain unknown, and with respect to epicardial contribution, further technical advances in lineage tracing understanding the precise derivation of ECs from the epicardium during development are warranted, as this may have therapeutic applications in the context of adult heart injury and EPDC-based neovascularization (see Sect. 3.4.2).

3.2.2 Epicardial Contribution to Cardiomyocytes

A significant area of contention regarding the fate of the developing epicardium relates to a putative contribution of EPDCs to cardiomyocytes. Studies using proepicardial explants derived from chick embryos first reported the potential of the proepicardium to give rise to cardiomyocytes (Guadix et al. 2006). However, this was inconsistent with previous published observations arising from quail-chick chimeras transplant models (Manner 1999). Proepicardium differentiation to cardiomyocytes has been also shown in mice ex vivo (Ruiz-Villalba et al. 2013; del Monte et al. 2011), but other studies using epicardial explants failed to recapitulate a cardiomyocyte fate (Zamora et al. 2007; Austin et al. 2008; Ruiz-Villalba et al. 2013). Of significance, Cre-based reporter lines revealed a putative contribution of EPDCs to cardiomyocytes. Two reports published in 2008 showed cardiomyocytes arising from EPDCs using both *Tbx18* and *Wt1* as Cre drivers (Cai et al. 2008; Zhou et al. 2008). Fate mapping using the *Tbx18Cre* line revealed an epicardial contribution to cardiomyocytes in the interventricular septum and left ventricle of the embryonic heart (Cai et al. 2008). However, in a subsequent study by Christoffels and colleagues (Christoffels et al. 2009), endogenous expression of Tbx18 was detected in cardiomyocytes in the left ventricle, raising concerns about the conclusions from the Tbx18-based lineage tracing (Cai et al. 2008). The parallel use of a *Wt1EGFP-Cre* line (Zhou et al. 2008) to establish a cardiomyocyte contribution of Wt1-expressing EPDCs has also been challenged on the basis that Wt1 expression could be detected in cardiomyocytes and that available *Wt1-Cre* lines do not faithfully recapitulate Wt1 endogenous expression (Rudat and Kispert 2012). Nevertheless, it is worth mentioning that subsequent lineage-tracing studies using the proepicardial drivers *Scx-Cre* and *Sema3D-Cre* have also shown a modest contribution of the (pro)epicardium to the cardiomyocyte lineage, 6.6% and 0.36%, respectively (Katz et al. 2012). However, the tissue specificity in the expression of Scx and Sema3D remains to be addressed. Collectively, these studies indicate a potential (albeit modest) epicardial contribution to the myocardial lineage, but equally they highlight the requirement for improved lineage-tracing approaches to further define the extent of EPDC conversion into cardiomyocytes during vertebrate heart development.

3.2.3 Epicardial Contribution to Fat Cells

Visceral fat, unlike subcutaneous fat, has a mesothelial origin (Chau et al. 2014), and recently, the epicardium has been implicated in adipose tissue development in the forming of the heart. A lineage-tracing study, based on the use of the *Wt1EGFP-Cre* line, revealed that EPDCs within the interventricular groove co-localized with adipocyte markers, suggesting that the epicardium contributed to adipose cells through a process coined epicardial-to-fat transition (EFT) (Liu et al. 2014). These observations were confirmed by using a viral vector driving Cre expression under the control of *mesothelin*, which encodes for a mesothelium-restricted membrane protein (MSLN). In addition, the authors reported a reactivation of the epicardial contribution to fat cells upon myocardial injury in the mouse adult heart (Liu et al. 2014). The latter was also described in a recent lineage-tracing study using the *Tbx18Cre* line (Yamaguchi et al. 2015), and importantly Tbx18 has not been reported to be endogenously expressed in fat cells. Specifically, murine EPDCs adopted an adipose fate after mesenchymal transformation, via a mechanism dependent on the expression of the peroxisome proliferator-activated receptor gamma (PPARγ), a nuclear receptor and master regulator of adipocyte differentiation and lipid metabolism (Yamaguchi et al. 2015). As such, these studies raise questions about the possible role of fat of epicardial origin and support the multipotent progenitor status of EPDCs. Whether the process of EFT is exclusive to mouse or conserved across other vertebrates, including humans, is yet to be elucidated.

In summary, a contribution of EPDCs to coronary smooth muscle, mesenchymal cells in the valve primordium, and fibroblasts, both interstitial and perivascular, seems to be widely accepted by the cardiovascular research field. Additional studies are warranted in order to fully determine the potential and extent of EPDC contribution to the different cardiovascular lineages discussed, in particular the coronary endothelium, myocardium, and adipose tissue. Of significance, the proepicardium is now regarded as a heterogeneous source of precursor cells. As described by Katz and colleagues (Katz et al. 2012), the PE reveals a heterogeneous expression of markers that appear to direct their commitment to different cell types. Whether this is predetermined in the proepicardium or via specification of PE cells from mesoderm progenitors (Fig. 3.2), and how this process occurs, requires further investigation.

3.3 The Epicardium as a Signaling Center in the Developing Heart

Cardiac organogenesis relies on interactions via paracrine signaling between the different cell types that form the heart. As such, cardiomyocytes promote EMT of endocardial (Potts and Runyan 1989; Potts et al. 1991; Boyer et al. 1999) and epicardial cells (Zamora et al. 2007; von Gise et al. 2011), whereas epicardial and endocardial signaling to the myocardium promotes cardiomyocyte proliferation and differentiation (Lee et al. 1995; Hertig et al. 1999; Lavine et al. 2005). Likewise, the epicardium promotes coronary artery formation (Chen et al. 2014), while the coronary endothelium and EPDCs regulate formation of the conduction system (Eralp et al. 2006; Gourdie et al. 1998). Importantly, all of these coordinated events are essential and mediate the maturation of the heart, leading to a functional organ. Recent findings have begun to elucidate the role of epicardium-specific signaling in heart development.

3.3.1 Epicardial Signals Promoting Cardiomyocyte Proliferation

A non-cell-autonomous role for the epicardium was first evident in studies using chick heart explants where physical removal of the epicardial layer resulted in a decrease in cardiomyocyte proliferation (Stuckmann et al. 2003; Gittenberger-de Groot et al. 2000; Perez-Pomares et al. 2002b; Pennisi et al. 2003). This finding was supported by mouse models defective for the vascular cell adhesion molecule (VCAM) or integrin alpha 4, key molecules mediating epicardial attachment to the myocardium. In these mouse mutants, the epicardium and myocardium were detached from each other and epicardial signaling was lost, causing arrest of heart growth and hypoplasia of the myocardial layer, which in turn led to embryonic lethality (Kwee et al. 1995; Yang et al. 1995). Moreover, epicardial cells in culture were shown to secrete growth factors that induced a mitogenic response on primary embryonic cardiomyocytes (Chen et al. 2002; Kang and Sucov 2005), suggesting that epicardial cells could promote cardiomyocyte proliferation through secretion of diffusible molecules. One secreted factor suggested to induce myocyte proliferation was retinoic acid (RA), which is synthetized in the epicardium by the enzyme retinaldehyde dehydrogenase 2 (Raldh 2)(Xavier-Neto et al. 2000) and exerts its effects via heterodimeric retinoic acid receptors, promoting epicardial release of further mitogenic soluble signals (Chen et al. 2002; Merki et al. 2005). Retinoic acid signaling is controlled by Wt1 (Guadix et al. 2011) and drives the expression of several members of the fibroblast growth factor (FGF) family in the epicardium. Accordingly, it has been shown that FGFs −2, −4, −9, −16, and −20 are expressed by EPDCs in response to RA and convey mitogenic signaling to the myocardium via FGF receptors −1 and −2 (Pennisi et al. 2003; Lavine et al. 2005). Moreover, RA regulates myocardial proliferation by a parallel signaling pathway, whereby it promotes secretion of the hormone erythropoietin (EPO) from the liver (Brade et al. 2011). EPO then induces epicardial expression and secretion of the insulin growth factor

(IGF), which in turn promotes cardiomyocyte proliferation by activation of the MAPK/ERK (mitogen-activated protein kinase, originally termed extracellular signal-regulated kinase) signaling pathway (Brade et al. 2011).

More recently, follistatin-like 1 (Fstl1) has been identified as an epicardium-secreted factor that holds therapeutic potential in the context of adult heart injury (Wei et al. 2015) (see Sect. 3.4.2). In the developing heart, the expression of Fstl1 in the epicardium suggests a potential paracrine mode of action, but its interplay with the other components of the epicardial secretome (e.g., RA, FGFs) is yet to be determined. Likewise, the molecular mechanism underpinning Flstl1-mediated cardiomyocyte proliferation remains poorly understood.

Epicardial expression of Crim1, the cysteine-rich transmembrane BMP (bone morphogenetic protein) regulator 1 (chordin-like) known to bind a variety of growth factors, has been shown to be required for myocardial development (Iyer et al. 2016). In detail, epicardial-restricted deletion of Crim1, via the use of Wt1-Cre and Wt1-CreER drivers, led to abnormal heart development, with an increase in epicardial EMT and thinned compact myocardial layer, suggesting an autocrine and paracrine mode of action for Crim 1. The molecular mechanism by which Crim 1 exerts its cell-autonomous effects is not known. Similarly, it is unclear whether the myocardial phenotype is secondary to a defect in epicardial EMT or in mitogen secretion by the epicardium and whether Crim 1 signaling is required in cardiomyocytes to promote proliferation and growth of the compact layer.

3.3.2 Epicardial Signals Directing Coronary Angiogenesis

In addition to contributing components of the coronary vasculature, the epicardium signals to regulate coronary vessel formation. Indeed, in vivo loss-of-function studies have revealed that epicardial cells secrete growth factors and cytokines that act directly on the forming coronary vasculature. For example, epicardial-derived VEGF-C targets the sinus venosus endothelium, inducing vessel sprouting and subsequent formation of coronary vessels (Chen et al. 2014). However, such secreted molecules may also act on cardiomyocytes, which in turn secrete angiogenic signals, such as VEGF-A, further promoting angiogenesis (Majesky 2004; Lavine et al. 2006, 2008; Tomanek et al. 2001).

To better understand and exploit, in a disease setting, the role of epicardial signaling underlying coronary angiogenesis, further mechanistic studies at a molecular level are warranted. Hypoxia and hypoxia-induced factor (HIF) signaling are widely accepted to play an essential role in heart development (Bishop and Ratcliffe 2015). This oxygen-sensing mechanism allows for direct transcriptional regulation via the heterodimeric complex HIF composed of alpha and beta subunits, the latter being constitutively expressed (Bishop and Ratcliffe 2015). Three HIF-alpha subunits have been described (HIF-1alpha, HIF-2alpha, and HIF-3alpha), and these are post-transcriptionally modified, being targeted for proteasomal degradation in normal oxygen tension. This enzymatic process is oxygen-dependent, and, therefore, under hypoxia HIF-alpha escapes destruction, translocates into the nucleus, and regulates its downstream targets through direct DNA binding. More than 100 direct HIF target genes have been identified and implicated in a variety of cellular responses (Bishop and Ratcliffe 2015). Importantly, HIF stimulates angiogenesis by activating the transcription of growth factors, such as VEGF (reviewed in Liu et al. (2012)). Recent studies have also implicated hypoxia in cardiomyocyte homeostasis, suggesting that HIF signaling is required for proliferation and regulation of cardiomyocyte metabolism (Guimaraes-Camboa et al. 2015). Moreover, it has been shown that environmental oxygen regulates postnatal cell cycle arrest in cardiomyocytes through DNA damage response (Puente et al. 2014) and that hypoxic cardiomyocytes in the adult heart have proliferative ability (Kimura et al. 2015). However, little is known about the role of hypoxia in the epicardium in the developing and adult heart. In the developing heart, hypoxic regions are localized within the epicardium and subepicardial space, and low oxygen tension has been shown to regulate EPDC migration (Kocabas et al. 2012). In addition, constitutive overexpression of HIF1-alpha in chick embryos prevents migration and differentiation of EPDCs through inhibition of the VEGF signaling pathway. In this context, epicardial VEGF acts in a paracrine manner, as a target of HIF1-alpha, promoting epicardial cell invasion and EMT (Tao et al. 2013). Further studies are needed in order to ascertain whether HIF signaling in the epicardium also extends to regulation of EPDC cycling and metabolism, as has been described in cardiomyocytes.

3.3.3 Other Epicardial Signals Contributing to Heart Development

The Purkinje fibers are specialized cardiac cells that control the propagation of electric activity across the heart, modulating the heart rate through an autonomic response. Due to the close spatial relationship between differentiating EPDCs and Purkinje fibers within the developing ventricular myocardium, epicardial cells have been suggested to play a role in supporting the formation of the cardiac conduction system (Eralp et al. 2006). Accordingly, mechanical or genetic disruption of epicardial development in the chick embryo, by suppressing epicardial EMT and subsequent EPDC migration, resulted in hypoplasia of the Purkinje fiber network (Eralp et al. 2006), although no specific signaling mechanism for this paracrine process was described. More recently, it has been reported that in the absence of the epicardial layer, the heart is also unable to increase its rate of contraction in response to adrenaline (Kelder et al. 2015), a mediator of the autonomic response, which modulates heart rate in response to environmental stimuli. Taken together, these two studies highlight additional roles for epicardium-derived signaling during heart development; however, the implications of these findings remain unclear. The mechanism by which the epicardium regulates autonomic response and Purkinje cell development is still unexplored, as is the extrapolation to the adult epicardium and contractile function.

3.4 Adult Epicardium

3.4.1 Physiological Role of the Epicardium in Homeostatic Conditions

The adult epicardium has been generally assumed to be a quiescent tissue restricted to a protective role covering the myocardium. This was supported by the fact that adult epicardial cells in homeostatic conditions show a downregulation of embryonic genes (Smart et al. 2011), low proliferation rates (Wu et al. 2010), and no active EMT (Zhou et al. 2011). In adult zebrafish, despite the fact that the epicardium has a low proliferation rate, ablation of the epicardium through the expression of a cytotoxin resulted in increased proliferation rates and complete restoration of the epicardial layer in 2 weeks. Cues secreted from the outflow tract, including the diffusible morphogen sonic hedgehog (SHH), were essential in the recovery of the epicardium integrity, as shown ex vivo using whole-heart transplant assays (Wang et al. 2015). Likewise, pulse-chase and lineage-tracing experiments revealed that in teleosts the epicardium contributes to adult heart function by continuously supplementing the ventricular wall with new EPDCs in a process dependent on FGF signaling (Wills et al. 2008). Thus, the epicardium may act as a reservoir of progenitor cells in the adult heart evoking its embryonic function. This hypothesis extends to mammals and is further supported by the observation that in the adult murine heart, the epicardium and subepicardium form a hypoxic microenvironment (Kocabas et al. 2012) similar to that described in bona fide organ-specific stem cell niches, such as in the bone marrow or the subventricular zone in the brain (Parmar et al. 2007) (reviewed in Mohyeldin et al. (2010)). Cells in the epicardium-subepicardium displayed a glycolysis-based metabolism, rather than relying on mitochondrial oxidative phosphorylation, and expressed markers of cardiac progenitor cells including Nkx2.5, Gata4, Wt1, and Tbx18 (Kocabas et al. 2012). These hypoxic cells also exhibited a phenotype with intermediate features between a stem cell and a cardiomyocyte, e.g., disorganized bundles of filaments and intracytoplasmic dense bodies similar to primordial Z-lines (Gherghiceanu and Popescu 2010), as detailed by electron microscopy. More recently, cluster-like ultrastructures have been described in the adult epicardium, challenging the idea of the epicardium being a single mesothelial cell layer and further supporting the notion of the epicardium and subepicardium space as a progenitor cell "niche" (Balmer et al. 2014). The epicardial cell "clusters" were established postnatally, during the neonatal period ranging between postnatal day (P)1 and P7, and responded dynamically upon myocardial infarction (MI), displaying cluster breakdown with subsequent epicardial expansion prior to reformation of the ultrastructure at later stages after injury (Balmer et al. 2014). Moreover, these cluster-like structures contained CD45+ hematopoietic progenitors, which were replenished throughout adult life by the bone marrow (BM), and were implicated in EPDC turnover, which was decreased with age, suggesting the adult epicardium is active even at baseline (Balmer et al. 2014).

Despite significant effort devoted to understanding cardiac myocyte turnover both in mouse and human in vitro and in vivo models (Bergmann et al. 2009; Senyo et al. 2013), epicardial cell turnover remains relatively unexplored. It has been shown that epicardial cells reach their proliferative potential by E11.5, and then proliferation progressively decreases, leading to quiescence by the end of the first postnatal week (Wu et al. 2010). Of significance, dividing epicardial cells display two distinct mitotic spindle orientations, and a model has been put forward by Wu and co-workers to explain EPDC "asymmetric division." If the mitotic spindle is established parallel to the basal membrane, two epicardial cells are formed that remain on the surface of the heart, whereas epicardial cells undergoing perpendicular cell division give rise to one cell that invades the subepicardial space, entering the myocardial compartment, and another that remains in the surface, dependent on beta-catenin regulation of adherens junctions (Wu et al. 2010). Whether these two distinct mitotic spindle orientations in epicardial cell division are maintained in the adult heart or whether the ability of the adult epicardium to supplement cells to the heart in homeostatic conditions is dependent on asymmetrical cell division is yet to be explored.

Studies based on carbon (C)-14 integration during the Cold War nuclear bomb testing to date cardiomyocytes in the human adult heart, have reported a 1% turnover of cardiomyocytes that decreased with age (Bergmann et al. 2009). More recently, this approach has been applied to mesenchymal and endothelial cell populations to reveal that while cardiomyocyte number varies very little with age, the number of endothelial and mesenchymal cells significantly increases during the adult life (Bergmann et al. 2015). Specifically, endothelial cells have a very fast turnover that is irrespective of age (around 5.8 years), and mesenchymal cells have a slower turnover, which declines with age (17.3 years on average) (Bergmann et al. 2015). However, neither the turnover of epicardial cells nor a possible epicardial origin for the newly formed cells during adult life in the human heart was explored in this model. As such, similar approaches are warranted to gain insight into epicardial cell behavior under homeostatic conditions in the adult heart, in order to harness its potential for self-renewal, leading to development of resident cell-based therapies to regenerate the failing heart.

3.4.2 The Epicardium to Regenerate the Adult Heart

The first study suggesting that the adult epicardium can respond to heart injury was performed in zebrafish. Upon amputation of 20% of the ventricle, the epicardial layer became activated, reexpressing markers associated with embryonic development, e.g., Raldh2 and Tbx18, and increasing its proliferation rate (Lepilina et al. 2006). This response was initially organ-wide and then became restricted to the injury site, whereby the adult epicardium contributes EPDCs in response to FGF signaling to induce neovascularization. More recently, epicardial ablation in the zebrafish has shown to impair heart regeneration by preventing cardiomyocyte proliferation (Wang et al. 2015).

As opposed to the zebrafish, the mammalian adult heart is unable to regenerate. Upon ischemic injury, there is permanent cardiomyocyte cell loss, inflammatory response and scar formation, downstream of fibroblast proliferation, and subsequent extracellular matrix (ECM) deposition (Li et al. 2014; Frangogiannis et al. 2002; Frangogiannis 2014). The epicardial layer has been shown to respond to injury (Zhou et al. 2011; Smart et al. 2007a, 2011; van Wijk et al. 2012), where epicardial cells reexpress embryonic markers such as Wt1, Tbx18, and Raldh2, followed by thickening of the epicardium with increased proliferation rates (van Wijk et al. 2012) in an organ-wide manner, similar to what had been described for zebrafish. This activation of epicardial cells induced expression of EMT markers (e.g., Snail/Slug and α-smooth muscle actin) and mobilization of EPDCs, which remained within the surface of the heart (van Wijk et al. 2012). Fluorescence-activated cellular sorting (FACS) and transcriptional profiling of these EPDCs revealed expression of fibroblast and smooth muscle markers. Notably, reactivation of embryonic genes in EPDCs also recapitulated secretion of paracrine factors (Zhou et al. 2011). Accordingly, conditioned medium from (post cardiac injury) ex vivo cultures of adult EPDCs was shown to improve blood vessel formation in vitro and reduce infarct size, improving heart function in vivo (Zhou et al. 2011). Despite the reactivation of embryonic genes and generation of mesenchymal cells through EMT, these newly formed cells remained within the subepicardial region and did not migrate into the myocardial layer, limiting their paracrine protective effect and progenitor cell potential. However, this response to injury in the adult heart was enhanced by priming the epicardium with the G-actin monomer-binding peptide thymosin β4 (Tβ4) (Smart et al. 2011). Tβ4 has pleiotropic effects in the adult heart, promoting, among others, cell survival in cardiomyocytes (Bock-Marquette et al. 2009) and modulation of the inflammatory response post-injury (Sosne et al. 2007; Evans et al. 2013) (reviewed in (Bollini et al. 2015)). Moreover, Tβ4 enhanced the ability of the adult epicardium to reactivate an embryonic gene program and increased EPDC migration, differentiation, and signaling in vitro (Smart et al. 2007a, 2010). Likewise, in vivo-based studies (Bock-Marquette et al. 2009; Smart et al. 2007b, 2010) revealed thickening of the epicardial layer and neovascularization after treatment (priming) with Tβ4 in a murine model of MI. In addition to de novo vessels (Smart et al. 2007b, 2010), the adult epicardium has also been shown to contribute new cardiomyocytes, albeit at a low yield (Smart et al. 2011), in response to priming with Tβ4 prior to MI. Noteworthy, treatment with Tβ4 post-MI was not sufficient to induce cardiomyogenesis (Zhou et al. 2012).

Despite the ability of Tβ4 to induce a reexpression of embryonic epicardial genes, a recent study has highlighted that these adult Tβ4-primed EPDCs are molecularly distinct from their embryonic counterpart (Bollini et al. 2014). Briefly, EPDCs were isolated from adult hearts after Tβ4 treatment followed by MI, using FACS, and compared with embryonic EPDCs. The number of EPDCs detected after MI was higher at 4 days post-injury (dpi), and these exhibited a highly heterogeneous phenotype, with variable expression of progenitor and mesenchymal cell markers (e.g., Sca1, c-kit, CD29, CD90, CD44, and CD105) and a phenotypical signature that differed from the embryonic EPDCs, e.g., higher expression of the stem cell marker Sca-1 and the mesenchymal markers CD90, CD44, and CD105 (Bollini et al. 2014). Specifically, the Sca1-/Wt1-expressing population showed a strong progenitor cell trait, expressing high levels of embryonic markers, such as Tbx18, Islet-1, or Gata4, which suggested that this pool of EPDCs might retain the potential to give rise to cardiovascular derivatives including cardiomyocytes. Of significance, follistatin-1 (Fstl-1) has been recently identified as a further potential therapeutic factor in cardiac repair (Wei et al. 2015). Fstl-1 is expressed throughout the embryonic heart, but becomes restricted to the epicardium in the adult heart, being reexpressed in cardiomyocytes after injury. Fstl-1 was detected in conditioned medium from the rat-derived epicardial cell line EMC (Wada et al. 2003), using mass spectrometry. Coculture of these cells with embryonic stem cell (ESC)-derived cardiomyocytes increased their number, which was recapitulated when treating cardiomyocytes with EMC-conditioned medium. Interestingly, treatment with recombinant follistatin-1 generated a similar response in vitro and, when administered via patches during MI, improved the post-injury outcome by increasing neovascularization, reducing fibrosis, and promoting cardiomyocyte proliferation (Wei et al. 2015).

It has been shown that the neonatal mouse heart has a potential to regenerate after ventricular amputation via cardiomyocyte proliferation (Porrello et al. 2011) in an analogous way to that observed in the adult zebrafish; however, this process is only active during the first week of life (from P1 to P7). Of significance, epicardial-associated embryonic gene expression declines within this period, but upon injury between stages P1 and P4, the postnatal epicardium reactivates supporting myocardium regeneration (Porrello et al. 2011). Whether epicardial signals underpin cardiomyocyte proliferation in response to injury remains to be determined; as does whether loss of epicardium signaling by P7 may contribute to cardiomyocyte cell cycle exit and terminal differentiation.

3.5 Human Epicardium

Almost all of the existing knowledge on the biology of the epicardium originates from the study of animal models. The number of studies performed using human samples is limited, due to the difficulty in sourcing human material. Therefore, it has generally been assumed that the findings arising from studies using the mouse model are likely to be applicable to humans. However, even if most processes regarding epicardial formation and function seem to be conserved across species, some differences have been reported (see Sect. 3.1), highlighting the need for experimental characterization of the human epicardium.

It was initially thought that the epicardium starts covering the surface of the human heart at Carnegie stage (CS) 12 (equivalent to E9.5 in mouse), forming a squamous epithelial layer that completely envelops the heart by the late CS 16 (equivalent to E11.5 in mouse). This was assumed from an early descriptive study based on light microscopy observation of sections of human embryonic hearts at different stages of development (Hirakow 1992) and from extrapolation of what had been previously described in chick and rodent models (Sylva et al. 2014). A recent characterization of the human embryonic and fetal (F) epicardium between CS11 and F3 (equivalent to E8.5–9.0 and E17.5–18.5 in mouse, respectively) using three-dimensional (3D) optical projection tomography (OPT), however, revealed that the epicardium is formed by day CS14, significantly earlier than previously thought (Risebro et al. 2015). Interestingly, it was revealed that the human embryonic and fetal epicardium displayed a more complex structure; confocal microscopy revealed that the ventricular, but not the atrial, epicardium comprised regions containing several layers of cells, as opposed to being a stereotypical squamous monolayered epithelium, as was initially thought. Of significance, this heterogeneity also affected epicardial cell behavior, and it was shown in explants that the ventricular epicardium undergoes EMT spontaneously, whereas the atrial epicardium retains an epithelial morphology (Risebro et al. 2015). Whether the differences between the epicardium of different chambers are retained or lost in the postnatal heart remains to be explored, as does the implications of these differences regarding the potential of the epicardium to give rise to different cardiovascular lineages. Likewise, it will be important to understand their relevance in the context of adult heart remodeling after myocardial infarction. Recently, a contribution of the human embryonic epicardium to the coronary vasculature has been experimentally investigated (Tomanek 2015). By imaging sections from hearts at different developmental stages, it was reported that the formation of the coronary system in the human embryo is similar to what had been described in other animal models, i.e., the epicardium supplies spindle-like cells that integrate into the coronary capillaries. The descriptive nature of the study highlights the need for further experimental validation and the development of new models.

Two recent studies have reported in vitro differentiation of epicardial-like cells from human embryonic stem cells (hESC) and induced pluripotent stem cells (hiPSC) (Iyer et al. 2015; Witty et al. 2014). As such, these studies provide an alternative experimental approach to overcome the limited supply of human samples for primary cultures and establish new models to study human epicardial developmental cell biology. Of significance, both studies reported expression of key epicardial factors in epithelial-like cells with cobblestone morphology, after a stepwise process of cell differentiation. There are, however, some differences in the experimental approach to generate epicardial cells. Iyer and colleagues reported generation of epicardial cells from the lateral mesoderm using a monolayer differentiation approach, whereas Witty and colleagues used combined embryoid bodies and monolayer differentiation. Also, even though both protocols included BMP and WNT signaling as necessary to promote differentiation, the former also used retinoic acid (RA) to enhance expression of epicardial proteins. While both studies reported TGFβ-driven epicardial EMT and contribution to smooth muscle and fibroblasts, Iyer and colleagues reported that human epicardial-like cells can be transplanted into the chick embryo, differentiating and integrating within the developing coronary vasculature (Iyer et al. 2015). Derivation of epicardial cells from pluripotent cells may therefore represent a valuable tool to study the human epicardium. Generation of epicardial cells from human inducible pluripotent stem cells (hiPSCs) could be useful to experimentally investigate patient-specific mutations or variants with increased susceptibility to coronary disease, specifically since single nucleotide polymorphisms (SNPs) in the *TCF21* locus have been associated with coronary

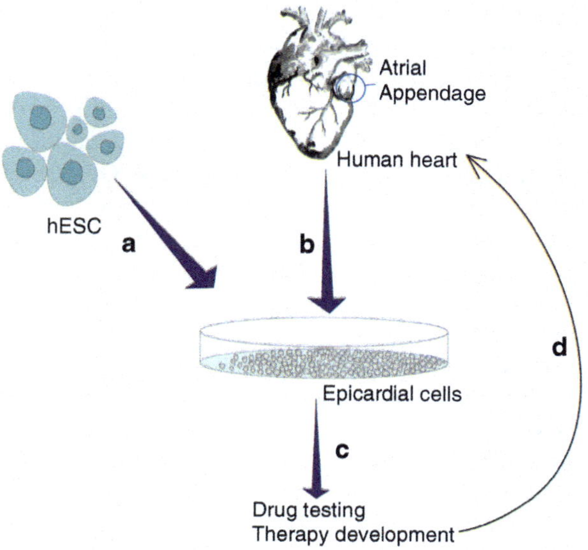

Fig. 3.3 Tools to study the human epicardium. (**a**) Epicardial cells can be generated from human pluripotent stem cells and patient-derived atrial appendages in vitro. (**b**) In vitro expansion of epicardial cells can be used to gain insight into human epicardial cell development and basic physiology, and to develop new therapies based on high-throughput screens (**c**), as a potential treatment for the failing heart (**d**)

heart disease in population studies (Miller et al. 2013; Sazonova et al. 2015). Besides helping to provide understanding on the genetic basis of cardiac disease, in vitro differentiation of human epicardial cells could also be a powerful tool in the development of new small molecules for future pharmacological therapies. The system allows the generation of large cell numbers that are required for high-throughput screenings and subsequent identification of molecular targets to harness the regenerative potential of the adult heart (Fig. 3.3).

The human adult epicardium has been studied in vivo using postmortem samples from patients with different cardiac pathologies, such as MI, congestive heart failure, and hypertension (Braitsch et al. 2013). The epicardial layer reportedly undergoes thickening and reexpression of the embryonic markers WT1, TBX18, and TCF21 during heart failure, in a similar way to what had been shown in mouse models of myocardial infarction. However, a major difference between human and mouse diseased hearts is the higher accumulation of adipose tissue in the epicardial layer of human hearts. As such, the human adult epicardium may show a higher tendency to give rise to adipose tissue, a feature of the epicardium that was recently reported in murine heart development and in response to injury in the adult heart (Liu et al. 2014). Further validation of these observations is needed to gain insight into the role of epicardial-to-fat transformation in the human heart, since the origin of the fat tissue may have important implications in the human response to cardiac injury and disease.

The human adult epicardium has also been studied ex vivo by deriving EPDCs from right atrial appendage biopsies obtained during coronary artery bypass graft (CABG) surgery (Bax et al. 2011; van Tuyn et al. 2007; Winter et al. 2009; Clunie-O'Connor et al. 2015). These cells can be maintained in culture for several passages and exhibit a cobblestone morphology (van Tuyn et al. 2007) expressing markers such as WT1 and TBX18 (Bax et al. 2011). Moreover, atrial appendage-derived epicardial cells can undergo EMT upon treatment with the transforming growth factor beta (TGFβ) and differentiate to different cell types, including smooth muscle cells and fibroblasts (Bax et al. 2011). Interestingly, when cocultured with human cardiomyocytes in hypoxic conditions, biopsy-derived cells increased their proliferation and secreted several growth factors including VEGF, placental growth factor, and platelet-derived growth factor (PDGF). Importantly, this study showed how both cell types can act synergistically in vitro, promoting an increase in cell proliferation (Winter et al. 2009). Moreover, primary human EPDCs were shown to improve cardiac function in a mouse model of MI. Recently, this ex vivo model to study human epicardial cells has been proposed to be a valuable resource to develop high-throughput phenotypic screens for potential drug discovery (Clunie-O'Connor et al. 2015) (Fig. 3.3).

From the early studies of the epicardium using the chick embryo moving forward to studies using the state-of-the-art experimental approaches (e.g., ESC-derived epicardial-like cells), our understanding of the role of the epicardium in mammalian heart development and repair has increased considerably. Nevertheless, there are still many unanswered questions to be addressed, such as the functional implications of proepicardium and epicardium cell heterogeneity, the role of the epicardium in adult homeostatic conditions, and how to fully harness the therapeutic potential of EPDCs to promote an efficient response following cardiac injury, acting as a resident cell source to promote regeneration. As a consequence, further research is warranted to meet these answers. The next years, therefore, offer significant potential regarding the exploitation of the therapeutic potential of the epicardium in cardiovascular regenerative medicine.

Compliance with Ethical Standards

Conflict of Interest The authors declare that they have no conflict of interest.

Ethical Approval This article does not contain any studies with human participants performed by any of the authors.

References

Acharya A, Baek ST, Huang G, Eskiocak B, Goetsch S, Sung CY, Banfi S, Sauer MF, Olsen GS, Duffield JS, Olson EN, Tallquist MD (2012) The bHLH transcription factor Tcf21 is required for lineage-specific EMT of cardiac fibroblast progenitors. Development 139(12):2139–2149. doi:10.1242/dev.079970

Austin AF, Compton LA, Love JD, Brown CB, Barnett JV (2008) Primary and immortalized mouse epicardial cells undergo differentiation in response to TGFbeta. Dev Dyn 237(2):366–376. doi:10.1002/dvdy.21421

Balmer GM, Bollini S, Dube KN, Martinez-Barbera JP, Williams O, Riley PR (2014) Dynamic haematopoietic cell contribution to the developing and adult epicardium. Nat Commun 5:4054. doi:10.1038/ncomms5054

Bax NA, van Oorschot AA, Maas S, Braun J, van Tuyn J, de Vries AA, Groot AC, Goumans MJ (2011) In vitro epithelial-to-mesenchymal transformation in human adult epicardial cells is regulated by TGFbeta-signaling and WT1. Basic Res Cardiol 106(5):829–847. doi:10.1007/s00395-011-0181-0

Bergmann O, Bhardwaj RD, Bernard S, Zdunek S, Barnabe-Heider F, Walsh S, Zupicich J, Alkass K, Buchholz BA, Druid H, Jovinge S, Frisen J (2009) Evidence for cardiomyocyte renewal in humans. Science 324(5923):98–102. doi:10.1126/science.1164680

Bergmann O, Zdunek S, Felker A, Salehpour M, Alkass K, Bernard S, Sjostrom SL, Szewczykowska M, Jackowska T, Dos Remedios C, Malm T, Andra M, Jashari R, Nyengaard JR, Possnert G, Jovinge S, Druid H, Frisen J (2015) Dynamics of cell generation and turnover in the human heart. Cell 161(7):1566–1575. doi:10.1016/j.cell.2015.05.026

Bishop T, Ratcliffe PJ (2015) HIF hydroxylase pathways in cardiovascular physiology and medicine. Circ Res 117(1):65–79. doi:10.1161/CIRCRESAHA.117.305109

Bock-Marquette I, Shrivastava S, Pipes GCT, Thatcher JE, Blystone A, Shelton JM, Galindo CL, Melegh B, Srivastava D, Olson EN, DiMaio JM (2009) Thymosin β4 mediated PKC activation is essential to initiate the embryonic coronary developmental program and epicardial progenitor cell activation in adult mice in vivo. J Mol Cell Cardiol 46(5):728–738. doi:10.1016/j.yjmcc.2009.01.017

Bollini S, Vieira JM, Howard S, Dube KN, Balmer GM, Smart N, Riley PR (2014) Re-activated adult epicardial progenitor cells are a heterogeneous population molecularly distinct from their embryonic counterparts. Stem Cells Dev 23(15):1719–1730. doi:10.1089/scd.2014.0019

Bollini S, Riley PR, Smart N (2015) Thymosin beta4: multiple functions in protection, repair and regeneration of the mammalian heart. Expert Opin Biol Ther 15(Suppl 1):S163–S174. doi:10.1517/14712598.2015.1022526

Boyer AS, Ayerinskas II, Vincent EB, McKinney LA, Weeks DL, Runyan RB (1999) TGFbeta2 and TGFbeta3 have separate and sequential activities during epithelial-mesenchymal cell transformation in the embryonic heart. Dev Biol 208(2):530–545. doi:10.1006/dbio.1999.9211

Brade T, Kumar S, Cunningham TJ, Chatzi C, Zhao X, Cavallero S, Li P, Sucov HM, Ruiz-Lozano P, Duester G (2011) Retinoic acid stimulates myocardial expansion by induction of hepatic erythropoietin which activates epicardial Igf2. Development 138(1):139–148. doi:10.1242/dev.054239

Braitsch CM, Kanisicak O, van Berlo JH, Molkentin JD, Yutzey KE (2013) Differential expression of embryonic epicardial progenitor markers and localization of cardiac fibrosis in adult ischemic injury and hypertensive heart disease. J Mol Cell Cardiol 65:108–119. doi:10.1016/j.yjmcc.2013.10.005

Cai CL, Martin JC, Sun Y, Cui L, Wang L, Ouyang K, Yang L, Bu L, Liang X, Zhang X, Stallcup WB, Denton CP, McCulloch A, Chen J, Evans SM (2008) A myocardial lineage derives from Tbx18 epicardial cells. Nature 454(7200):104–108. doi:10.1038/nature06969

Cano E, Carmona R, Ruiz-Villalba A, Rojas A, Chau YY, Wagner KD, Wagner N, Hastie ND, Munoz-Chapuli R, Perez-Pomares JM (2016) Extracardiac septum transversum/proepicardial endothelial cells pattern embryonic coronary arterio-venous connections. Proc Natl Acad Sci U S A 113(3):656–661. doi:10.1073/pnas.1509834113

Chau YY, Bandiera R, Serrels A, Martinez-Estrada OM, Qing W, Lee M, Slight J, Thornburn A, Berry R, McHaffie S, Stimson RH, Walker BR, Chapuli RM, Schedl A, Hastie N (2014) Visceral and subcutaneous fat have different origins and evidence supports a mesothelial source. Nat Cell Biol 16(4):367–375. doi:10.1038/ncb2922

Chen T, Chang TC, Kang JO, Choudhary B, Makita T, Tran CM, Burch JB, Eid H, Sucov HM (2002) Epicardial induction of fetal cardiomyocyte proliferation via a retinoic acid-inducible trophic factor. Dev Biol 250(1):198–207

Chen HI, Sharma B, Akerberg BN, Numi HJ, Kivela R, Saharinen P, Aghajanian H, McKay AS, Bogard PE, Chang AH, Jacobs AH, Epstein JA, Stankunas K, Alitalo K, Red-Horse K (2014) The sinus venosus contributes to coronary vasculature through VEGFC-stimulated angiogenesis. Development 141(23):4500–4512. doi:10.1242/dev.113639

Christoffels VM, Grieskamp T, Norden J, Mommersteeg MT, Rudat C, Kispert A (2009) Tbx18 and the fate of epicardial progenitorsNature 458 (7240):E8-E9.; discussion E9-10. doi:10.1038/nature07916

Clunie-O'Connor C, Smits AM, Antoniades C, Russell AJ, Yellon DM, Goumans MJ, Riley PR (2015) The derivation of primary human epicardium-derived cells. Curr Protoc Stem Cell Biol 35:2C 5 1–2C 5 12. doi:10.1002/9780470151808.sc02c05s35 del Monte G, Casanova JC, Guadix JA, MacGrogan D, Burch JB, Perez-Pomares JM, de la Pompa JL (2011) Differential notch signaling in the epicardium is required for cardiac inflow development and coronary vessel morphogenesis. Circ Res 108(7):824–836. doi:10.1161/CIRCRESAHA.110.229062

Dettman RW, Denetclaw W Jr, Ordahl CP, Bristow J (1998) Common epicardial origin of coronary vascular smooth muscle, perivascular fibroblasts, and intermyocardial fibroblasts in the avian heart. Dev Biol 193(2):169–181. doi:10.1006/dbio.1997.8801

Duim SN, Kurakula K, Goumans MJ, Kruithof BP (2015) Cardiac endothelial cells express Wilms' tumor-1: Wt1 expression in the developing, adult and infarcted heart. J Mol Cell Cardiol 81:127–135. doi:10.1016/j.yjmcc.2015.02.007

Eralp I, Lie-Venema H, Bax NA, Wijffels MC, Van Der Laarse A, Deruiter MC, Bogers AJ, Van Den Akker NM, Gourdie RG, Schalij MJ, Poelmann RE, Gittenberger-De Groot AC (2006) Epicardium-derived cells are important for correct development of the Purkinje fibers in the avian heart. Anat Rec A Discov Mol Cell Evol Biol 288(12):1272–1280. doi:10.1002/ar.a.20398

Evans MA, Smart N, Dube KN, Bollini S, Clark JE, Evans HG, Taams LS, Richardson R, Levesque M, Martin P, Mills K, Riegler J, Price AN, Lythgoe MF, Riley PR (2013) Thymosin beta4-sulfoxide attenuates inflammatory cell infiltration and promotes cardiac wound healing. Nat Commun 4:2081. doi:10.1038/ncomms3081

Frangogiannis NG (2014) The inflammatory response in myocardial injury, repair, and remodelling. Nat Rev Cardiol 11(5):255–265. doi:10.1038/nrcardio.2014.28

Frangogiannis NG, Smith CW, Entman ML (2002) The inflammatory response in myocardial infarction. Cardiovasc Res 53(1):31–47

Gherghiceanu M, Popescu LM (2010) Cardiomyocyte precursors and telocytes in epicardial stem cell niche: electron microscope images. J Cell Mol Med 14(4):871–877. doi:10.1111/j.1582-4934.2010.01060.x

Gittenberger-de Groot AC, Vrancken Peeters MP, Mentink MM, Gourdie RG, Poelmann RE (1998) Epicardium-derived cells contribute a novel population to the myocardial wall and the atrioventricular cushions. Circ Res 82(10):1043–1052

Gittenberger-de Groot AC, Vrancken Peeters MP, Bergwerff M, Mentink MM, Poelmann RE (2000) Epicardial outgrowth inhibition leads to compensatory mesothelial outflow tract collar and abnormal cardiac septation and coronary formation. Circ Res 87(11):969–971

Gourdie RG, Wei Y, Kim D, Klatt SC, Mikawa T (1998) Endothelin-induced conversion of embryonic heart muscle cells into impulse-conducting Purkinje fibers. Proc Natl Acad Sci U S A 95(12):6815–6818

Guadix JA, Carmona R, Munoz-Chapuli R, Perez-Pomares JM (2006) In vivo and in vitro analysis of the vasculogenic potential of avian proepicardial and epicardial cells. Dev Dyn 235(4):1014–1026. doi:10.1002/dvdy.20685

Guadix JA, Ruiz-Villalba A, Lettice L, Velecela V, Munoz-Chapuli R, Hastie ND, Perez-Pomares JM, Martinez-Estrada OM (2011) Wt1 controls retinoic acid signalling in embryonic epicardium through transcriptional activation of Raldh2. Development 138(6):1093–1097. doi:10.1242/dev.044594

Guimaraes-Camboa N, Stowe J, Aneas I, Sakabe N, Cattaneo P, Henderson L, Kilberg MS, Johnson RS, Chen J, McCulloch AD, Nobrega MA, Evans SM, Zambon AC (2015) HIF1 alpha represses cell stress pathways to allow proliferation of hypoxic fetal cardiomyocytes. Dev Cell 33(5):507–521. doi:10.1016/j.devcel.2015.04.021

Hertig CM, Kubalak SW, Wang Y, Chien KR (1999) Synergistic roles of neuregulin-1 and insulin-like growth factor-I in activation of the phosphatidylinositol 3-kinase pathway and cardiac chamber morphogenesis. J Biol Chem 274(52):37362–37369

Hirakow R (1992) Epicardial formation in staged human embryos. Kaibogaku Zasshi J Anat 67(5):616–622

Iyer D, Gambardella L, Bernard WG, Serrano F, Mascetti VL, Pedersen RA, Talasila A, Sinha S (2015) Robust derivation of epicardium and its differentiated smooth muscle cell progeny from human pluripotent stem cells. Development 142(8):1528–1541. doi:10.1242/dev.119271

Iyer S, Chou FY, Wang R, Chiu HS, Raju VK, Little MH, Thomas WG, Piper M, Pennisi DJ (2016) Crim1 has cell-autonomous and paracrine roles during embryonic heart development. Sci Rep 6:19832. doi:10.1038/srep19832

Kang JO, Sucov HM (2005) Convergent proliferative response and divergent morphogenic pathways induced by epicardial and endocardial signaling in fetal heart development. Mech Dev 122(1):57–65. doi:10.1016/j.mod.2004.09.001

Katz TC, Singh MK, Degenhardt K, Rivera-Feliciano J, Johnson RL, Epstein JA, Tabin CJ (2012) Distinct compartments of the proepicardial organ give rise to coronary vascular endothelial cells. Dev Cell 22(3):639–650. doi:10.1016/j.devcel.2012.01.012

Kelder TP, Duim SN, Vicente-Steijn R, Vegh AM, Kruithof BP, Smits AM, van Bavel TC, Bax NA, Schalij MJ, Gittenberger-de Groot AC, DeRuiter MC, Goumans MJ, Jongbloed MR (2015) The epicardium as modulator of the cardiac autonomic response during early development. J Mol Cell Cardiol. doi:10.1016/j.yjmcc.2015.10.025

Kikuchi K, Gupta V, Wang J, Holdway JE, Wills AA, Fang Y, Poss KD (2011) tcf21+ epicardial cells adopt non-myocardial fates during zebrafish heart development and regeneration. Development 138(14):2895–2902. doi:10.1242/dev.067041

Kimura W, Xiao F, Canseco DC, Muralidhar S, Thet S, Zhang HM, Abderrahman Y, Chen R, Garcia JA, Shelton JM, Richardson JA, Ashour AM, Asaithamby A, Liang H, Xing C, Lu Z, Zhang CC, Sadek HA (2015) Hypoxia fate mapping identifies cycling cardiomyocytes in the adult heart. Nature 523(7559):226–230. doi:10.1038/nature14582

Kocabas F, Mahmoud AI, Sosic D, Porrello ER, Chen R, Garcia JA, DeBerardinis RJ, Sadek HA (2012) The hypoxic epicardial and subepicardial microenvironment. J Cardiovasc Transl Res 5(5):654–665. doi:10.1007/s12265-012-9366-7

Komiyama M, Ito K, Shimada Y (1987) Origin and development of the epicardium in the mouse embryo. Anat Embryol 176(2):183–189

Kwee L, Baldwin HS, Shen HM, Stewart CL, Buck C, Buck CA, Labow MA (1995) Defective development of the embryonic and extraembryonic circulatory systems in vascular cell adhesion molecule (VCAM-1) deficient mice. Development 121(2):489–503

Lavine KJ, Yu K, White AC, Zhang X, Smith C, Partanen J, Ornitz DM (2005) Endocardial and epicardial derived FGF signals regulate myocardial proliferation and differentiation in vivo. Dev Cell 8(1):85–95. doi:10.1016/j.devcel.2004.12.002

Lavine KJ, White AC, Park C, Smith CS, Choi K, Long F, Hui CC, Ornitz DM (2006) Fibroblast growth factor signals regulate a wave of hedgehog activation that is essential for coronary vascular development. Genes Dev 20(12):1651–1666. doi:10.1101/gad.1411406

Lavine KJ, Long F, Choi K, Smith C, Ornitz DM (2008) Hedgehog signaling to distinct cell types differentially regulates coronary artery and vein development. Development 135(18):3161–3171. doi:10.1242/dev.019919

Lee KF, Simon H, Chen H, Bates B, Hung MC, Hauser C (1995) Requirement for neuregulin receptor erbB2 in neural and cardiac development. Nature 378(6555):394–398. doi:10.1038/378394a0

Lepilina A, Coon AN, Kikuchi K, Holdway JE, Roberts RW, Burns CG, Poss KD (2006) A dynamic epicardial injury response supports progenitor cell activity during zebrafish heart regeneration. Cell 127(3):607–619. doi:10.1016/j.cell.2006.08.052

Li N, Wang C, Jia L, Du J (2014) Heart regeneration, stem cells, and cytokines. Regener Med Res 2(1):6. doi:10.1186/2050-490X-2-6

Liu J, Stainier DY (2010) Tbx5 and Bmp signaling are essential for proepicardium specification in zebrafish. Circ Res 106(12):1818–1828. doi:10.1161/CIRCRESAHA.110.217950

Liu W, Shen SM, Zhao XY, Chen GQ (2012) Targeted genes and interacting proteins of hypoxia inducible factor-1. Int J Biochem Mol Biol 3(2):165–178

Liu Q, Huang X, Oh JH, Lin RZ, Duan S, Yu Y, Yang R, Qiu J, Melero-Martin JM, Pu WT, Zhou B (2014) Epicardium-to-fat transition in injured heart. Cell Res 24(11):1367–1369. doi:10.1038/cr.2014.125

Majesky MW (2004) Development of coronary vessels. Curr Top Dev Biol 62:225–259. doi:10.1016/S0070-2153(04)62008-4

Manner J (1992) The development of pericardial villi in the chick embryo. Anat Embryol 186(4):379–385

Manner J (1999) Does the subepicardial mesenchyme contribute myocardioblasts to the myocardium of the chick embryo heart? A quail-chick chimera study tracing the fate of the epicardial primordium. Anat Rec 255(2):212–226

Merki E, Zamora M, Raya A, Kawakami Y, Wang J, Zhang X, Burch J, Kubalak SW, Kaliman P, Izpisua Belmonte JC, Chien KR, Ruiz-Lozano P (2005) Epicardial retinoid X receptor alpha is required for myocardial growth and coronary artery formation. Proc Natl Acad Sci U S A 102(51):18455–18460. doi:10.1073/pnas.0504343102

Mikawa T, Fischman DA (1992) Retroviral analysis of cardiac morphogenesis: discontinuous formation of coronary vessels. Proc Natl Acad Sci U S A 89(20):9504–9508

Mikawa T, Gourdie RG (1996) Pericardial mesoderm generates a population of coronary smooth muscle cells migrating into the heart along with ingrowth of the epicardial organ. Dev Biol 174(2):221–232. doi:10.1006/dbio.1996.0068

Miller CL, Anderson DR, Kundu RK, Raiesdana A, Nurnberg ST, Diaz R, Cheng K, Leeper NJ, Chen CH, Chang IS, Schadt EE, Hsiung CA, Assimes TL, Quertermous T (2013) Disease-related growth factor and embryonic signaling pathways modulate an enhancer of TCF21 expression at the 6q23.2 coronary heart disease locus. PLoS Genet 9(7):e1003652. doi:10.1371/journal.pgen.1003652

Mohyeldin A, Garzon-Muvdi T, Quinones-Hinojosa A (2010) Oxygen in stem cell biology: a critical component of the stem cell niche. Cell Stem Cell 7(2):150–161. doi:10.1016/j.stem.2010.07.007

Parmar K, Mauch P, Vergilio JA, Sackstein R, Down JD (2007) Distribution of hematopoietic stem cells in the bone marrow according to regional hypoxia. Proc Natl Acad Sci U S A 104(13):5431–5436. doi:10.1073/pnas.0701152104

Pennisi DJ, Ballard VL, Mikawa T (2003) Epicardium is required for the full rate of myocyte proliferation and levels of expression of myocyte mitogenic factors FGF2 and its receptor, FGFR-1, but not for transmural myocardial patterning in the embryonic chick heart. Dev Dyn 228(2):161–172. doi:10.1002/dvdy.10360

Peralta M, Steed E, Harlepp S, Gonzalez-Rosa JM, Monduc F, Ariza-Cosano A, Cortes A, Rayon T, Gomez-Skarmeta JL, Zapata A, Vermot J, Mercader N (2013) Heartbeat-driven pericardiac fluid forces contribute to epicardium morphogenesis. Curr Biol 23(18):1726–1735. doi:10.1016/j.cub.2013.07.005

Perez-Pomares JM, Macias D, Garcia-Garrido L, Munoz-Chapuli R (1997) Contribution of the primitive epicardium to the subepicardial mesenchyme in hamster and chick embryos. Dev Dyn 210(2):96–105. doi:10.1002/(SICI)1097-0177(199710)210:2<96::AID-AJA3>3.0.CO;2-4

Perez-Pomares JM, Macias D, Garcia-Garrido L, Munoz-Chapuli R (1998) The origin of the subepicardial mesenchyme in the avian embryo: an immunohistochemical and quail-chick chimera study. Dev Biol 200(1):57–68

Perez-Pomares JM, Carmona R, Gonzalez-Iriarte M, Atencia G, Wessels A, Munoz-Chapuli R (2002a) Origin of coronary endothelial cells from epicardial mesothelium in avian embryos. Int J Dev Biol 46(8):1005–1013

Perez-Pomares JM, Phelps A, Sedmerova M, Carmona R, Gonzalez-Iriarte M, Munoz-Chapuli R, Wessels A (2002b) Experimental studies on the spatiotemporal expression of WT1 and RALDH2 in the embryonic avian heart: a model for the regulation of myocardial and valvuloseptal development by epicardially derived cells (EPDCs). Dev Biol 247(2):307–326

Poelmann RE, Gittenberger-de Groot AC, Mentink MM, Bokenkamp R, Hogers B (1993) Development of the cardiac coronary vascular endothelium, studied with antiendothelial antibodies, in chicken-quail chimeras. Circ Res 73(3):559–568

Porrello ER, Mahmoud AI, Simpson E, Hill JA, Richardson JA, Olson EN, Sadek HA (2011) Transient regenerative potential of the neonatal mouse heart. Science 331(6020):1078–1080. doi:10.1126/science.1200708

Potts JD, Runyan RB (1989) Epithelial-mesenchymal cell transformation in the embryonic heart can be mediated, in part, by transforming growth factor beta. Dev Biol 134(2):392–401

Potts JD, Dagle JM, Walder JA, Weeks DL, Runyan RB (1991) Epithelial-mesenchymal transformation of embryonic cardiac endothelial cells is inhibited by a modified antisense oligodeoxynucleotide to transforming growth factor beta 3. Proc Natl Acad Sci U S A 88(4):1516–1520

Puente BN, Kimura W, Muralidhar SA, Moon J, Amatruda JF, Phelps KL, Grinsfelder D, Rothermel BA, Chen R, Garcia JA, Santos CX, Thet S, Mori E, Kinter MT, Rindler PM, Zacchigna S, Mukherjee S, Chen DJ, Mahmoud AI, Giacca M, Rabinovitch PS, Aroumougame A, Shah AM, Szweda LI, Sadek HA (2014) The oxygen-rich postnatal environment induces cardiomyocyte cell-cycle arrest through DNA damage response. Cell 157(3):565–579. doi:10.1016/j.cell.2014.03.032

Red-Horse K, Ueno H, Weissman IL, Krasnow MA (2010) Coronary arteries form by developmental reprogramming of venous cells. Nature 464(7288):549–553. doi:10.1038/nature08873

Risebro CA, Vieira JM, Klotz L, Riley PR (2015) Characterisation of the human embryonic and foetal epicardium during heart development. Development 142(21):3630–3636. doi:10.1242/dev.127621

Rodgers LS, Lalani S, Runyan RB, Camenisch TD (2008) Differential growth and multicellular villi direct proepicardial translocation to the developing mouse heart. Dev Dyn 237(1):145–152. doi:10.1002/dvdy.21378

Rudat C, Kispert A (2012) Wt1 and epicardial fate mapping. Circ Res 111(2):165–169. doi:10.1161/CIRCRESAHA.112.273946

Ruiz-Villalba A, Ziogas A, Ehrbar M, Perez-Pomares JM (2013) Characterization of epicardial-derived cardiac interstitial cells: differentiation and mobilization of heart fibroblast progenitors. PLoS One 8(1):e53694. doi:10.1371/journal.pone.0053694

Sauer B, Henderson N (1988) Site-specific DNA recombination in mammalian cells by the Cre recombinase of bacteriophage P1. Proc Natl Acad Sci U S A 85(14):5166–5170

Sazonova O, Zhao Y, Nurnberg S, Miller C, Pjanic M, Castano VG, Kim JB, Salfati EL, Kundaje AB, Bejerano G, Assimes T, Yang X, Quertermous T (2015) Characterization of TCF21 downstream target regions identifies a transcriptional network linking multiple independent coronary artery disease loci. PLoS Genet 11(5):e1005202. doi:10.1371/journal.pgen.1005202

Schulte I, Schlueter J, Abu-Issa R, Brand T, Manner J (2007) Morphological and molecular left-right asymmetries in the development of the proepicardium: a comparative analysis on mouse and chick embryos. Dev Dyn 236(3):684–695. doi:10.1002/dvdy.21065

Senyo SE, Steinhauser ML, Pizzimenti CL, Yang VK, Cai L, Wang M, Wu TD, Guerquin-Kern JL, Lechene CP, Lee RT (2013) Mammalian heart renewal by pre-existing cardiomyocytes. Nature 493(7432):433–436. doi:10.1038/nature11682

Serluca FC (2008) Development of the proepicardial organ in the zebrafish. Dev Biol 315(1):18–27. doi:10.1016/j.ydbio.2007.10.007

Smart N, Risebro CA, Melville AA, Moses K, Schwartz RJ, Chien KR, Riley PR (2007a) Thymosin beta4 induces adult epicardial progenitor mobilization and neovascularization. Nature 445(7124):177–182. doi:10.1038/nature05383

Smart N, Risebro CA, Melville AA, Moses K, Schwartz RJ, Chien KR, Riley PR (2007b) Thymosin beta-4 is essential for coronary vessel development and promotes neovascularization via adult epicardium. Ann N Y Acad Sci 1112:171–188. doi:10.1196/annals.1415.000

Smart N, Risebro CA, Clark JE, Ehler E, Miquerol L, Rossdeutsch A, Marber MS, Riley PR (2010) Thymosin beta4 facilitates epicardial neovascularization of the injured adult heart. Ann N Y Acad Sci 1194:97–104. doi:10.1111/j.1749-6632.2010.05478.x

Smart N, Bollini S, Dube KN, Vieira JM, Zhou B, Davidson S, Yellon D, Riegler J, Price AN, Lythgoe MF, Pu WT, Riley PR (2011) De novo cardiomyocytes from within the activated adult heart after injury. Nature 474(7353):640–644. doi:10.1038/nature10188

Sosne G, Qiu P, Christopherson PL, Wheater MK (2007) Thymosin beta 4 suppression of corneal NFkappaB: a potential anti-inflammatory pathway. Exp Eye Res 84(4):663–669. doi:10.1016/j.exer.2006.12.004

Stuckmann I, Evans S, Lassar AB (2003) Erythropoietin and retinoic acid, secreted from the epicardium, are required for cardiac myocyte proliferation. Dev Biol 255(2):334–349

Sylva M, van den Hoff MJ, Moorman AF (2014) Development of the human heart. Am J Med Genet A 164A(6):1347–1371. doi:10.1002/ajmg.a.35896

Tao J, Doughman Y, Yang K, Ramirez-Bergeron D, Watanabe M (2013) Epicardial HIF signaling regulates vascular precursor cell invasion into the myocardium. Dev Biol 376(2):136–149. doi:10.1016/j.ydbio.2013.01.026

Tian X, Hu T, Zhang H, He L, Huang X, Liu Q, Yu W, He L, Yang Z, Yan Y, Yang X, Zhong TP, Pu WT, Zhou B (2014) Vessel formation. De novo formation of a distinct coronary vascular population in neonatal heart. Science 345(6192):90–94. doi:10.1126/science.1251487

Tomanek RJ (2015) Developmental progression of the coronary vasculature in human embryos and fetuses. Anat Rec. doi:10.1002/ar.23283

Tomanek RJ, Sandra A, Zheng W, Brock T, Bjercke RJ, Holifield JS (2001) Vascular endothelial growth factor and basic fibroblast growth factor differentially modulate early postnatal coronary angiogenesis. Circ Res 88(11):1135–1141

van Tuyn J, Atsma DE, Winter EM, van der Velde-van Dijke I, Pijnappels DA, Bax NA, Knaan-Shanzer S, Gittenberger-de Groot AC, Poelmann RE, van der Laarse A, van der Wall EE, Schalij MJ, de Vries AA (2007) Epicardial cells of human adults can undergo an epithelial-to-mesenchymal transition and obtain characteristics of smooth muscle cells in vitro. Stem Cells 25(2):271–278. doi:10.1634/stemcells.2006-0366

van Wijk B, Gunst QD, Moorman AF, van den Hoff MJ (2012) Cardiac regeneration from activated epicardium. PLoS One 7(9):e44692. doi:10.1371/journal.pone.0044692

von Gise A, Zhou B, Honor LB, Ma Q, Petryk A, Pu WT (2011) WT1 regulates epicardial epithelial to mesenchymal transition through beta-catenin and retinoic acid signaling pathways. Dev Biol 356(2):421–431. doi:10.1016/j.ydbio.2011.05.668

Wada AM, Smith TK, Osler ME, Reese DE, Bader DM (2003) Epicardial/mesothelial cell line retains vasculogenic potential of embryonic epicardium. Circ Res 92(5):525–531. doi:10.1161/01.RES.0000060484.11032.0B

Wagner KD, Wagner N, Bondke A, Nafz B, Flemming B, Theres H, Scholz H (2002) The Wilms' tumor suppressor Wt1 is expressed in the coronary vasculature after myocardial infarction. FASEB J 16(9):1117–1119. doi:10.1096/fj.01-0986fje

Wang J, Cao J, Dickson AL, Poss KD (2015) Epicardial regeneration is guided by cardiac outflow tract and hedgehog signalling. Nature 522(7555):226–230. doi:10.1038/nature14325

Wei K, Serpooshan V, Hurtado C, Diez-Cunado M, Zhao M, Maruyama S, Zhu W, Fajardo G, Noseda M, Nakamura K, Tian X, Liu Q, Wang A, Matsuura Y, Bushway P, Cai W, Savchenko A, Mahmoudi M, Schneider MD, van den Hoff MJ, Butte MJ, Yang PC, Walsh K, Zhou B, Bernstein D, Mercola M, Ruiz-Lozano P (2015) Epicardial FSTL1 reconstitution regenerates the adult mammalian heart. Nature 525(7570):479–485. doi:10.1038/nature15372

Wessels A, van den Hoff MJ, Adamo RF, Phelps AL, Lockhart MM, Sauls K, Briggs LE, Norris RA, van Wijk B, Perez-Pomares JM, Dettman RW, Burch JB (2012) Epicardially derived fibroblasts preferentially contribute to the parietal leaflets of the atrioventricular valves in the murine heart. Dev Biol 366(2):111–124. doi:10.1016/j.ydbio.2012.04.020

Wills AA, Holdway JE, Major RJ, Poss KD (2008) Regulated addition of new myocardial and epicardial cells fosters homeostatic cardiac growth and maintenance in adult zebrafish. Development 135(1):183–192. doi:10.1242/dev.010363

Wilm B, Ipenberg A, Hastie ND, Burch JB, Bader DM (2005) The serosal mesothelium is a major source of smooth muscle cells of the gut vasculature. Development 132(23):5317–5328. doi:10.1242/dev.02141

Winter EM, van Oorschot AA, Hogers B, van der Graaf LM, Doevendans PA, Poelmann RE, Atsma DE, Gittenberger-de Groot AC, Goumans MJ (2009) A new direction for cardiac regeneration therapy: application of synergistically acting epicardium-derived cells and cardiomyocyte progenitor cells. Circ Heart Fail 2(6):643–653. doi:10.1161/CIRCHEARTFAILURE.108.843722

Witty AD, Mihic A, Tam RY, Fisher SA, Mikryukov A, Shoichet MS, Li RK, Kattman SJ, Keller G (2014) Generation of the epicardial lineage from human pluripotent stem cells. Nat Biotechnol 32(10):1026–1035. doi:10.1038/nbt.3002

Wu M, Smith CL, Hall JA, Lee I, Luby-Phelps K, Tallquist MD (2010) Epicardial spindle orientation controls cell entry into the myocardium. Dev Cell 19(1):114–125. doi:10.1016/j.devcel.2010.06.011

Wu B, Zhang Z, Lui W, Chen X, Wang Y, Chamberlain AA, Moreno-Rodriguez RA, Markwald RR, O'Rourke BP, Sharp DJ, Zheng D, Lenz J, Baldwin HS, Chang CP, Zhou B (2012) Endocardial cells form the coronary arteries by angiogenesis through myocardial-endocardial VEGF signaling. Cell 151(5):1083–1096. doi:10.1016/j.cell.2012.10.023

Xavier-Neto J, Shapiro MD, Houghton L, Rosenthal N (2000) Sequential programs of retinoic acid synthesis in the myocardial and epicardial layers of the developing avian heart. Dev Biol 219(1):129–141. doi:10.1006/dbio.1999.9588

Yamaguchi Y, Cavallero S, Patterson M, Shen H, Xu J, Kumar SR, Sucov HM (2015) Adipogenesis and epicardial adipose tissue: a novel fate of the epicardium induced by mesenchymal transformation and PPARgamma activation. Proc Natl Acad Sci U S A 112(7):2070–2075. doi:10.1073/pnas.1417232112

Yang JT, Rayburn H, Hynes RO (1995) Cell adhesion events mediated by alpha 4 integrins are essential in placental and cardiac development. Development 121(2):549–560

Zamora M, Manner J, Ruiz-Lozano P (2007) Epicardium-derived progenitor cells require beta-catenin for coronary artery formation. Proc Natl Acad Sci U S A 104(46):18109–18114. doi:10.1073/pnas.0702415104

Zhou B, Ma Q, Rajagopal S, Wu SM, Domian I, Rivera-Feliciano J, Jiang D, von Gise A, Ikeda S, Chien KR, Pu WT (2008) Epicardial progenitors contribute to the cardiomyocyte lineage in the developing heart. Nature 454(7200):109–113. doi:10.1038/nature07060

Zhou B, von Gise A, Ma Q, Hu YW, Pu WT (2010) Genetic fate mapping demonstrates contribution of epicardium-derived cells to the annulus fibrosis of the mammalian heart. Dev Biol 338(2):251–261. doi:10.1016/j.ydbio.2009.12.007

Zhou B, Honor LB, He H, Ma Q, Oh JH, Butterfield C, Lin RZ, Melero-Martin JM, Dolmatova E, Duffy HS, Gise A, Zhou P, Hu YW, Wang G, Zhang B, Wang L, Hall JL, Moses MA, McGowan FX, Pu WT (2011) Adult mouse epicardium modulates myocardial injury by secreting paracrine factors. J Clin Invest 121(5):1894–1904. doi:10.1172/JCI45529

Zhou B, Honor LB, Ma Q, Oh JH, Lin RZ, Melero-Martin JM, von Gise A, Zhou P, Hu T, He L, Wu KH, Zhang H, Zhang Y, Pu WT (2012) Thymosin beta 4 treatment after myocardial infarction does not reprogram epicardial cells into cardiomyocytes. J Mol Cell Cardiol 52(1):43–47. doi:10.1016/j.yjmcc.2011.08.020

Generation and Application of Human Pluripotent Stem Cell-Derived Cardiomyocytes

4

Adam J.T. Schuldt, Marisol Romero-Tejeda, and Paul W. Burridge

Abstract

Human induced pluripotent stem cell-derived cardiomyocytes (hiPSC-CMs) are a powerful technology with established applications in regenerative medicine, disease modeling, drug testing and discovery, and developmental biology. The in vitro generation of human cardiomyocytes has progressed rapidly over the last 15 years, and cardiac differentiation of human pluripotent stem cells (hPSC) is now a simple and routine process for many laboratories, generating contracting fetal-like cardiomyocytes that recapitulate a variety of known cardiovascular disease phenotypes. Here we discuss the history and existing knowledge in the field on the generation of cardiomyocytes from human pluripotent stem cells. We cover the potential clinical and investigational applications of hPSC-CMs, as well as the role of the pluripotent state in differentiation, and current methodologies for differentiation, purification, directed cardiomyocyte subtype specification and maturation, and large-scale production. We conclude with a discussion on the future of the field.

A.J. Schuldt
Department of Pharmacology, Northwestern University Feinberg School of Medicine, Chicago, IL 60611, USA

Center for Pharmacogenomics, Northwestern University Feinberg School of Medicine, Chicago, IL 60611, USA

Division of Cardiology, Northwestern University Feinberg School of Medicine, Chicago, IL 60611, USA

M. Romero-Tejeda • P.W. Burridge (✉)
Department of Pharmacology, Northwestern University Feinberg School of Medicine, Chicago, IL 60611, USA

Center for Pharmacogenomics, Northwestern University Feinberg School of Medicine, Chicago, IL 60611, USA
e-mail: paul.burridge@northwestern.edu

M. Ieda, W.-H. Zimmermann (eds.), *Cardiac Regeneration*, Cardiac and Vascular Biology 4, DOI 10.1007/978-3-319-56106-6_4

4.1 Introduction

The last two decades have witnessed a sea change in cardiology, in which the heart has gone from being perceived as incapable of regenerating new cardiomyocytes to an organ which contains potential for cardiomyocyte replacement by proliferation of existing cardiomyocytes, differentiation of resident or circulating stem and progenitor cells, the transdifferentiation of endogenous cardiac fibroblasts, or the engraftment of exogenous human pluripotent stem cell-derived cardiomyocytes (hPSC-CMs). Heart disease remains the leading cause of death in both men and women in the USA (Mozaffarian et al. 2016). It has been estimated that 1–1.7 billion cardiomyocytes have been lost in a failing heart (Murry et al. 2006; Beltrami et al. 1994). Yet, in the face of such a large contractile cell deficit, the level of in situ cardiomyocyte renewal in humans remains low. While medical therapies for coronary heart disease and heart failure have improved mortality related to heart disease since 1980 (Mozaffarian et al. 2016), these therapies usually do not represent cures. Furthermore, a number of genetic cardiac diseases have limited treatment options available. Progress in treating these diseases is hindered by the limitations inherent to studying animal models that are often inadequate mimics of human disease and may involve different pathophysiology despite a similar phenotype.

Several cell types have been proposed as candidates for use in cardiac regenerative therapies, and most of these have demonstrated limited success in generating new cardiomyocytes in vivo, particularly in large animal studies (Bolli et al. 2013; Johnston et al. 2009; Leiker et al. 2008; Hatzistergos et al. 2010), with observed improvements in cardiac parameters being largely attributed to paracrine effects (Gallina et al. 2015; Broughton and Sussman 2016; Khanabdali et al. 2016; Tang et al. 2010). Among these cells, hPSCs have proven to have the greatest potential for cardiomyocyte differentiation and myocardial repair. Initially, this option was limited to embryonic stem cells (ESC); however, their use for therapeutic purposes entails an allogenic therapy with concerns over immune rejection and the need for immunosuppression. In 2006, Takahashi and Yamanaka opened the field of induced pluripotent stem cells (iPSC) with their creation of cells that behaved like mouse embryonic stem cells (mESC) from mouse fibroblasts that had been retrovirally transduced with four genes: *Pou5f1 (Oct4)*, *Sox2*, *Klf4*, and *Myc* (Takahashi and Yamanaka 2006). These cells showed gene expression profiles similar to mESCs, potential to form teratomas and differentiate into cells from all three germ layers, and had clonogenic potential. Subsequently, germline competence was established by selecting for *Nanog*- or native *Oct4*-expressing miPSCs (Okita et al. 2007; Wernig et al. 2007; Maherali et al. 2007). In 2007 and 2008, reports of iPSCs derived from human cells (hiPSC) were published using the same *OCT4*, *SOX2*, *KLF4*, and *MYC* combination of factors (Takahashi et al. 2007; Park et al. 2008). Despite the identification of numerous other reprogramming gene combinations, the use of the four "OSKM" Yamanaka factors with non-integrating Sendai viruses for transduction (Fusaki et al. 2009) remains the most common approach for generation of hiPSCs today.

The advances in hiPSCs make autologous regenerative cardiac therapies more plausible and also introduce a platform upon which to model genetic cardiac diseases by way of creating disease-specific and even patient-specific models for investigation. Despite this progress, there remains a great deal of work to be done to improve our understanding of the genetics and epigenetics of human heart development, the generation of de novo cardiomyocytes both in vitro and in vivo, and factors involved in cardiac repair before safe and effective approaches for human cardiac regenerative medicine can be developed for large-scale use.

4.2 Applications

The creation of hiPSC-CMs has opened the floodgates to possibilities for therapeutic applications of these cells as well as never-before-available models for investigation of development, disease processes, and drug and toxicity screening. The ability to easily and reproducibly generate large numbers of human cardiomyocytes from hiPSCs has placed a powerful tool in the hands of investigators to accelerate progress in a number of areas, several of which are outlined here.

4.2.1 Regenerative Medicine

While regenerating over a billion cardiomyocytes in a failing heart is a lofty goal, current evidence would support its pursuit. Multiple stem and progenitor cell populations have been proposed as candidates to meet this challenge, including bone marrow-derived stem cells (e.g., mesenchymal stem cells, c-KIT$^+$ cells, CD34$^+$ cells, Hoechst exclusion or "side population cells"), resident cardiac progenitor cells (e.g., c-KIT$^+$, CD34$^+$, ISL1$^+$, Hoechst exclusion, and cardiosphere-derived cells), cord blood cells, and fat-derived stem cells. Indeed, several of these cell types have shown modest improvements in cardiac function in post-injury models, as measured by parameters such as left ventricular ejection fraction (LVEF), regional systolic function, or scar reduction following myocardial infarction (MI). However, scant evidence supports clinically meaningful levels of cardiomyocyte differentiation of any of these cell types in vivo (Bolli et al. 2013; Johnston et al. 2009; Leiker et al. 2008; Hatzistergos et al. 2010; Gallina et al. 2015; Broughton and Sussman 2016; Khanabdali et al. 2016; Tang et al. 2010).

In contrast, ESCs and iPSCs have demonstrated far greater capacity to generate de novo cardiomyocytes in vitro (Kattman et al. 2011; Lian et al. 2012, 2013; Hudson et al. 2012; Burridge et al. 2014), as well as in vivo (Chong et al. 2014; Carpenter et al. 2012). Despite the successes with in vitro cardiomyocyte differentiation, in vivo data demonstrating significant regeneration of cardiac tissue is less plentiful. Nevertheless, there is a reason for optimism. Since a defining characteristic of PSCs is the ability to form teratomas, most in vivo studies have focused on delivery of PSC-CMs, rather than PSCs themselves. Delivery of 2×10^6 hiPSC-CMs to a rat heart post-MI demonstrated the presence of cardiomyocyte markers in engrafted cells as late as 10 weeks after injection, although there was no significant improvement in cardiac function by MRI (Carpenter et al. 2012). However,

Ong and colleagues delivered the same dose of hiPSC-CMs to an immunodeficient mouse heart post-MI and observed poor engraftment at 35 days using bioluminescent imaging despite improvements in cardiac function by MRI and invasive hemodynamic monitoring (Ong et al. 2015). Further analyses suggested that release of proangiogenic and antiapoptotic factors by the transplanted cells might underlie the observed improvement in cardiac function after injury (Ong et al. 2015). In the first study in nonhuman primates, delivery of 1 billion hESC-CMs to post-MI macaque hearts resulted in extensive remuscularization with human cardiomyocytes with evidence of electrical integration with host myocardium (Chong et al. 2014). This study demonstrated the feasibility of generation of large numbers of hPSC-CMs in addition to providing strong evidence of myocardial regeneration.

A growing body of evidence supports the notion that combination cell therapy may improve transplanted cell retention and cardiac repair. Ye et al. (2014) delivered a combination of hiPSC-derived cardiomyocytes, endothelial cells, and smooth muscle cells to a porcine heart following MI. In addition, they placed a fibrin patch containing IGF1-releasing gelatin microspheres over the site of injury prior to cell injection. The use of the fibrin patch resulted in improved cell retention (~9% of injected cells vs. ~4% without the patch, as assessed by quantitative PCR for human Y chromosome in tissue digests). Histological analyses confirmed hiPSC-CMs within muscle fibers in the heart. This treatment resulted in significant improvements in LVEF, wall stress, apoptosis, myocardial energetics, and arteriole density.

4.2.2 Biological Pacemaker Development

In addition to treating disorders of cardiac mechanical pump function, there has also been interest in using cell-based therapies to treat electrical heart problems. Patients who suffer from sinoatrial node dysfunction and bradycardia currently rely on electronic pacemakers to take over pacing of the heart. While these devices are highly effective, they do have shortcomings which include limited battery life necessitating generator changes, the potential for devastating infections, lack of autonomic responsiveness, pacing-induced cardiomyopathy, and lead fracture or dislodgement, which is particularly problematic in the pediatric population as they grow and attempt to carry out a normal active life (Rosen et al. 2011). The ability to biologically replace a failing sinoatrial node might alleviate several of these problems. In the case of genetic causes of sinus node dysfunction, nodal cells could be created from autologous hiPSCs that have been genetically modified to correct the defect and restore normal pacing function. For example, overexpression of HCN4 (the ion channel responsible for the pacemaker current, I_f, which has been implicated in sick sinus syndrome) in mESC-CMs increased their spontaneous beating rate and allowed them to pace hiPSC-CMs in co-culture (Saito et al. 2015). These cells also demonstrated β-adrenergic responsiveness.

To date, all efforts at cardiomyocyte generation from hiPSCs have produced a mixed population of cardiac myocytes which resemble ventricular, atrial, and nodal cells, based on their action potential characteristics. On the one hand, this heterogeneity raises concerns about potential arrhythmia generation in clinical applications (see "Cardiomyocyte Subtype Specification" below). However, this also highlights the fact that the creation of nodal cardiomyocytes from hPSCs appears to be possible in vitro, and ongoing efforts to control the subtype of cardiomyocytes produced will likely allow for the generation of pure ventricular as well as nodal cells for their respective applications. Kehat and colleagues were the first to demonstrate the principle that hPSC-CMs could provide cardiac pacing by delivering hESC-CMs to porcine hearts with complete atrioventricular block (Kehat et al. 2004). The delivered cells integrated into the cardiac syncytium and paced the hearts, as demonstrated by 3D electrical mapping. Similar results might be expected from hiPSC-derived cells, although Lee et al. (2011) provide reason for caution. They investigated the calcium-handling characteristics of hiPSC- and hESC-derived cells and found that the calcium handling of hiPSC-CMs was relatively immature compared to hESC-CMs (Lee et al. 2011). This could limit their potential as biological pacemakers (Barbuti and Robinson 2015), although efforts to improve maturation of hiPSC-CMs cells are currently a major focus of research by several groups.

4.2.3 Understanding of Human Cardiac Development

While the use of hiPSCs for regenerative therapies may be their most obvious application in the cardiovascular space, there are multiple other areas in which these cells are already proving valuable. Improving the efficiency of hiPSC differentiation into cardiomyocytes has relied upon a thorough understanding of normal cardiac development. Human heart development is uniquely difficult to study, as embryonic and fetal heart tissue is difficult to obtain and raises obvious ethical concerns. Even adult primary cardiomyocytes are difficult to isolate and can only be maintained for days in active cell culture. These facts have forced the research community to utilize animal models to understand cardiac developmental biology, focusing largely on murine, avian, and frog cardiogenesis. The discovery and successful culture of ESCs opened new possibilities to study cardiac development in a human cell model. At present, much of what we know about human cardiac development has been discovered or verified by investigating hESC signaling and gene expression profiles. This knowledge has guided approaches to cardiomyocyte differentiation of hPSCs, while hPSC-CMs in turn provide ever improving models of cardiogenesis. Furthermore, hiPSCs provide a valuable model for studying cardiac development in both normal and diseased states. Human iPSCs can be created from patients with congenital or developmental heart disease and used to study abnormal development or the effects of various environmental toxins or pharmacologic agents on development in the setting of various patient-specific genetic backgrounds that are known to produce a phenotype, thus greatly enhancing our understanding of these diseases.

4.2.4 Disease Modeling

Human iPSCs provide a platform of unprecedented breadth for the study of virtually any disease process with a genetic association. The creation of hiPSC-CMs derived from patients allows for the study of their specific pathophysiology, whether it involves a genetic or acquired cardiomyopathy, arrhythmia, defect in development, or metabolic disorder. Several cardiac disease phenotypes have already been modeled in hiPSC-CMs from affected patients.

Familial cardiomyopathies have been associated with several known mutations, often involving genes that encode for proteins that comprise the contractile apparatus of the cell. In one of the first examples of the use of patient-derived hiPSCs to study cardiac disease, Sun et al. created hiPSC-CMs from seven members of a family affected by familial dilated cardiomyopathy (DCM), including controls and individuals carrying a point mutation (R173W) in the cardiac troponin T (*TNNT2*) gene (Sun et al. 2012). The hiPSC-CMs from patients with the mutation showed impaired myofilament regulation, contractility, and calcium handling compared to controls, although the cardiomyocytes were notably immature. Subsequent work from this group demonstrated abnormal β-adrenergic signaling and contractile function in these DCM hiPSC-CMs and a possible role for nuclear localization of the mutated TNNT2 in epigenetic modifications resulting in upregulation of phosphodiesterase 2A and 3A, which related to compromised β-adrenergic signaling and contractility (Wu et al. 2015).

Other mutations involved in DCMs have since been modeled in hiPSC-CMs as well, including a mutation in the gene encoding lamin A/C (*LMNA*), demonstrating increased electrical stimulation-induced nuclear senescence and apoptosis in hiPSC-CMs, which could be reduced or eliminated by pharmacologic blockade of the MEK1/ERK1/2 pathway (Siu et al. 2012). A mutation in the gene for desmin (*DES*) was identified in another DCM patient, and hiPSC-CMs derived from this patient demonstrated structural and functional abnormalities due to protein misfolding that recapitulated findings from pathologic specimens (Tse et al. 2013). Sarcomere insufficiency was identified as a cause of DCM in patients who carry truncating mutations in the gene for titin (*TTN*), a major structural protein in myofibrils (Hinson et al. 2015). In another example, hiPSC-CMs from a DCM patient carrying a phospholamban (*PLN*) mutation demonstrated abnormal Ca^{2+} handling, electrical instability, and abnormal cytoplasmic distribution of phospholamban protein, recapitulating findings in primary cardiomyocytes of patients with *PLN* mutations and DCM (Karakikes et al. 2015).

These efforts highlight the fact that there are now over 50 genes which have been implicated in familial DCMs (Skrzynia et al. 2015), and these various mutated genes confer nearly indistinguishable clinical phenotypes via very different mechanisms. The use of hiPSCs has been instrumental in defining these causative factors and may aid in identifying unique treatments for individual patients carrying specific mutations. As such, the ability to create cardiomyocytes from hiPSCs represents one of the greatest steps toward the realization of personalized medicine, in which treatments for diseases can be tailored to an individual patient based on their genetics or the specifics of their pathophysiology.

Several other cardiac diseases have now been modeled in hiPSCs, including hypertrophic cardiomyopathy (Lan et al. 2013; Han et al. 2014; Birket et al. 2015a; Tanaka et al. 2014; Ojala et al. 2016); arrhythmogenic right ventricular cardiomyopathy (Kim et al. 2013; Ma et al. 2013a; Wen et al. 2015); diabetic cardiomyopathy (Drawnel et al. 2014); familial long QT syndrome type 1 (Moretti et al. 2010; Egashira et al. 2012; Ma et al. 2015), type 2 (Itzhaki et al. 2011; Matsa et al. 2011; Lahti et al. 2012; Matsa et al. 2014; Jouni et al. 2015; Mehta et al. 2014), type 3 (Ma et al. 2013b; Fatima et al. 2013; Terrenoire et al. 2013; Malan et al. 2016), and type 8 (Timothy syndrome) (Yazawa et al. 2011); catecholaminergic polymorphic ventricular tachycardia types 1 (CPVT1) (Fatima et al. 2011; Jung et al. 2012; Itzhaki et al. 2012; Kujala et al. 2012; Zhang et al. 2013; Di Pasquale et al. 2013; Penttinen et al. 2015; Novak et al. 2015) and 2 (CPVT2) (Novak et al. 2012, 2015); hypoplastic left heart syndrome (Jiang et al. 2014; Kobayashi et al. 2014; Theis et al. 2015); and the cardiac phenotypes of Pompe disease (Huang et al. 2011; Raval et al. 2015; Sato et al. 2015) and Duchenne muscular dystrophy (Dick et al. 2013; Guan et al. 2014; Lin et al. 2015; Hashimoto et al. 2016).

Human iPSC-CMs also allow investigators to go beyond simply recreating disease physiology in human cells and tissues in vitro to demonstrating causality of a given mutation or signaling pathway by providing a platform for human gene knockdown and overexpression techniques or the introduction of mutations into control human cardiomyocytes to demonstrate reproducibility of a given phenotype. This is particularly important when one considers that observed differences between diseased hiPSC-CMs and control cells could be due to off-target genetic differences. For example, selective expression of a mutated titin gene in control hiPSC-CMs resulted in reduced contractile function, recapitulating the DCM phenotype seen in diseased cells (Hinson et al. 2015), and knockdown of LMNA in control hiPSC-CMs using shRNA recreated the phenotypic changes and susceptibility to electrical stress seen in *LMNA* mutant hiPSC-CMs (Siu et al. 2012). Similarly, normal genes can be introduced into diseased hiPSC-CMs to demonstrate rescue of a disease phenotype. Karakikes et al. used both a TALEN-mediated gene correction and a combination strategy of knockdown of the mutant *PLN* and addition of the normal gene into hiPSCs prior to cardiomyocyte differentiation to confirm the role of the *PLN* mutation in the DCM phenotype (Karakikes et al. 2015). This approach rescued Ca^{2+} handling and PLN distribution and fully restored normal cardiac phenotype in hiPSC-CMs.

4.2.5 Drug Discovery, Pharmacogenomics, and Cardiotoxicity Screening

Human iPSC-CMs provide an unprecedented ability to study cellular signaling, metabolism, and ion channel function in human cardiomyocytes. Therefore, it is not surprising that several of the hiPSC-CM disease models outlined above have resulted in the identification of novel drugs or potential therapeutic targets which may correct or improve the cardiac defects in question. Studying type 1 long QT syndrome (LQT1), Ma et al. observed decreased slowly activating delayed rectifier potassium channel current (I_{Ks}) in hiPSC-CMs from an LQT1 patient with a novel loss of function mutation in *KCNQ1* (Ma et al. 2015). This mutation resulted in prolonged action potential duration, which confers an increased risk of fatal arrhythmias in these patients. Application of the novel drug ML277 increased I_{Ks} in the LQT1 hiPSC-CMs and shortened action potential duration, identifying this as a potential therapy for LQT1. In a hiPSC-CM model of CPVT1, dantrolene was shown to suppress arrhythmogenic spontaneous Ca^{2+} release events and abolish delayed afterdepolarizations and spontaneous action potentials in CPVT1 hiPSC-CMs carrying an N-terminal mutation in the cardiac ryanodine receptor (Jung et al. 2012). And while recombinant enzyme replacement therapy effectively corrects lysosomal storage in Pompe disease, its effect is incomplete. The use of hiPSC-CMs from a patient with Pompe disease allowed investigators to screen for additional therapies, identifying L-carnitine as having potential benefit due to its ability to rescue mitochondrial function in diseased cardiomyocytes (Huang et al. 2011).

Just as hiPSCs can aid in the discovery of new drugs and drug targets, they also have great potential to improve the efficiency of bringing new therapies to market. Recent estimates suggest that the average cost to bring one new drug to market in the USA, including the cost of failures, is $1.24 billion (Kaitin 2010). Considering that only one in six drugs that made it into clinical trials in the USA actually made it to market (DiMasi et al. 2010), while the costs of clinical trials have been rising in recent years (Kaitin 2010), the wasted R&D expenditures involved are alarming. The high drug attrition rate is at least in part due to unforeseen toxicities encountered during clinical trials that were not predicted by suboptimal screening assays which rely on the use of nonhuman cells such as Chinese hamster ovary (CHO) cells or immortalized transgenically modified human cell lines, such as human embryonic kidney (HEK) cells, during drug development. Of particular concern is the potential for pharmacologically induced cardiac arrhythmias. Human iPSCs provide a model for testing drug toxicity in virtually any human tissue or cell type without having to rely on animal models which may produce very different results. As an example of such use, a recent initiative called the Comprehensive In Vitro Proarrhythmia Assay (CiPA) was begun in an effort to more accurately predict the risk of the potentially fatal cardiac arrhythmia, torsades de pointes (TdP). TdP has been associated with a long list of medications based on surrogate markers for TdP risk, which include inhibition of the potassium channel hERG and prolongation of the QT interval on an electrocardiogram (Fermini et al. 2016). Common practice has been to test an investigational drug on non-cardiomyocyte cells such as hERG transgenic CHO cells (Danker and Moller 2014) and monitor QT intervals in clinical trials. While this method is sensitive for TdP risk, it is not very specific and ignores potential for effects on other ion channels (Fermini et al. 2016). As a result, several drugs that may not truly increase risk of TdP have been forced to go through additional phase IIb and III trials to assess risk of QT prolongation and TdP, at a cost of approximately $1 billion (Fermini et al. 2016). The CiPA initiative was designed to assess drug effects on seven ion channels (not just hERG), generate in silico models of cardiac repolarization from these data, and utilize standardized hiPSC-CM protocols to assess effects on repolarization in an effort to screen out drugs with unacceptable risks of toxicity prior to reaching clinical trials. Combining such an approach with current state-of-the-art technology such as commercially available high-throughput robotic patch-clamp systems and microelectrode arrays could aid in rapid screening of multiple drugs at multiple doses using a panel of hiPSC-CMs from subjects with varying genetic backgrounds (Mordwinkin et al. 2013).

Importantly, patients with underlying cardiac disease or rare genetic mutations may not be included in investigational drug trials, making identification of cardiotoxicity risk virtually impossible until post-market analyses are performed. Furthermore, Liang et al. demonstrated increased susceptibility to cardiotoxicity in diseased cardiomyocytes by creating hiPSC-CMs from healthy subjects and patients with hereditary long QT syndrome, familial HCM, and familial DCM and performing assessments of the effects of known cardiotoxic drugs on ion channel currents in these cells (Liang et al. 2013). Diseased hiPSC-CMs demonstrated increased propensity for arrhythmias, which mimicked that of the patients from whom they were derived, and were superior to hERG assays alone. This work illustrates how high-throughput technologies could be valuable in identifying cardiotoxicities in patients with varying cardiac conditions *before* they are given a new drug.

As an example, hiPSC-CMs have demonstrated benefit in modeling, and even predicting, chemotherapy-induced cardiotoxicity. Anthracyclines are effective in treating a wide range of malignancies, but a subset of patients go on to develop cardiomyopathy and heart failure. Burridge et al. showed that hiPSC-CMs from breast cancer patients who had developed anthracycline-induced cardiomyopathy actually recapitulated the increased predilection toward cardiotoxicity relative to patients who received similar treatment but did not develop cardiotoxicity, suggesting that hiPSCs from cancer patients could be used to screen for risks of serious treatment side effects (Burridge et al. 2016). Such an in vitro model of chemotherapeutic-induced toxicity could be useful in screening cells from large numbers of individuals to identify particular gene variants or single nucleotide polymorphisms that might identify patients at risk for toxicity, moving healthcare closer to the goal of personalized medicine.

4.3 Cardiomyogenesis during Mammalian Heart Development

The heart is the first functioning organ to develop in an embryo, bearing the responsibility for pumping oxygen and nutrients to the other developing tissues. The heart forms through multiple stages of morphogenesis which begin with development of cardiac progenitor cells, formation of the linear heart tube, cardiac looping, chamber formation, septation, and finally maturation. Cardiac development begins following the formation of the blastocyst, when the inner cell mass (from which ESCs are obtained) undergoes gastrulation to form the three primitive germ layers:

endoderm, mesoderm, and ectoderm. Initiation of gastrulation and induction of the mesoderm layer is heavily dependent on NODAL, bone morphogenetic protein (BMP), Wnt, and fibroblast growth factors (FGF) signaling (Kimelman 2006; Noseda et al. 2011; Brade et al. 2013). Mesodermal induction is marked by expression of the T-box transcription factor Brachyury (T), which is itself a target of Wnt/β-catenin signaling. The mesoderm contains precursors of the first and second heart fields (FHF and SHF), which will ultimately give rise to the majority of the atria, ventricles, and the outflow tract myocardium (Brade et al. 2013).

The commitment and differentiation of mesodermal cells to a cardiac lineage is dependent on a complex interaction of signaling pathways. The T^+ cells require NOTCH-mediated inhibition of canonical Wnt/β-catenin signaling and activation of noncanonical Wnt pathways for induction of cardiac transcription factors (Fig. 4.1, upper panel), leading to expression of KDR (FLK1) (Brade et al. 2013; Kwon et al. 2009; Gessert and Kuhl 2010). These T^+/KDR^+ cells give rise to the $MESP1^+$ cardiogenic mesoderm, and finally the FHF and SHF, which form the cardiac crescent. The more anterolaterally located FHF cells are exposed to BMP and FGF signals as well as canonical Wnt pathway inhibitors which further drives their differentiation, and they begin to express NKX2-5, GATA4, and TBX5, as well as contractile proteins myosin light chain-2a (MYL7) and sarcomeric myosin heavy chain (Brade et al. 2013). The FHF cells of the cardiac crescent give rise to the linear heart tube, while the SHF cells migrate into the heart tube later, where FGF signaling maintains progenitor cell proliferation, while Sonic hedgehog-mediated signals from the endoderm and canonical Wnt signaling from the midline neural tube inhibit differentiation and stimulate proliferation (Brade et al. 2013; Kelly 2012). As

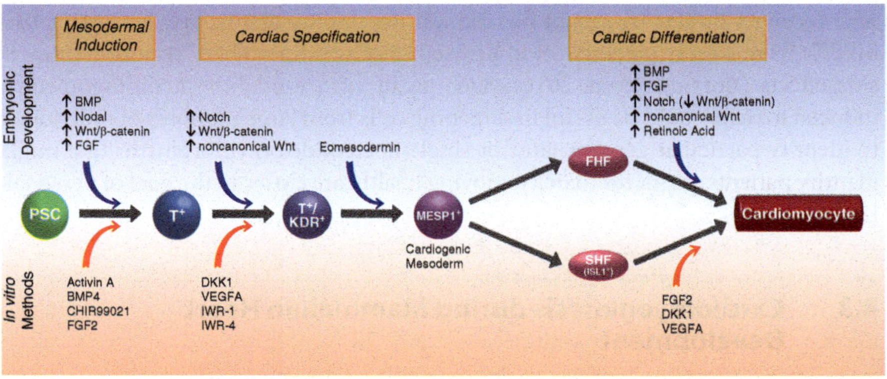

Fig. 4.1 Signaling pathways involved in embryonic cardiomyocyte development (*top*) and factors shown to influence differentiation in vitro (*bottom*). Increases or decreases in the pathways indicated have been shown to promote advancement of cell differentiation during embryonic development. In many cases, successful in vitro cardiomyocyte differentiation strategies have recapitulated these embryonic events. Abbreviations: *BMP* bone morphogenetic protein, *DKK1* Dickkopf-related protein-1, *FGF* fibroblast growth factor, *FHF* first heart field, *IWR* inhibitor of Wnt response, *PSC* pluripotent stem cell, *SHF* second heart field, *VEGFA* vascular endothelial growth factor

with the FHF, BMPs, NOTCH, and noncanonical Wnt signals promote cardiac differentiation of the SHF cells. The later stages of cardiac development include cardiac looping and chamber formation, as well as formation/septation of the outflow tract, which are reviewed elsewhere (Schleich et al. 2013).

4.4 Control of the Pluripotent State

Though all pluripotent cells harbor the ability to differentiate into somatic cell lineages, it is now increasingly evident that pluripotency can exist in a small number of diverse stable states. Of these, two pluripotent phases corresponding to successive stages in development have been described: the naïve state, which corresponds to the pre-implantation blastocyst in mice, and the primed state, which remains pluripotent but has undergone epigenetic modifications.

Historically, mESCs have been acquired from embryonic day 3.5 (E3.5) pre-implantation embryos at the blastocyst stage by explanting cells from the inner cell mass (ICM) and allowing for subsequent outgrowth (Evans and Kaufman 1981). These mESCs, still in the naïve state, form dome-shaped colonies and were classically cultured in media containing fetal bovine serum (FBS) and leukemia inhibitory factor (LIF) (Smith et al. 1988) or bone morphogenetic protein 4 (BMP4) and LIF (Ying et al. 2003). These cells are now commonly maintained by a combination of MAPK/ERK kinase (MEK) and glycogen synthase kinase 3 beta (GSK3B) inhibition plus LIF, commonly referred to as "2i/LIF" (Nichols and Smith 2009). In contrast, mESCs acquired from E5.5–7.5 post-implantation epiblasts (EpiSC) (Tesar et al. 2007; Brons et al. 2007) represent the primed state. They grow in monolayer colonies, proliferate in response to Activin/NODAL/TGFβ and FGF2 signaling (Brons et al. 2007), and have diverse propensities for differentiation (Bernemann et al. 2011). The discernable differences in morphology and varying dependence on exogenous factors between these two populations have been ascribed to the developmental stages from which these cells were derived.

In contrast, human PSCs, whether hESCs derived from blastocysts or hiPSCs reprogrammed from adult somatic cells, show morphological, epigenetic, and growth factor requirements reminiscent of the primed EpiSC-like state in mice. Early attempts at capturing a naïve state of human pluripotency similar to that of mESCs relied on the expression of transgenes, such as sustained doxycycline-dependent expression of *OCT4, SOX2, KLF4,* and *MYC* (De Los et al. 2012). However, recent transgene-free methods of deriving human naïve pluripotent cells have been described, and methods for transferring human pluripotent cells between the naïve and primed states are being developed (Guo et al. 2016; Gafni et al. 2013; Theunissen et al. 2014).

In a recent study, Hanna and colleagues (Gafni et al. 2013) used an *OCT4*-GFP reporter with a doxycycline-inducible system to express reprogramming factors previously shown to convert cells to a naïve-like state. A subsequent screen for conditions that allowed for sustained GFP expression in the absence of doxycycline found that a combination of 2i/LIF, TGFB1, FGF2, along with JNK and p38 inhibitors was sufficient to maintain reporter construct expression and sustain a naïve-like state in primed hiPSCs. Human ESCs and hiPSCs grown under these conditions exhibited molecular and epigenetic features similar to those of naïve mESCs, including utilization of the *OCT4* distal enhancer and a lack of X chromosome inactivation. Furthermore, these cells generated cross-species chimeric mouse embryos and displayed a distinct gene expression pattern when compared to primed hESCs and hiPSCs, underscoring their pluripotency potential and suggesting a distinct pluripotent state.

In another study, a screen for factors that maintain the expression of a reporter of *OCT4* distal enhancer activity identified five kinase inhibitors (of MEK, GSK3B, ROCK, BRAF, and SRC) that, in combination with Activin A and LIF, allow for the maintenance of hESCs in a naïve state as well as for the interconversion between primed and naïve states (Theunissen et al. 2014). Human ESCs cultured under these conditions exhibited an upregulation of transcription factors associated with naïve pluripotency in mice, including *STELLA*, *DPPA2*, *DPPA5*, *REX1*, *KLF4*, *KLF5*, *TFCP2L1*, and *NANOG*. However, they expressed a distinct gene expression pattern when compared to hiPSCs and previously described naïve hESC states. Contrary to findings in mESCs, these cells showed an upregulation of the X-chromosome inactivator *XIST* and inactivation of X-linked gene expression. While this may indicate that the relationship between X inactivation and the naïve pluripotent state may not be evolutionarily conserved, the authors point out that it may also be attributable to X chromosome erosion in their late passage cells and thus warrants further investigation. These reports illustrate the challenges in capturing a naïve state in hPSCs and the lack of consensus regarding what this state might look like.

Refining the relationship between the naïve and primed state provides the opportunity to explore the epigenetic and transcription factor networks governing the stabilization of pluripotency (Guo et al. 2016; Gafni et al. 2013; Theunissen et al. 2014). Notably, a number of pathways appear to play contradictory roles in maintaining the naïve versus primed state. FGF2 signaling, classically used to maintain the pluripotent state of primed hESCs and hiPSCs, plays an antagonistic role in maintaining the naïve state cells (Guo et al. 2016; Gafni et al. 2013). BMP4, which can be used to support naïve cells (Ying et al. 2003), results in the mesoderm or trophectoderm differentiation of primed cells (Xu et al. 2002; Yu et al. 2011), while BMP inhibitors promote self-renewal of hPSCs (Frank et al. 2012). Likewise, JNK and p38 inhibition results in the stabilization of the naïve state but is not required for maintenance of primed cells (Gafni et al. 2013) (Fig. 4.2).

Further investigation of the distinct mechanisms by which these two pluripotent states are sustained may help explain subtle differences in their susceptibility to differentiation and could allow for the derivation of pluripotent states with increased stability and wider differentiation potential. For example, in mice, BMP signaling through SMAD has been shown to be dispensable for pluripotency and instead has been implicated in lineage priming predisposition via control of DNA methylation (Gomes Fernandes et al. 2016). Because the elucidation of signaling networks governing pluripotency is complicated by the use of different media formulations and cell attachment matrices across studies, systematic design of experiments (DOE) approaches using chemically defined media will help better characterize the naïve state. If successfully captured and described, the naïve state of human pluripotency

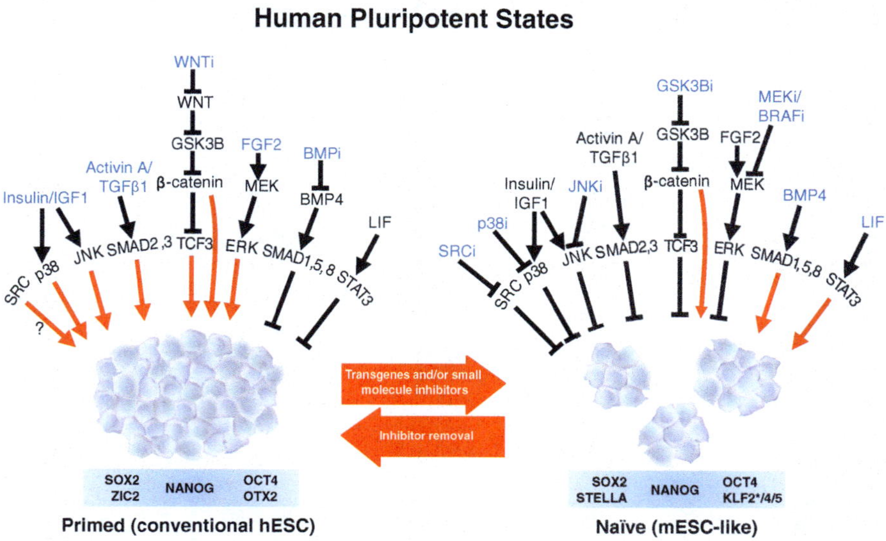

Fig. 4.2 Signaling pathways and reprogramming factors involved in deriving and sustaining the primed and naïve pluripotent states in humans. Multiple signaling pathways influence the primed and naïve states of hPSC. Notably, most of these pathways have opposite effects on stabilization of the primed vs. naïve state. *Blue* text indicates factors that have been utilized in vitro to induce or maintain the respective pluripotent state. *Red arrows* indicate pathways that promote the indicated pluripotent state. Genes in *blue boxes* have been shown to be expressed in the respective state. *KLF2, while not expressed in human pre-implantation epiblast cells (Blakeley et al. 2015), can promote the naïve pluripotent state when overexpressed

may supply a source of unobstructed cells in a stable pluripotent state with the capacity for unrestricted and unbiased differentiation.

4.5 Human Pluripotent Growth Media

The use of patient-derived hiPSCs in regenerative medicine will require the homogeneous production of large quantities of cells cultured in defined and xeno-free environments (Fig. 4.3). Initial conditions for culturing hESCs relied heavily on factors known to support pluripotency in mESCs and involved the use of DMEM supplemented with fetal bovine serum (FBS), glutamine, β-mercaptoethanol (BME), and nonessential amino acids. The use of a mouse embryonic fibroblast (MEF) feeder layer as a source for cell adhesion and of soluble factors was deemed necessary for maintaining pluripotency in both the presence and absence of leukemia inhibitory factor (LIF) (Thomson et al. 1998). Although these culture conditions provided a source of cells suitable for studying human development, the use of xenogenic factors rendered the cultures inadequate for clinical applications.

The elimination of FBS by KnockOut serum replacement (KSR), as well as the addition of human recombinant FGF2, improved the cloning efficiency of cells cultured in KnockOut DMEM (KO-DMEM) (Amit et al. 2000). Likewise, the use of a Matrigel matrix eliminated the need for an MEF feeder layer and established an important role

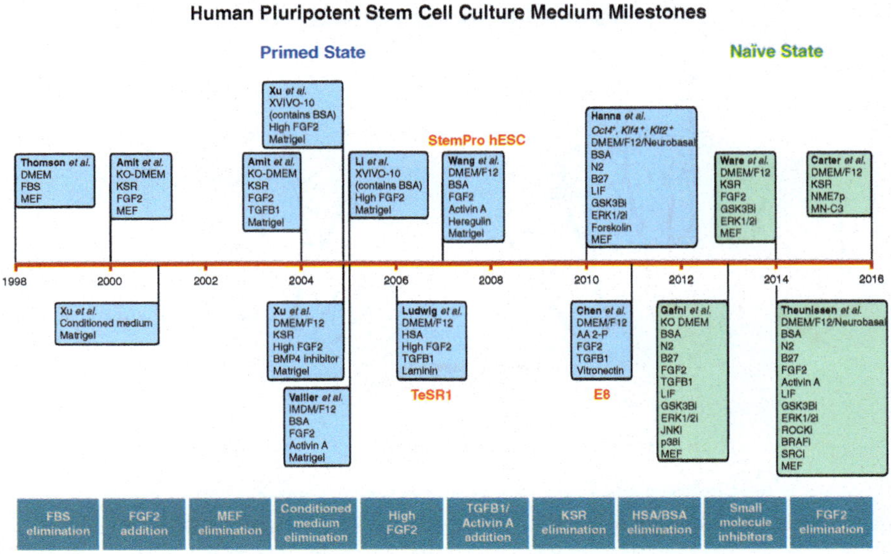

Fig. 4.3 Milestones in human pluripotent stem cell culture. Important publications in the evolution of pluripotent stem cell media. Included are key media components, growth factors, and small molecules employed to sustain pluripotency. *Red text* indicates the name of the media formulation when commercialized. *Blue boxes* indicate media derived for primed hPSCs, with naïve hPSC media in *green boxes*. Boxes below the timeline depict important milestones. Abbreviations: *AA 2-P* L-ascorbic acid 2-phosphate, *BMP4* bone morphogenetic protein 4, *BSA* bovine serum albumin, *FBS* fetal bovine serum, *FGF2* basic fibroblast growth factor, *HSA* human serum albumin, *KO-DMEM*, KnockOut DMEM, *KSR* KnockOut serum replacement (Xu et al. 2001, 2005a, b; Amit et al. 2004; Hanna et al. 2010; Ware et al. 2014; Carter et al. 2016)

for laminin in mediating cell-matrix interactions but required the use of conditioned media from heterologous sources (Xu et al. 2001). This obstacle was overcome by the addition of transforming growth factor β1 (TGFB1) (Amit et al. 2004) as well as the inclusion of higher levels of FGF2 (Xu et al. 2005a) and suppression of BMP signaling using NOGGIN (Xu et al. 2005b). While these factors allowed for the feeder-free maintenance of hESCs in basal media, the use of KSR and Matrigel remained problematic, as both contain undefined components of animal origin. Nonetheless, the shift away from conditioned media spurred the identification of key factors necessary to maintain pluripotency, such as the cooperation of the Activin/Nodal and FGF2 pathways. This allowed for the replacement of KSR and BMP inhibition with bovine serum albumin (BSA) and insulin (Vallier et al. 2005), as well as for the culture of hESCs in serum-free medium (X-VIVO 10) supplemented with high levels of human recombinant FGF2 (Li et al. 2005). The optimized development of defined medium containing DMEM/F12, human serum albumin (HSA), high FGF2, and TGFB1 (TeSR1) by Ludwig et al. marked the first time hESCs had been both derived and maintained in defined conditions free of nonhuman products (Ludwig et al. 2006), and its commercialization laid the groundwork for establishing standardized methods of culturing hESCs. A focus on highly expressed receptor tyrosine kinases in hESCs led to the development of pluripotency media containing FGF2 and Activin A as well as the ERBB2/ERBB3 ligand heregulin1β (commercialized as StemPro hESC) and established an important role for insulin signaling in maintaining pluripotency in the absence of KSR (Wang et al. 2007).

However, the use of human or bovine serum albumin in these media limited their large-scale use for therapeutic applications. The careful examination of basal (DMEM/F12) and TeSR components allowed for the formulation of albumin-free media by revealing BME-mediated toxicity in the absence of BSA and allowed for the identification of components necessary to maintain pluripotency (Chen et al. 2011). The resulting formulation (E8) consists of chemically defined components including FGF2, TGFB1, insulin, L-ascorbic acid 2-phosphate, selenium, and transferrin and is free of xenogenic factors, making it suitable for use in clinical applications (Chen et al. 2011). Although hESCs and hiPSCs can now be maintained in chemically defined conditions, the further optimization of components using DOE approaches may result in formulations that allow for improved culture conditions (Marinho et al. 2015).

4.6 Priming for Cardiac Differentiation

Just as optimization of culture conditions can improve maintenance of the pluripotent state, the right conditions can also bias pluripotency toward cardiac cells. This approach requires the stabilization of a pluripotent cell in a state that is poised to become mesoderm. The derivation of such a cell state must consider determinants of fate, including the relative expression of pluripotency factors, as these can play both inductive and inhibitive roles in lineage determination.

Pluripotency has proven to be quite diverse. Various studies have noted differences in directed differentiation efficiency among hiPSC lines as well as in the distribution of cell types within teratomas formed by these cells. Furthermore, lineage-associated genes and pluripotency factors can be co-expressed and can fluctuate within a single culture (Montserrat et al. 2013; Cahan and Daley 2013). While it is unclear whether the diversity in gene expression and differentiation potential arises due to donor- or clone-specific gene expression, time in culture, variability in culture conditions, or is simply a reflection of pluripotency in vivo, these observations lend support for the idea that cells can take on various stable states that are primed, yet undifferentiated. Therefore, the challenge in deriving a cell line primed specifically for cardiac differentiation lies in minimizing the heterogeneity observed and driving pluripotency toward a specific state that is exclusively poised for differentiation into the cell of interest.

Establishing a primed mesodermal state will require a delicate balance of expressed pluripotency factors and mesodermal genes. Pluripotency factors, including *OCT4*, *SOX2*, and *NANOG*, were identified based on their expression in the pluripotent state. However, it is now clear that these factors also play a role in specifying lineage commitment. OCT4 and SOX2 have been associated with the induction of mesendoderm and neuroectoderm fates, respectively (Thomson et al. 2011); therefore, their relative expression will need to be regulated in a way that allows for both pluripotency and facilitated entry into mesodermal lineages.

Recent reports show that pluripotency can also be achieved by overexpression of genes associated with embryonic lineages, such as the mesendodermal lineage specifier *GATA3* (Montserrat et al. 2013; Shu et al. 2013). This suggests a model in which transcription factors can act as lineage specifiers and repressors rather than as direct inducers of pluripotency (Montserrat et al. 2013; Thomson et al. 2011; Shu et al. 2013). The successful reprogramming of human fibroblasts using lineage determinants to replace core pluripotency factors *OCT4* and *SOX2* (Montserrat et al. 2013) implies that a mesodermally primed state may be achieved by sustaining pluripotency via controlled expression of optimal relative levels of mesodermal and endodermal lineage specifiers. Although cells reprogrammed via lineage specifiers appear indistinguishable from traditionally derived hiPSCs by genome-wide transcription analysis (Montserrat et al. 2013), it is possible that differences in gene expression may become evident only upon differentiation, as is observed in different hESC lines with varied differentiation propensities (Osafune et al. 2008). Comparisons of directed differentiation success in hiPSC lines stabilized via different lineage determinants could address this issue and aid in the elucidation of conditions best suited for cardiac differentiation. Likewise, the development of computational methods for predicting transcription regulatory networks and lineage specifiers for subpopulations within heterogeneous cultures (Okawa and del Sol 2015) has the potential to increase our understanding of how the pluripotent state can be stabilized and expedite the discovery of favorable conditions for cardiac priming.

4.7 Maintenance of hiPSC-Derived Mesodermal Intermediates and Cardiac Progenitor Cells

A remaining barrier in the application of hiPSC-CMs is the variability in differentiation success when using large numbers of hiPSC lines at varying passages. The development of an expandable intermediate mesodermal cell line to provide a renewable source of cells fated to become mesodermally derived tissues could overcome this obstacle. An hiPSC-derived cardiac progenitor cell (CPC) line could also potentially be used directly in regenerative therapies. Although the stabilization of such cell lines has been challenging, several groups have recently reported diverse methods for effectively capturing and expanding the mesoderm and cardiac progenitor states.

Cao and colleagues described the stabilization of self-renewing MESP1/MESP2$^+$, SSEA1$^+$, GATA4$^+$, MEF2C$^+$, ISL1$^+$ CPCs through BMP4, Wnt, and MEK-ERK activation in defined conditions (Cao et al. 2013). In this approach, hESCs were cultured in medium containing BMP4, CHIR99021, and ascorbic acid and showed a loss in expression of pluripotency markers followed by an upregulation of early mesodermal genes within 3 days of differentiation. Cells were subsequently stabilized and propagated for >15 passages with Activin/NODAL and BMP inhibition along with Wnt activation (CHIR99021). The resulting CPCs lacked in vivo tumorigenicity and readily differentiated into smooth muscle cells, endothelial cells, and beating, cross-striated NKX2-5$^+$/TNNT2$^+$ cardiomyocytes. However, while promising, the success of this protocol has still yet to be replicated.

More recently, a high-throughput screening method was used to identify a combination of small molecules and growth factors that allow for the sustained expression of T in mesoderm cells (Kumar et al. 2015). Cells cultured in RPMI/B27 with a combination of FGF2 and the GSK3B inhibitor CHIR98014 on an extracellular matrix composed of fibronectin, vitronectin, and collagens I, III, IV, and V retained the ability to expand and stably express the mesodermal markers MESP1, MIXL1, and LHX1. Although the mesodermal progenitor cells successfully differentiated into renal cells, they failed to differentiate into other mesodermally derived lineages, including cardiac. However, these conditions could help increase our understanding of development along mesodermal lineages and possibly improve differentiation into cardiac cells. Using a similar approach, Birket and colleagues demonstrated the sustained culture of PDGFRA$^+$, pre–NKX2-5$^+$ CPCs using controlled *MYC* expression in the presence of IGF1 and hedgehog signaling (Birket et al. 2015b). In this study, MYC expression at differentiation day 4.75 blocked CPCs from transitioning to an NKX2-5$^+$ state for over 40 population doublings (Birket et al. 2015b). While this discovery will aid in an increased understanding of cardiac development, the use of transgenic elements currently precludes its use in clinical applications.

In contrast, two very recent studies focused on reprogramming mouse fibroblasts directly to an expandable CPC state in defined conditions. A candidate gene approach to identify transcription factors and chromatin remodelers that allow for reprogramming of mouse adult cardiac fibroblasts into CPCs revealed that five cardiac factors (*Mesp1*, *Tbx5*, *Gata4*, *Nkx2-5*, and *Smarcd3* (*Baf60c*)) along with JAK/STAT and Wnt signaling (LIF and BIO) gave rise to *Nkx2-5*$^+$ CPCs (Lalit et al. 2016). Although these CPCs could be passaged and successfully differentiated into cardiomyocytes, smooth muscle cells, and endothelial cells, they did not spontaneously contract, suggesting that further optimization is necessary. Similarly, a second group has reported that the use of BMP4, Activin A, and Wnt activation (CHIR99021), along with FGF, VEGFA, and PDGF inhibition, allows for the sustained culture of KDR$^+$, PDGFRA$^+$ CPCs derived from murine fibroblasts (Zhang et al. 2016).

The approaches outlined above all result in the stabilization of a mesodermal precardiac state; however, each employs varied combinations of growth factors and small molecules, resulting in populations defined by different markers. This suggests that the resulting CPCs represent different stages of mesodermal differentiation, and continued analysis is necessary to define optimal conditions for deriving CPCs restricted to differentiation into intended cardiac subtypes. While transgene-free and chemically defined methods of capturing the mesodermal or cardiac precursor state in humans remain to be elucidated, the studies above are an auspicious beginning.

4.8 Directed Cardiac Differentiation of Human Pluripotent Stem Cells

The current efficiency of hiPSC-CM generation owes in large part to previous efforts to understand the pathways involved in cardiac development, including modulation of the TGFβ/Activin/NODAL, BMP, Wnt, NOTCH, FGF, vascular endothelial growth factor (VEGF), and Dickkopf-related protein-1 (DKK1) pathways (Fig. 4.1). Many of the methods used in differentiating hiPSCs along various lineages stems from earlier work using hESCs. Early on, growth of hESCs in suspension was found to produce spherical aggregates of cells termed embryoid bodies (EB) (Itskovitz-Eldor et al. 2000) and was shown by Kehat et al. in 2001 to result in spontaneous contractions in 8–10% of the EBs (Kehat et al. 2001). Cardiac-specific gene expression and calcium transients were observed in cells from the contracting EBs. Efforts to improve on this technique soon followed (Table 4.1). EBs of differing sizes resulted in inconsistent differentiation efficiencies, so attempts were made to control the size of EBs using forced aggregation techniques, in which a known number of hESCs were plated in U- or V-bottomed wells (Ng et al. 2005; Burridge et al. 2007). This method significantly improved the

Table 4.1 Methods for static cardiac differentiation of human pluripotent stem cells

Pink cells indicate positive inducers (growth factors, FBS, and Wnt-inducing small molecules). Yellow cells indicate inhibitors. Green cells indicate media used. Abbreviations: *BME* β-mercaptoethanol, *BMP4* bone morphogenetic protein 4, *DKK1* Dickkopf-related protein-1, *EB* embryoid body, *END-2* endoderm-like cell line, *FBS* fetal bovine serum, *FGF2* basic fibroblast growth factor, *HF* human fibroblasts, *IWR-1* inhibitor of Wnt response-1, *KO-DMEM*, KnockOut DMEM, *KSR* KnockOut Serum Replacement, *MEF* mouse embryonic fibroblasts, *MEF-CM* MEF conditioned medium, *MTG* α-monothioglycerol, *NEAA* non-essential amino acids, *N/S* not stated, *PFHM* Protein-Free Hybridoma Medium, *PVA* polyvinyl alcohol, *SCF* stem cell factor, *VEGFA* vascular endothelial growth factor (Kattman et al. 2011; Lian et al. 2012; Hudson et al. 2012; Burridge et al. 2014; Kehat et al. 2001; Burridge et al. 2007; Mummery et al. 2003; Yang et al. 2008; Laflamme et al. 2007; Zhang et al. 2011; Zhang et al. 2012; Elliott et al. 2011; Uosaki et al. 2011; Zhang et al. 2015; Graichen et al. 2008; Xu et al. 2008; Takei et al. 2009; Tran et al. 2009; Burridge et al. 2011; Ren et al. 2011; Willems et al. 2011; Cao et al. 2012; Gonzalez et al. 2011; Willems et al. 2012; Minami et al. 2012; Fonoudi et al. 2013; Lian et al. 2015; Aguilar et al. 2015; van den Berg et al. 2016; Lin et al. 2016)

Milestone	Pluripotency Media	Differentiation protocol, growth factors, small molecules and media by day of protocol (d0–d12): factors	Differentiation protocol: media	Efficiency	Reference
Spontaneous differentiation	KO-DMEM, 20% FBS	20% FBS	KO-DMEM	8% OF EBs	Kehat et al., 2001 [135]
END-2 co-culture	DMEM, 20% FBS	20% FBS, END-2 cells	DMEM	N/S	Mummery et al., 2003 [138]
Growth factor-directed mesoderm induction, forced aggregation	MEF-CM	Activin A, FGF2 / 20% FBS	CDM-PVA, ascorbic acid / DMEM	24% of EBs	Burridge et al., 2007 [137]
Growth factor monolayer	MEF-CM	Activin A / BMP4	RPMI+B27	30% MYH7+	Laflamme et al., 2007 [141]
First application of a small molecule (p38 MAPKi)	KO-DMEM, 20% KSR, FGF2	SB203580	END-2 conditioned: DMEM, NEAA, insulin, transferrin, sodium selenite, BME	22% MYH6+	Graichen et al., 2008 [192]
First chemically-defined medium, insulin has negative effect on differentiation	KO-DMEM, 20% KSR, FGF2	PGI2, SB203580	DMEM, NEAA, transferrin, sodium selenite, BME	11% MYH6+	Xu et al., 2008 [193]
High efficiency, growth factor optimization, Wnt inhibition	DMEM/F12, 20% KSR, FGF2	BMP4 / BMP4, Activin A, FGF2 / DKK1, VEGFA / DKK1, VEGFA, FGF2	StemPro-34, ascorbic acid	50% TNNT2+	Yang et al., 2008 [140]
Use of BMP4	DMEM, 20% KSR, FGF2	BMP4, 20% FBS / 20% FBS	DMEM, NEAA, BME	98% of EBs	Takei et al., 2009 [194]
Serum and insulin inhibit differentiation	DMEM/F12, 20% KSR, FGF2	WNT3A, 15% FBS / 15% FBS / 5% FBS / 0.2% BSA	KO-DMEM, NEAA, BME	N/S	Tran et al., 2009 [195]
Developmental stage-specific signaling requirements	DMEM/F12, 20% KSR, FGF2	BMP4 / BMP4, Activin A, FGF2 / DKK1, VEGFA / VEGFA, FGF2	StemPro-34, ascorbic acid	50-70% TNNT2+	Kattman et al., 2011 [19]
High efficiency, thorough optimization, chemically defined media	MEF-CM	BMP4, FGF2 / 20% FBS or HSA	RPMI+PVA / RPMI, ascorbic acid / RPMI, insulin, lipids	94% of EBs	Burridge et al., 2011 [196]
Small molecule Wnt inhibition	DMEM/F12, 20% KSR, FGF2	20% FBS, BMP4 / 20% FBS / 20% FBS, IWR-1 / 20% FBS / 2.5% FBS	KO-DMEM, NEAA, BME	16% TNNT2+	Ren et al., 2011 [197]

Small molecule Wnt inhibition, serum-free	KO-DMEM, 20% KSR, FGF2	BMP4 | BMP4, Activin A, FGF2 | IWR-1 | IWR-1, FGF2, VEGFA | FGF2, VEGFA | 2% FBS	StemPro-34 | DMEM	30% MYH6+	Willems et al., 2011 [198]
First successful monolayer for multiple lines	MEF-CM	Activin A | BMP4 | DKK1	RPMI+B27 without insulin	54% TNNT2+	Uosaki et al., 2011 [164]
EB and monolayer protocol	DMEM/F12, 20% KSR, FGF2	BMP4, WNT3A, Activin A, VEGFA, SCF	LI-APEL (IMDM/F12, PFHM-II, PVA, BSA, lipids, ITS, MTG) ascorbic acid | LI-AEL (IMDM/F12, PFHM-II, BSA, lipids, ITS, MTG), ascorbic acid	~25% GFP+	Elliott et al., 2011 [163]
Cardiomyocyte subtype specification	MEF-CM	BMP4, | Activin A | | Noggin | DKK1, BMS-189453 or retinoic acid | DKK1	RPMI+B27	83% V or 94% A	Zhang et al., 2011 [142]
Defined pluripotency media, monolayer small molecule Wnt inhibition	mTeSR1	BMP4, Activin A | IWP-4	RPMI+B27	60% MYH6+	Hudson et al., 2012 [22]
Mechanism of ascorbic acid	DMEM/F12, 20% KSR, FGF2	| 20% FBS | 5% FBS	DMEM/F12, KSR | DMEM, ascorbic acid	90-100% of EBs	Cao et al., 2012 [199]
Small molecule only, monolayer	MEF-CM	CHIR99021 | IWR-1, SB431542, purmorphamine	DMEM/F12+N2+B27 | RPMI+B27	70-80% NKX2-5+	Gonzalez et al., 2011 [200]
Small molecule only, monolayer	mTeSR1	CHIR99021 | | IWP-4	RPMI+B27 without insulin | RPMI+B27	85% TNNT2+	Lian et al., 2012 [20]
Matrigel sandwich	mTeSR1	Activin A | BMP4	RPMI+B27 without insulin | RPMI+B27	98% TNNT2+	Zhang et al., 2012 [150]
TGFB1 inhibition	KO-DMEM, 20% KSR, FGF2	BMP4 | Activin A, BMP4, FGF2, ITD-1 | VEGFA, DKK1, FGF2	StemPro34, NEAA, BME, ascorbic acid	60% TNNT2+	Willems et al., 2012 [201]
Novel Wnt inhibitor	DMEM/F12, 20% KSR	CHIR99021, BIO | KY02111, XAV939	IMDM, NEAA, BME, HSA or BSA	98% TNNT2+	Minami et al., 2012 [202]
Transcription factor transduction	MEF-CM	Activin A | ISL1, BMP4 | ISL1	RPMI+B27	75% TNNT2+	Fonoudi et al., 2013 [203]
Small molecule, chemically defined	E8	CHIR99012 | Wnt-C59	RPMI, recombinant HSA, ascorbic acid 2-phosphate	95% TNNT2+	Burridge et al., 2014 [23]
Without albumin, chemically defined	FTDA	BMP FGF CHIR | | IWP-2 or Wnt-C59	KO-DMEM, ascorbic acid, transferrin, sodium selenite (insulin and Y27632 for first day)	80-95% TNNT2+	Zhang et al., 2015 [172]
Small molecule, without albumin, chemically defined	E8 or mTeSR1	CHIR99021 | | IWP-2	RPMI, putrescine, progesterone, sodium selenite, | RPMI, insulin, ascorbic acid 2-phosphate	88-98% TNNT2+	Lian et al., 2015 [204]
Use of BMP inhibitor	mTeSR1	CHIR99021 | DMH1	DMEM/F12+B27	75% TNNT2+	Aguilar et al., 2015 [205]
Multiple pluripotent culture methods	Any	BMP4, Activin A, CHIR99021 | XAV939	IMDM/F12, PFHM-II, BSA, lipids, low insulin, transferrin, sodium selenite, MTG	84% TNNT2+	van den Berg et al., 2016 [206]
E8-based media	E8	CHIR99021 | | IWP-2	DMEM/F12, ascorbic acid, lipids, transferrin, sodium selenite, sodium bicarbonate (heparin from d1-d7)	94% TNNT2+	Lin et al., 2016 [207]

reproducibility of EB formation within a given hESC line but highlighted the variability in CM differentiation between hESC lines (Burridge et al. 2007).

Turning to developmental biology, modulators of key signaling pathways involved in mesodermal induction and cardiomyocyte differentiation were added to the culture conditions. Considering the importance of signals from neighboring endodermal cells during cardiac development, Mummery et al. co-cultured hESCs with a visceral endoderm-like cell line (END-2 cells), which resulted in cardiomyocyte differentiation, although the efficiency remained low (Mummery et al. 2003). The combination of Activin A and BMP improved cardiomyocyte differentiation of hEB-derived cells (Burridge et al. 2007; Yao et al. 2006). As protocols incorporated additional signaling pathways and improvements in the timing of their modulation, the efficiency of differentiation greatly improved. Using a staged approach, differentiation could be boosted by mesodermal induction with Activin A and BMP4, followed by canonical Wnt inhibition with DKK1, with subsequent FGF2 addition to promote expansion of the developing cardiomyocytes (Yang et al. 2008). This approach improved the frequency of contracting cardiomyocytes to over 50%.

Although successful, the method of EB formation followed by selection of beating EBs and separation of cardiomyocytes was very labor intensive. Therefore, attempts to differentiate hESCs in monolayer cultures using defined media were undertaken. Human ESCs grown on Matrigel and treated with Activin A for 24 h, followed by BMP4 for 4 days, yielded >30% CMs (Laflamme et al. 2007). As with the EB approach, additional cytokines including FGF2 and DKK1 improved monolayer yields as well (Zhang et al. 2011). The small molecules CHIR99021, an inhibitor of the downstream canonical Wnt pathway inhibitor glycogen synthase kinase 3 (GSK3) (Lian et al. 2012; Lian et al. 2013), and inhibitor of Wnt Production-4 (IWP4) and -2 (IWP2) (Lian et al. 2012, 2013; Hudson et al. 2012) were also shown to be effective and in sequential combination could drive differentiation efficiency as high as 98%.

The efforts with hiPSCs began in 2007, when Takahashi et al. first demonstrated evidence of spontaneous beating and cardiac gene expression in hiPSCs (Takahashi et al. 2007). Zhang et al. followed with a more thorough characterization of hiPSC-CMs in vitro (Zhang et al. 2009). They observed spontaneous beating in up to ~10% of hiPSC-EBs, comparable to hESC-EBs (Zhang et al. 2009). From there, investigations of hiPSCs largely followed the work being done on hESCs. The role of Activin/Nodal and BMP signaling in hiPSCs was confirmed (Kattman et al. 2011), and small molecule Wnt and GSK3B inhibitors proved effective in hiPSCs as well (Lian et al. 2012).

In keeping with the move away from serum-containing media, effective cardiac differentiation protocols were established using chemically defined media. Burridge et al. developed a protocol for highly efficient and reproducible differentiation using a low-cost fully chemically defined medium (Burridge et al. 2014). Supplementing the basal medium RPMI 1640 with recombinant human albumin and L-ascorbic acid 2-phosphate and use of sequential CHIR99021 and the Wnt inhibitor Wnt-C59, cardiomyocyte yields as high as 80–95% could be produced. Metabolic selection techniques could further enrich the cells to >95% pure cardiomyocytes (Burridge et al. 2014).

Rao et al. provided further insight into the molecular basis of cardiomyocyte differentiation when they demonstrated that early BMP and Wnt activation in hESCs drives mesodermal induction via cooperative rapid silencing of *SOX2* (Rao et al.

2016). Further, they demonstrated that Wnt inhibition is necessary at the mesodermal stage to restrict MSX1 and CDX2/CDX1 upregulation and allow for cardiac lineage induction. These findings suggest additional targets for modulation to more precisely control hPSC fate in vitro.

Another approach to inducing cardiac differentiation involves the use of cardiac transcription factors to "forward program" the cells. In 2011, Dixon et al. demonstrated that lentivirus-mediated expression of the four transcription factors *GATA4*, *TBX5*, *NKX2–5*, and *BAF60C* (collectively known as "GTNB") in hESCs and hiPSCs could generate 6–12% cardiomyocytes in culture conditions that typically maintain pluripotency (Dixon et al. 2011). Subsequently, similar results were achieved with only the transcription factors *BAF60C*, *GATA4*, and *MESP1* using an electroporation method or even with further reduction to just *GATA4* and *MESP1*, although with lower efficiency (Hartung et al. 2013). Demonstrating potential for cardiomyocyte subtype specification, forward programming using the human *TBX3* gene in mouse ESCs combined with *Myh6*-based antibiotic selection resulted in generation of >80% pure nodal-like cells (Jung et al. 2014), although a similar result has yet to be achieved with hPSCs. These efforts demonstrate a completely separate, and perhaps complementary, avenue for controlling cardiomyocyte differentiation, and it is conceivable that optimal control of cardiomyocyte differentiation and even subtype specification may involve some combination of cytokines, small molecules, and transient transcription factor expression.

4.9 Gene Expression during Differentiation

Human iPSC-CMs exhibit properties of isolated primary cardiomyocytes including expression of contractile proteins such as cardiac troponin T (Takahashi et al. 2007; Kattman et al. 2011; Lian et al. 2012; Burridge et al. 2014; Ye et al. 2014; Zhang et al. 2009), sarcomeric α-actinin (Lian et al. 2012; Burridge et al. 2014; Ye et al. 2014; Zhang et al. 2009), and myosin heavy and light chains (Lian et al. 2012; Burridge et al. 2014; Ye et al. 2014; Zhang et al. 2009); sarcomeric organization (Lian et al. 2012; Burridge et al. 2014; Ye et al. 2014; Zhang et al. 2009); spontaneous action potential generation (Lian et al. 2012; Burridge et al. 2014; Zhang et al. 2009); contraction/electromechanical coupling (Takahashi et al. 2007; Lian et al. 2012; Ye et al. 2014; Zhang et al. 2009); calcium transients; β-adrenergic signaling/responsiveness (Zhang et al. 2009); and intracellular connections via connexin 43 (Ye et al. 2014). Differentiation of cardiomyocytes from hiPSCs progresses in a stepwise fashion consistent with patterns seen in embryonic development and differentiation of hESCs. On induction of differentiation, downregulation of the pluripotency genes *OCT4* (Lian et al. 2012; Burridge et al. 2014; Zhang et al. 2009), *NANOG* (Lian et al. 2012; Zhang et al. 2009), and *SOX2* (Lian et al. 2012; Rao et al. 2016) occurs along with upregulation of mesodermal genes, including *T* and *MIXL1* (Lian et al. 2012; Burridge et al. 2014). Expression of the cardiac mesoderm marker *MESP1* soon follows (Burridge et al. 2014). At the mesodermal stage, *MSX1* and *CDX2/CDX1* are also restricted to allow further progression (Rao et al. 2016). Differentiating hiPSCs have been shown to progress through a cardiac progenitor state (Lian et al. 2012), as indicated by expression of *ISL1* (Lian et al. 2012; Burridge et al. 2014) and *WT1* (Lian et al. 2012). Other early cardiac genes also become activated around this time, including *KDR* (Burridge et al. 2014) and *GATA4* (Lian et al. 2012; Burridge et al. 2014). The later cardiac markers *NKX2-5* (Takahashi et al. 2007; Zhang et al. 2009; Lian et al. 2012; Burridge et al. 2014), *TBX2* (Lian et al. 2012), *TBX5* (Lian et al. 2012; Burridge et al. 2014), *MEF2C* (Takahashi et al. 2007; Lian et al. 2012; Burridge et al. 2014), and *PLN* (Lian et al. 2012; Zhang et al. 2009) appear between days 3 and 8. Finally, cardiac myofilament mRNA levels increase after day 8. Transcripts for *TNNT2* (Burridge et al. 2014; Zhang et al. 2009), *TNNI2* and *TNNI3* (Lian et al. 2012), *MYH6* (Burridge et al. 2014), *MLC2A* (*MYL7*) (Takahashi et al. 2007; Lian et al. 2012; Zhang et al. 2009), *MLC2V* (*MYL2*) (Lian et al. 2012; Burridge et al. 2014; Zhang et al. 2009), and *ACTN2* (Zhang et al. 2009) have all been demonstrated. Human iPSC-CMs also express the cardiac ion channel genes *HCN1*, *HCN4*, *KCNQ1*, and *KCNH2* (Burridge et al. 2014). In general agreement with these patterns, a recent study using gene ontology on hESCs evaluated 2917 gene transcripts with significant expression changes at four specified time points: hESCs, primitive mesoderm, cardiac mesoderm, and differentiated CMs (Tompkins et al. 2016). The data demonstrated an upregulation of genes for cardiac development, cardiac structural proteins, glycolysis, and mitochondrial oxidative phosphorylation, while mitotic and organelle fission genes were suppressed with differentiation. This finding is consistent with known characteristics of cardiomyocytes. Corresponding with mRNA transcripts, cardiac protein translation has also been confirmed in differentiated hiPSCs, as expected (Takahashi et al. 2007; Kattman et al. 2011; Lian et al. 2012; Burridge et al. 2014; Ye et al. 2014; Zhang et al. 2009).

As myofilament protein expression stabilizes and the cells mature, *ISL1* expression extinguishes by day 30 post-induction (Lian et al. 2012). Likewise, it has been noted that both myosin light chain mRNA and protein expression shifts from predominantly MLC2A early after differentiation to MLC2V by 45–60 days after induction, consistent with progression of the cells from an unspecified or atrial cardiac phenotype to a more mature ventricular phenotype (Lian et al. 2012; Burridge et al. 2014). Markers of the various cardiomyocyte subtypes have been identified in populations of hiPSC-CMs as well, with expression of the atrial-specific genes *NPPA*, *CX40*, and *SLN*, ventricle-specific genes *MLC2V* and *IRX4*, and the nodal gene *TBX18* (Burridge et al. 2014) all observed. This indicates that hiPSCs have the potential to form all of these cardiomyocyte subtypes, while efforts to more precisely control subtype specification are ongoing (see below).

4.10 Large-Scale Production of hiPSC-Derived Cardiomyocytes

Beginning with the earliest hESC cardiac differentiation protocols, there was a consideration for scale-up, with the knowledge that EBs are a suitable format for large-scale bioreactors commonly used in commercial cell production (Burridge et al. 2007). Despite this, monolayer protocols have proved popular as they vastly simplify media changing regimes, do not have the complex requirements of EB formation, and allow for simple assessment of differentiation success (Laflamme et al. 2007). It is now well accepted in the field that monolayer differentiation, commonly performed in six-well plates, does not scale up well past 100 cm^2 surface area. It is theorized that this is due to the small molecule methodologies currently in use being dependent on paracrine effects, with the CHIR99021-induced GSK3B inhibition allowing Wnt signaling to occur and forming a positive feedback loop with Activin/NODAL/TGFβ, BMP, and FGF production and signaling (Burridge et al. 2012, 2014). In addition, a major barrier to scale up has been the widespread use of prohibitively expensive media, such as mTeSR1 for pluripotent growth (Lian et al. 2012; Zhang et al. 2012) and StemPro-34 (Yang et al. 2008) or RPMI + B27 for differentiation (Lian et al. 2012; Zhang et al. 2012). There are now multiple efforts to replace these media with cost-effective alternatives such as chemically defined E8 for pluripotent growth, which can be made in-house for less than $50 per liter, and various chemically defined media for cardiac differentiation, such as CDM3 (Burridge et al. 2014) costing as little as $11 per liter. Another cost consideration has been the use of growth factors, which become prohibitively expensive at large volumes, although at least for cardiac differentiation, these have largely now been replaced with small molecules. In each case, there has been an obligation to prove that novel media have the minimal number of components necessary to simplify the formulation and optimize results.

There has been significant progress with cardiac differentiation in stirred suspension vessels (Kempf et al. 2016), as hiPSCs easily adapt to suspension-based culture using either cell aggregates (hEBs) or attachment to matrix-coated microcarriers (Lam et al. 2016; Lecina et al. 2010). Both techniques have resulted in efficient differentiation, although microcarriers substantially increase the complexity of handling (Lecina et al. 2010; Lam et al. 2014; Leahy et al. 1999). Early techniques began with FBS-based spontaneous differentiation (Niebruegge et al. 2009), progressed through growth factor-based systems (Chen et al. 2012), and most recently, protocols have been demonstrated adopting the small molecule-based approach using CHIR99021 for 24 h followed by a Wnt inhibitor such as IWP-2 or IWP-4 and media volumes of ~100 ml (Chen et al. 2015; Kempf et al. 2014; Kempf et al. 2015). These techniques primarily involve simple Corning spinner flasks (Chen et al. 2015) or full stirred bioreactors such as the DASGIP platform (Kempf et al. 2014). These methods commonly have a pluripotent culture step of 3–4 days for hEB formation, with seeding at ~$0.3–0.5 \times 10^6$ cells/ml, and result in cardiomyocyte production with purities of ~50–90% TNNT2$^+$ in the 1.0×10^6 cells/ml range (Chen et al. 2015; Kempf et al. 2014, 2015). In particular, bioreactors allow constant perfusion of media, with online monitoring of key process parameters such as pH, dissolved oxygen, glucose consumption, and lactate accumulation potentially allowing a reduction in differentiation variability.

4.11 Purification of hiPSC-Derived Cardiomyocytes

Current cardiac differentiation methodologies can achieve purities of ~80–90% TNNT2$^+$ cardiomyocytes, yet there is still considerable variation in differentiation reproducibility and variability between hiPSC lines. Numerous techniques have been adopted to purify cardiomyocytes from contaminating cell types such as fibroblasts, endothelial cells, or smooth muscle cells. One of the first techniques was the use of the mitochondrial membrane potential sensor TMRM, based on the hypothesis that cardiomyocytes have a higher mitochondrial number per cell than the contaminating cell types (Hattori et al. 2010) although this was later demonstrated to be less effective in hESC-CM than animal-derived cardiomyocytes (Dubois et al. 2011). Also used have been genetically encoded sensors such as *MYH6-GFP* or *NKX2-5-GFP* (Elliott et al. 2011). Considerable effort has been expended to discover unique hiPSC-CM cell surface markers, which has led to the discovery of SIRPA (Dubois et al. 2011) and VCAM1 (Uosaki et al. 2011) as suitable markers, which can be very effective when used in combination (Elliott et al. 2011). Metabolic selection has proven particularly successful, as replacing D-glucose in common media (DMEM) with 4 mM L-lactic acid from around d20 to ~d27 of differentiation can achieve purities of up to 98.9% ACTN2$^+$ cells (Tohyama et al. 2013). This simple process is also well suited to suspension cultures (Hemmi et al. 2014). One common approach in commercial hiPSC-CM cell production has been the use of antibiotic selection (Anderson et al. 2007; Ma et al. 2011), using a cardiac-specific promoter such *MYH6* or *TNNT2* linked to a puromycin resistance gene or similar strategy (Kita-Matsuo et al. 2009). With these commercial cells becoming more ubiquitous, this genetic modification has now become acceptable within the industry.

4.12 Cardiomyocyte Subtype Specification

After cardiomyocyte generation, the next major step is determining how to produce the specific cardiomyocyte subtypes found in the heart, such as ventricular, atrial, and nodal. Subtype isolation could prove to be useful for transplantation purposes, as hPSC-derived ventricular cardiomyocytes could be used to repair ischemic damage to the ventricular myocardium following MI. The production of pure isolates of specific subtypes may be critical for the prevention of arrhythmias in therapeutic applications, as injection of a mixed population of hESC-CMs resulted in ventricular arrhythmias in a primate MI model (Chong et al. 2014). Likewise, hPSC-derived nodal cells could be used to restore proper cardiac rhythm and electrical conduction, and hPSC-derived atrial cells could be used to assess drugs which target the atria (Devalla et al. 2015). Common cardiac differentiation protocols produce predominantly ventricular cells with ~15–20% atrial cells and ~5% nodal cells when assessed by patch-clamp electrophysiology (Burridge et al. 2014). The divergence of atrial and ventricular markers between mouse and human models has slowed progress toward the production of specific subtypes. For example, in mice, MYL7 (MLC2A) is expressed in the atria during development, and MYL2 (MLC2V) is expressed in the ventricles, whereas in humans, MYL7 is expressed throughout the atria, ventricles, and outflow tract, and only MYL2 is specific for the ventricles (Chuva de Sousa Lopes et al. 2006). In hiPSC-CMs, the expression of MYL7 decreases from ~d15 to d60 of differentiation, while the expression of MYL2 increases (Burridge et al. 2014; Zhang et al. 2012; Zhang et al. 2015) with little change in the subtype ratios as detected by patch clamp. Reporter gene constructs such as *MLC2V*-GFP have been successfully employed to isolate ventricular cells from hPSCs (Bizy et al. 2013). Similarly, sarcolipin (*SLN*)-tdTomato has been used to isolate atrial cells (Josowitz et al. 2014). Other genes that are subtype specific include *IRX4* (ventricular), *HCN4* and *TBX18* (nodal), and *NPPA (ANF)* and *NPPB (BNF)* (ventricular and atrial). Modification of existing differentiation protocols to skew cardiomyocyte sub-populations has also been successful. Using factors such as the small molecule AG1478 to inhibit neuregulin signaling can increase the nodal population to ~50% (Zhu et al. 2010). Similarly, inhibiting retinoic acid signaling using the small molecule BMS-189453 increased the ventricular population to 83%, while additional retinoic acid increased the atrial population to 94% (Zhang et al. 2011). In mESCs, *Tbx3* overexpression has been used to increase the population of nodal cells to ~80% (Jung et al. 2014), and *Shox2* has been similarly employed (Ionta et al. 2015).

Measuring cardiomyocyte subtype by action potential morphology derived from patch clamping has been a mainstay of the hiPSC-CM field (Moretti et al. 2010). Recently the reliability of this technique in the cells has been called into question (Du et al. 2015), suggesting that the atrial, ventricular, and nodal designations are in fact based on the arbitrary clustering of continuous variables (Kane et al. 2016).

4.13 Cardiomyocyte Maturation

It is now well established that cardiomyocytes generated from hPSCs by any of the current differentiation protocols are of an immature, fetal type in terms of automaticity, proliferation, metabolism, gene expression, sarcomere organization, shape/aspect ratio, calcium handling, lack of T-tubules, and electrophysiological properties (Yang et al. 2014a) and that these hPSC-CMs can mature with extended time in culture (>80 days) (Lundy et al. 2013). Numerous media-based methodologies have been developed for improving cardiomyocyte maturation. The addition of triiodothyronine (T_3) (Yang et al. 2014b; Ivashchenko et al. 2013), which binds the THRA and THRB receptors, results in the downregulation of MYH7 and upregulation of MYH6 and SERCA2A (ATP2A2), along with numerous ion channels. In addition to T3, the application of dexamethasone has also been shown to increase maturation metrics (Kosmidis et al. 2015). The combination of insulin and dexamethasone (Kim et al. 2013; Wen et al. 2015) or IGF1, dexamethasone, and T3 (Birket et al. 2015a) have been shown to enhance the electrophysiologic properties, bioenergetics, and contractile force of hiPSC-CM.

The lack of Kir2.1 (I_{K1}), an inward-rectifier potassium ion channel, is thought to be a main contributor to the observed immature electrophysiological properties of hPSC-CMs, and forced expression *KCNJ2* that encodes Kir2.1 has been demonstrated to be effective in inducing maturation (Lieu et al. 2013). Other techniques such as adrenergic stimulation, electrical stimulation, stretch, and micropatterning have also been employed to successfully improve maturation (Zhu et al. 2014). In particular, soft surfaces have been demonstrated to hold potential as a simple method for maturation including thick layers of matrix such as the "Matrigel mattress" (Feaster et al. 2015) or a layer of PDMS below Matrigel (Herron et al. 2016).

Adult cardiomyocytes primarily derive their ATP from oxidative phosphorylation, breaking down fatty acids such as palmitate and linoleic acid to acetyl-CoA via beta oxidation. In contrast, hiPSC-CM and fetal cardiomyocytes derive their ATP from glycolysis. Numerous attempts have been made to switch hiPSC-CMs to non-glucose metabolism. One method that has been demonstrated is to replace the D-glucose in the medium with D-galactose, resulting in ATP net-neutral glycolysis and the production of pyruvate. Pyruvate can then be decarboxylated to acetyl-CoA, which can enter the citric acid cycle and cause a shift to mitochondrial metabolism (Rana et al. 2012; Wang et al. 2014). Another approach under active investigation is to force hPSC-CMs into a mature metabolic profile by combining the D-galactose usage with the presence of oleic and palmitic fatty acids (Rana et al. 2012), allowing the use of beta oxidation.

4.14 Future Directions and Challenges

Although the progress in generating cardiomyocytes from hPSCs in the past decade has been tremendous, a great deal of work remains. To achieve their full clinical and translational potential, the production of hPSC-CMs must be scaled up considerably while simultaneously improving the interline reproducibility of cardiac differentiation. This will no doubt necessitate the discovery of new techniques for growth and differentiation as well as identification of additional culture factors and fine-tuning of doses to control the pluripotent state and steer differentiation more precisely. Along these lines, greater control over cardiomyocyte subtype specification must also be achieved to maximize safety in cell-based therapeutic applications, as well as the reliability of developmental, disease modeling, and drug and toxicity screening data. The identification of ventricular arrhythmias in macaques that received non-subtype selected hESC-CMs post-MI may illustrate the importance of this issue (Chong et al. 2014). While Ye et al. saw no arrhythmias in a porcine MI model with delivery of a 100-fold lower dose of hiPSC-CMs (Ye et al. 2014), the use of smaller numbers of cells undermines the problem of regenerating large amounts of myocardium. Going one step beyond subtype specification, it will also be necessary to demonstrate that the electrophysiologic characteristics of a given hPSC-CM subtype closely resemble those of the intended delivery site (e.g., ventricle), as electrical heterogeneity could also predispose to arrhythmia generation. This leads us to the need to improve our ability to rapidly mature hPSC-CMs to an adult phenotype for most clinical applications. However, maturation is also important for in vitro studies to accurately model certain disease phenotypes, in particular those that present during adulthood such as ARVC and HCM, as well as for generating relevant results from drug and toxicity screening applications. While time in culture leads to hPSC-CM maturation, waiting over 80 days after hiPSC generation and subsequent differentiation is suboptimal for most applications. For all of these challenges, high-throughput screening assays will prove indispensable.

Finally, another area of great effort not touched upon in this chapter involves determination of the best method for delivery of hPSC-CMs to the diseased heart, as well as elucidation of ways to improve long-term cell retention. A constant theme in all cardiac cell therapy attempts to date is the poor long-term engraftment rates. Precise analyses of cell engraftment are difficult in vivo due to the difficulties in tracking all of the delivered cells. However, the few available estimates of PSC retention suggest that rates are as low as <1–2% as few as 2–4 weeks after basic intramyocardial injection (Chong et al. 2014; Ye et al. 2014; Xiong et al. 2013). As noted above, co-injection of hiPSC-derived cardiomyocytes, smooth muscle cells, and endothelial cells through an IGF1-releasing fibrin patch improved retention estimates to about 9% (Ye et al. 2014). While the efficacy of this method needs to be validated using more direct cell tracking techniques, it represents a step in the right direction. Certainly, improvement of cell engraftment warrants ongoing efforts. Considering how remarkably far this field has come in the last 10 years, the prospects of overcoming these challenges for cardiac regeneration and therapeutic discovery in the near future are very encouraging.

Acknowledgments We would like to acknowledge funding support from the US National Institutes of Health R00 HL121177, the Dixon Translational Research Grant Young Investigator Award, and Innovation Development Progress (IDP) Research Innovation Challenge Grant (P.W.B.). We apologize to those investigators whose work was omitted here due to space limitations.

Compliance with Ethical Standards

Conflict of Interest P.W.B. is on the scientific advisory board of Pluriomics B.V. and a shareholder in Stem Cell Theranostics.

Ethical Approval This article does not contain any studies with human participants performed by any of the authors.

References

Aguilar JS, Begum AN, Alvarez J, Zhang XB, Hong Y, Hao J (2015) Directed cardiomyogenesis of human pluripotent stem cells by modulating Wnt/beta-catenin and BMP signalling with small molecules. Biochem J 469(2):235–241

Amit M, Carpenter MK, Inokuma MS, Chiu CP, Harris CP, Waknitz MA, Itskovitz-Eldor J, Thomson JA (2000) Clonally derived human embryonic stem cell lines maintain pluripotency and proliferative potential for prolonged periods of culture. Dev Biol 227(2):271–278

Amit M, Shariki C, Margulets V, Itskovitz-Eldor J (2004) Feeder layer- and serum-free culture of human embryonic stem cells. Biol Reprod 70(3):837–845

Anderson D, Self T, Mellor IR, Goh G, Hill SJ, Denning C (2007) Transgenic enrichment of cardiomyocytes from human embryonic stem cells. Mol Ther 15(11):2027–2036

Barbuti A, Robinson RB (2015) Stem cell-derived nodal-like cardiomyocytes as a novel pharmacologic tool: insights from sinoatrial node development and function. Pharmacol Rev 67(2):368–388

Beltrami CA, Finato N, Rocco M, Feruglio GA, Puricelli C, Cigola E, Quaini F, Sonnenblick EH, Olivetti G, Anversa P (1994) Structural basis of end-stage failure in ischemic cardiomyopathy in humans. Circulation 89(1):151–163

van den Berg CW, Elliott DA, Braam SR, Mummery CL, Davis RP (2016) Differentiation of human pluripotent stem cells to cardiomyocytes under defined conditions. Methods Mol Biol 1353:163–180

Bernemann C, Greber B, Ko K, Sterneckert J, Han DW, Arauzo-Bravo MJ, Scholer HR (2011) Distinct developmental ground states of epiblast stem cell lines determine different pluripotency features. Stem Cells 29(10):1496–1503

Birket MJ, Ribeiro MC, Kosmidis G, Ward D, Leitoguinho AR, van de Pol V, Dambrot C, Devalla HD, Davis RP, Mastroberardino PG, Atsma DE, Passier R, Mummery CL (2015a) Contractile defect caused by mutation in MYBPC3 revealed under conditions optimized for human PSC-cardiomyocyte function. Cell Rep 13(4):733–745

Birket MJ, Ribeiro MC, Verkerk AO, Ward D, Leitoguinho AR, den Hartogh SC, Orlova VV, Devalla HD, Schwach V, Bellin M, Passier R, Mummery CL (2015b) Expansion and patterning of cardiovascular progenitors derived from human pluripotent stem cells. Nat Biotechnol 33(9):970–979

Bizy A, Guerrero-Serna G, Hu B, Ponce-Balbuena D, Willis BC, Zarzoso M, Ramirez RJ, Sener MF, Mundada LV, Klos M, Devaney EJ, Vikstrom KL, Herron TJ, Jalife J (2013) Myosin light chain 2-based selection of human iPSC-derived early ventricular cardiac myocytes. Stem Cell Res 11(3):1335–1347

Blakeley P, Fogarty NM, del Valle I, Wamaitha SE, Hu TX, Elder K, Snell P, Christie L, Robson P, Niakan KK (2015) Defining the three cell lineages of the human blastocyst by single-cell RNA-seq. Development 142(18):3151–3165

Bolli R, Tang XL, Sanganalmath SK, Rimoldi O, Mosna F, Abdel-Latif A, Jneid H, Rota M, Leri A, Kajstura J (2013) Intracoronary delivery of autologous cardiac stem cells improves cardiac function in a porcine model of chronic ischemic cardiomyopathy. Circulation 128(2):122–131

Brade T, Pane LS, Moretti A, Chien KR, Laugwitz KL (2013) Embryonic heart progenitors and cardiogenesis. Cold Spring Harb Perspect Med 3(10):a013847

Brons IG, Smithers LE, Trotter MW, Rugg-Gunn P, Sun B, Chuva de Sousa Lopes SM, Howlett SK, Clarkson A, Ahrlund-Richter L, Pedersen RA, Vallier L (2007) Derivation of pluripotent epiblast stem cells from mammalian embryos. Nature 448(7150):191–195

Broughton KM, Sussman MA (2016) Empowering adult stem cells for myocardial regeneration V2.0: success in small steps. Circ Res 118(5):867–880

Burridge PW, Anderson D, Priddle H, Barbadillo Munoz MD, Chamberlain S, Allegrucci C, Young LE, Denning C (2007) Improved human embryonic stem cell embryoid body homogeneity and cardiomyocyte differentiation from a novel V-96 plate aggregation system highlights interline variability. Stem Cells 25(4):929–938

Burridge PW, Thompson S, Millrod MA, Weinberg S, Yuan X, Peters A, Mahairaki V, Koliatsos VE, Tung L, Zambidis ET (2011) A universal system for highly efficient cardiac differentiation of human induced pluripotent stem cells that eliminates interline variability. PLoS One 6(4):e18293

Burridge PW, Keller G, Gold JD, Wu JC (2012) Production of de novo cardiomyocytes: human pluripotent stem cell differentiation and direct reprogramming. Cell Stem Cell 10(1):16–28

Burridge PW, Matsa E, Shukla P, Lin ZC, Churko JM, Ebert AD, Lan F, Diecke S, Huber B, Mordwinkin NM, Plews JR, Abilez OJ, Cui B, Gold JD, Wu JC (2014) Chemically defined generation of human cardiomyocytes. Nat Methods 11(8):855–860

Burridge PW, Li YF, Matsa E, Wu H, Ong SG, Sharma A, Holmstrom A, Chang AC, Coronado MJ, Ebert AD, Knowles JW, Telli ML, Witteles RM, Blau HM, Bernstein D, Altman RB, Wu JC (2016) Human induced pluripotent stem cell-derived cardiomyocytes recapitulate the predilection of breast cancer patients to doxorubicin-induced cardiotoxicity. Nat Med 22(5):547–556

Cahan P, Daley GQ (2013) Origins and implications of pluripotent stem cell variability and heterogeneity. Nat Rev Mol Cell Biol 14(6):357–368

Cao N, Liu Z, Chen Z, Wang J, Chen T, Zhao X, Ma Y, Qin L, Kang J, Wei B, Wang L, Jin Y, Yang HT (2012) Ascorbic acid enhances the cardiac differentiation of induced pluripotent stem cells through promoting the proliferation of cardiac progenitor cells. Cell Res 22(1):219–236

Cao N, Liang H, Huang J, Wang J, Chen Y, Chen Z, Yang HT (2013) Highly efficient induction and long-term maintenance of multipotent cardiovascular progenitors from human pluripotent stem cells under defined conditions. Cell Res 23(9):1119–1132

Carpenter L, Carr C, Yang CT, Stuckey DJ, Clarke K, Watt SM (2012) Efficient differentiation of human induced pluripotent stem cells generates cardiac cells that provide protection following myocardial infarction in the rat. Stem Cells Dev 21(6):977–986

Carter MG, Smagghe BJ, Stewart AK, Rapley JA, Lynch E, Bernier KJ, Keating KW, Hatziioannou VM, Hartman EJ, Bamdad CC (2016) A primitive growth factor, NME7AB, is sufficient to induce stable naive state human pluripotency; reprogramming in this novel growth factor confers superior differentiation. Stem Cells 34(4):847–859

Chen G, Gulbranson DR, Hou Z, Bolin JM, Ruotti V, Probasco MD, Smuga-Otto K, Howden SE, Diol NR, Propson NE, Wagner R, Lee GO, Antosiewicz-Bourget J, Teng JM, Thomson JA (2011) Chemically defined conditions for human iPSC derivation and culture. Nat Methods 8(5):424–429

Chen VC, Couture SM, Ye J, Lin Z, Hua G, Huang HI, Wu J, Hsu D, Carpenter MK, Couture LA (2012) Scalable GMP compliant suspension culture system for human ES cells. Stem Cell Res 8(3):388–402

Chen VC, Ye J, Shukla P, Hua G, Chen D, Lin Z, Liu JC, Chai J, Gold J, Wu J, Hsu D, Couture LA (2015) Development of a scalable suspension culture for cardiac differentiation from human pluripotent stem cells. Stem Cell Res 15(2):365–375

Chong JJ, Yang X, Don CW, Minami E, Liu YW, Weyers JJ, Mahoney WM, Van Biber B, Cook SM, Palpant NJ, Gantz JA, Fugate JA, Muskheli V, Gough GM, Vogel KW, Astley CA, Hotchkiss CE, Baldessari A, Pabon L, Reinecke H, Gill EA, Nelson V, Kiem HP, Laflamme MA, Murry CE

(2014) Human embryonic-stem-cell-derived cardiomyocytes regenerate non-human primate hearts. Nature 510(7504):273–277

Chuva de Sousa Lopes SM, Hassink RJ, Feijen A, van Rooijen MA, Doevendans PA, Tertoolen L, Brutel de la Riviere A, Mummery CL (2006) Patterning the heart, a template for human cardiomyocyte development. Dev Dyn 235(7):1994–2002

Danker T, Moller C (2014) Early identification of hERG liability in drug discovery programs by automated patch clamp. Front Pharmacol 5:203

De Los AA, Loh YH, Tesar PJ, Daley GQ (2012) Accessing naive human pluripotency. Curr Opin Genet Dev 22(3):272–282

Devalla HD, Schwach V, Ford JW, Milnes JT, El-Haou S, Jackson C, Gkatzis K, Elliott DA, Chuva de Sousa Lopes SM, Mummery CL, Verkerk AO, Passier R (2015) Atrial-like cardiomyocytes from human pluripotent stem cells are a robust preclinical model for assessing atrial-selective pharmacology. EMBO Mol Med 7(4):394–410

Di Pasquale E, Lodola F, Miragoli M, Denegri M, Avelino-Cruz JE, Buonocore M, Nakahama H, Portararo P, Bloise R, Napolitano C, Condorelli G, Priori SG (2013) CaMKII inhibition rectifies arrhythmic phenotype in a patient-specific model of catecholaminergic polymorphic ventricular tachycardia. Cell Death Dis 4:e843

Dick E, Kalra S, Anderson D, George V, Ritson M, Laval S, Barresi R, Aartsma-Rus A, Lochmuller H, Denning C (2013) Exon skipping and gene transfer restore dystrophin expression in hiPSC-cardiomyocytes harbouring DMD mutations. Stem Cells Dev 22(20):2714–2724

DiMasi JA, Feldman L, Seckler A, Wilson A (2010) Trends in risks associated with new drug development: success rates for investigational drugs. Clin Pharmacol Ther 87(3):272–277

Dixon JE, Dick E, Rajamohan D, Shakesheff KM, Denning C (2011) Directed differentiation of human embryonic stem cells to interrogate the cardiac gene regulatory network. Mol Ther 19(9):1695–1703

Drawnel FM, Boccardo S, Prummer M, Delobel F, Graff A, Weber M, Gerard R, Badi L, Kam-Thong T, Bu L, Jiang X, Hoflack JC, Kiialainen A, Jeworutzki E, Aoyama N, Carlson C, Burcin M, Gromo G, Boehringer M, Stahlberg H, Hall BJ, Magnone MC, Kolaja K, Chien KR, Bailly J, Iacone R (2014) Disease modeling and phenotypic drug screening for diabetic cardiomyopathy using human induced pluripotent stem cells. Cell Rep 9(3):810–821

Du DT, Hellen N, Kane C, Terracciano CM (2015) Action potential morphology of human induced pluripotent stem cell-derived cardiomyocytes does not predict cardiac chamber specificity and is dependent on cell density. Biophys J 108(1):1–4

Dubois NC, Craft AM, Sharma P, Elliott DA, Stanley EG, Elefanty AG, Gramolini A, Keller G (2011) SIRPA is a specific cell-surface marker for isolating cardiomyocytes derived from human pluripotent stem cells. Nat Biotechnol 29(11):1011–1018

Egashira T, Yuasa S, Suzuki T, Aizawa Y, Yamakawa H, Matsuhashi T, Ohno Y, Tohyama S, Okata S, Seki T, Kuroda Y, Yae K, Hashimoto H, Tanaka T, Hattori F, Sato T, Miyoshi S, Takatsuki S, Murata M, Kurokawa J, Furukawa T, Makita N, Aiba T, Shimizu W, Horie M, Kamiya K, Kodama I, Ogawa S, Fukuda K (2012) Disease characterization using LQTS-specific induced pluripotent stem cells. Cardiovasc Res 95(4):419–429

Elliott DA, Braam SR, Koutsis K, Ng ES, Jenny R, Lagerqvist EL, Biben C, Hatzistavrou T, Hirst CE, Yu QC, Skelton RJ, Ward-van Oostwaard D, Lim SM, Khammy O, Li X, Hawes SM, Davis RP, Goulburn AL, Passier R, Prall OW, Haynes JM, Pouton CW, Kaye DM, Mummery CL, Elefanty AG, Stanley EG (2011) NKX2-5(eGFP/w) hESCs for isolation of human cardiac progenitors and cardiomyocytes. Nat Methods 8(12):1037–1040

Evans MJ, Kaufman MH (1981) Establishment in culture of pluripotential cells from mouse embryos. Nature 292(5819):154–156

Fatima A, Xu G, Shao K, Papadopoulos S, Lehmann M, Arnaiz-Cot JJ, Rosa AO, Nguemo F, Matzkies M, Dittmann S, Stone SL, Linke M, Zechner U, Beyer V, Hennies HC, Rosenkranz S, Klauke B, Parwani AS, Haverkamp W, Pfitzer G, Farr M, Cleemann L, Morad M, Milting H, Hescheler J, Saric T (2011) In vitro modeling of ryanodine receptor 2 dysfunction using human induced pluripotent stem cells. Cell Physiol Biochem 28(4):579–592

Fatima A, Kaifeng S, Dittmann S, Xu G, Gupta MK, Linke M, Zechner U, Nguemo F, Milting H, Farr M, Hescheler J, Saric T (2013) The disease-specific phenotype in cardiomyocytes derived from induced pluripotent stem cells of two long QT syndrome type 3 patients. PLoS One 8(12):e83005

Feaster TK, Cadar AG, Wang L, Williams CH, Chun YW, Hempel JE, Bloodworth N, Merryman WD, Lim CC, Wu JC, Knollmann BC, Hong CC (2015) Matrigel mattress: a method for the generation of single contracting human-induced pluripotent stem cell-derived cardiomyocytes. Circ Res 117(12):995–1000

Fermini B, Hancox JC, Abi-Gerges N, Bridgland-Taylor M, Chaudhary KW, Colatsky T, Correll K, Crumb W, Damiano B, Erdemli G, Gintant G, Imredy J, Koerner J, Kramer J, Levesque P, Li Z, Lindqvist A, Obejero-Paz CA, Rampe D, Sawada K, Strauss DG, Vandenberg JI (2016) A new perspective in the field of cardiac safety testing through the comprehensive in vitro proarrhythmia assay paradigm. J Biomol Screen 21(1):1–11

Fonoudi H, Yeganeh M, Fattahi F, Ghazizadeh Z, Rassouli H, Alikhani M, Mojarad BA, Baharvand H, Salekdeh GH, Aghdami N (2013) ISL1 protein transduction promotes cardiomyocyte differentiation from human embryonic stem cells. PLoS One 8(1):e55577

Frank S, Zhang M, Scholer HR, Greber B (2012) Small molecule-assisted, line-independent maintenance of human pluripotent stem cells in defined conditions. PLoS One 7(7):e41958

Fusaki N, Ban H, Nishiyama A, Saeki K, Hasegawa M (2009) Efficient induction of transgene-free human pluripotent stem cells using a vector based on Sendai virus, an RNA virus that does not integrate into the host genome. Proc Jpn Acad Ser B Phys Biol Sci 85(8):348–362

Gafni O, Weinberger L, Mansour AA, Manor YS, Chomsky E, Ben-Yosef D, Kalma Y, Viukov S, Maza I, Zviran A, Rais Y, Shipony Z, Mukamel Z, Krupalnik V, Zerbib M, Geula S, Caspi I, Schneir D, Shwartz T, Gilad S, Amann-Zalcenstein D, Benjamin S, Amit I, Tanay A, Massarwa R, Novershtern N, Hanna JH (2013) Derivation of novel human ground state naive pluripotent stem cells. Nature 504(7479):282–286

Gallina C, Turinetto V, Giachino C (2015) A new paradigm in cardiac regeneration: the mesenchymal stem cell secretome. Stem Cells Int 2015:765846

Gessert S, Kuhl M (2010) The multiple phases and faces of wnt signaling during cardiac differentiation and development. Circ Res 107(2):186–199

Gomes Fernandes M, Dries R, Roost MS, Semrau S, de Melo BA, Davis RP, Ramakrishnan R, Szuhai K, Maas E, Umans L, Abon Escalona V, Salvatori D, Deforce D, Van Criekinge W, Huylebroeck D, Mummery C, Zwijsen A, Chuva de Sousa Lopes SM (2016) BMP-SMAD signaling regulates lineage priming, but is dispensable for self-renewal in mouse embryonic stem cells. Stem Cell Rep 6(1):85–94

Gonzalez R, Lee JW, Schultz PG (2011) Stepwise chemically induced cardiomyocyte specification of human embryonic stem cells. Angew Chem 50(47):11181–11185

Graichen R, Xu X, Braam SR, Balakrishnan T, Norfiza S, Sieh S, Soo SY, Tham SC, Mummery C, Colman A, Zweigerdt R, Davidson BP (2008) Enhanced cardiomyogenesis of human embryonic stem cells by a small molecular inhibitor of p38 MAPK. Differentiation 76(4):357–370

Guan X, Mack DL, Moreno CM, Strande JL, Mathieu J, Shi Y, Markert CD, Wang Z, Liu G, Lawlor MW, Moorefield EC, Jones TN, Fugate JA, Furth ME, Murry CE, Ruohola-Baker H, Zhang Y, Santana LF, Childers MK (2014) Dystrophin-deficient cardiomyocytes derived from human urine: new biologic reagents for drug discovery. Stem Cell Res 12(2):467–480

Guo G, von Meyenn F, Santos F, Chen Y, Reik W, Bertone P, Smith A, Nichols J (2016) Naive pluripotent stem cells derived directly from isolated cells of the human inner cell mass. Stem Cell Rep 6(4):437–446

Han L, Li Y, Tchao J, Kaplan AD, Lin B, Li Y, Mich-Basso J, Lis A, Hassan N, London B, Bett GC, Tobita K, Rasmusson RL, Yang L (2014) Study familial hypertrophic cardiomyopathy using patient-specific induced pluripotent stem cells. Cardiovasc Res 104(2):258–269

Hanna J, Cheng AW, Saha K, Kim J, Lengner CJ, Soldner F, Cassady JP, Muffat J, Carey BW, Jaenisch R (2010) Human embryonic stem cells with biological and epigenetic characteristics similar to those of mouse ESCs. Proc Natl Acad Sci U S A 107(20):9222–9227

Hartung S, Schwanke K, Haase A, David R, Franz WM, Martin U, Zweigerdt R (2013) Directing cardiomyogenic differentiation of human pluripotent stem cells by plasmid-based transient overexpression of cardiac transcription factors. Stem Cells Dev 22(7):1112–1125

Hashimoto A, Naito AT, Lee JK, Kitazume-Taneike R, Ito M, Yamaguchi T, Nakata R, Sumida T, Okada K, Nakagawa A, Higo T, Kuramoto Y, Sakai T, Tominaga K, Okinaga T, Kogaki S, Ozono K, Miyagawa S, Sawa Y, Sakata Y, Morita H, Umezawa A, Komuro I (2016) Generation of induced pluripotent stem cells from patients with duchenne muscular dystrophy and their induction to cardiomyocytes. Int Heart J 57(1):112–117

Hattori F, Chen H, Yamashita H, Tohyama S, Satoh YS, Yuasa S, Li W, Yamakawa H, Tanaka T, Onitsuka T, Shimoji K, Ohno Y, Egashira T, Kaneda R, Murata M, Hidaka K, Morisaki T, Sasaki E, Suzuki T, Sano M, Makino S, Oikawa S, Fukuda K (2010) Nongenetic method for purifying stem cell-derived cardiomyocytes. Nat Methods 7(1):61–66

Hatzistergos KE, Quevedo H, Oskouei BN, Hu Q, Feigenbaum GS, Margitich IS, Mazhari R, Boyle AJ, Zambrano JP, Rodriguez JE, Dulce R, Pattany PM, Valdes D, Revilla C, Heldman AW, McNiece I, Hare JM (2010) Bone marrow mesenchymal stem cells stimulate cardiac stem cell proliferation and differentiation. Circ Res 107(7):913–922

Hemmi N, Tohyama S, Nakajima K, Kanazawa H, Suzuki T, Hattori F, Seki T, Kishino Y, Hirano A, Okada M, Tabei R, Ohno R, Fujita C, Haruna T, Yuasa S, Sano M, Fujita J, Fukuda K (2014) A massive suspension culture system with metabolic purification for human pluripotent stem cell-derived cardiomyocytes. Stem Cells Transl Med 3(12):1473–1483

Herron TJ, Rocha AM, Campbell KF, Ponce-Balbuena D, Willis BC, Guerrero-Serna G, Liu Q, Klos M, Musa H, Zarzoso M, Bizy A, Furness J, Anumonwo J, Mironov S, Jalife J (2016) Extracellular matrix-mediated maturation of human pluripotent stem cell-derived cardiac monolayer structure and electrophysiological function. Circ Arrhythm Electrophysiol 9(4)

Hinson JT, Chopra A, Nafissi N, Polacheck WJ, Benson CC, Swist S, Gorham J, Yang L, Schafer S, Sheng CC, Haghighi A, Homsy J, Hubner N, Church G, Cook SA, Linke WA, Chen CS, Seidman JG, Seidman CE (2015) HEART DISEASE. Titin mutations in iPS cells define sarcomere insufficiency as a cause of dilated cardiomyopathy. Science 349(6251):982–986

Huang HP, Chen PH, Hwu WL, Chuang CY, Chien YH, Stone L, Chien CL, Li LT, Chiang SC, Chen HF, Ho HN, Chen CH, Kuo HC (2011) Human Pompe disease-induced pluripotent stem cells for pathogenesis modeling, drug testing and disease marker identification. Hum Mol Genet 20(24):4851–4864

Hudson J, Titmarsh D, Hidalgo A, Wolvetang E, Cooper-White J (2012) Primitive cardiac cells from human embryonic stem cells. Stem Cells Dev 21(9):1513–1523

Ionta V, Liang W, Kim EH, Rafie R, Giacomello A, Marban E, Cho HC (2015) SHOX2 overexpression favors differentiation of embryonic stem cells into cardiac pacemaker cells, improving biological pacing ability. Stem Cell Reports 4(1):129–142

Itskovitz-Eldor J, Schuldiner M, Karsenti D, Eden A, Yanuka O, Amit M, Soreq H, Benvenisty N (2000) Differentiation of human embryonic stem cells into embryoid bodies compromising the three embryonic germ layers. Mol Med 6(2):88–95

Itzhaki I, Maizels L, Huber I, Zwi-Dantsis L, Caspi O, Winterstern A, Feldman O, Gepstein A, Arbel G, Hammerman H, Boulos M, Gepstein L (2011) Modelling the long QT syndrome with induced pluripotent stem cells. Nature 471(7337):225–229

Itzhaki I, Maizels L, Huber I, Gepstein A, Arbel G, Caspi O, Miller L, Belhassen B, Nof E, Glikson M, Gepstein L (2012) Modeling of catecholaminergic polymorphic ventricular tachycardia with patient-specific human-induced pluripotent stem cells. J Am Coll Cardiol 60(11):990–1000

Ivashchenko CY, Pipes GC, Lozinskaya IM, Lin Z, Xiaoping X, Needle S, Grygielko ET, Hu E, Toomey JR, Lepore JJ, Willette RN (2013) Human-induced pluripotent stem cell-derived cardiomyocytes exhibit temporal changes in phenotype. Am J Physiol Heart Circ Physiol 305(6):H913–H922

Jiang Y, Habibollah S, Tilgner K, Collin J, Barta T, Al-Aama JY, Tesarov L, Hussain R, Trafford AW, Kirkwood G, Sernagor E, Eleftheriou CG, Przyborski S, Stojkovic M, Lako M, Keavney B, Armstrong L (2014) An induced pluripotent stem cell model of hypoplastic left heart syndrome (HLHS) reveals multiple expression and functional differences in HLHS-derived cardiac myocytes. Stem Cells Transl Med 3(4):416–423

Johnston PV, Sasano T, Mills K, Evers R, Lee ST, Smith RR, Lardo AC, Lai S, Steenbergen C, Gerstenblith G, Lange R, Marban E (2009) Engraftment, differentiation, and functional benefits of autologous cardiosphere-derived cells in porcine ischemic cardiomyopathy. Circulation 120(12):1075–1083

Josowitz R, Lu J, Falce C, D'Souza SL, Wu M, Cohen N, Dubois NC, Zhao Y, Sobie EA, Fishman GI, Gelb BD (2014) Identification and purification of human induced pluripotent stem cell-derived atrial-like cardiomyocytes based on sarcolipin expression. PLoS One 9(7):e101316

Jouni M, Si-Tayeb K, Es-Salah-Lamoureux Z, Latypova X, Champon B, Caillaud A, Rungoat A, Charpentier F, Loussouarn G, Baro I, Zibara K, Lemarchand P, Gaborit N (2015) Toward personalized medicine: using cardiomyocytes differentiated from urine-derived pluripotent stem cells to recapitulate electrophysiological characteristics of type 2 long QT syndrome. J Am Heart Assoc 4(9):e002159

Jung CB, Moretti A, Mederos Y, Schnitzler M, Iop L, Storch U, Bellin M, Dorn T, Ruppenthal S, Pfeiffer S, Goedel A, Dirschinger RJ, Seyfarth M, Lam JT, Sinnecker D, Gudermann T, Lipp P, Laugwitz KL (2012) Dantrolene rescues arrhythmogenic RYR2 defect in a patient-specific stem cell model of catecholaminergic polymorphic ventricular tachycardia. EMBO Mol Med 4(3):180–191

Jung JJ, Husse B, Rimmbach C, Krebs S, Stieber J, Steinhoff G, Dendorfer A, Franz WM, David R (2014) Programming and isolation of highly pure physiologically and pharmacologically functional sinus-nodal bodies from pluripotent stem cells. Stem Cell Rep 2(5):592–605

Kaitin KI (2010) Deconstructing the drug development process: the new face of innovation. Clin Pharmacol Ther 87(3):356–361

Kane C, Du DT, Hellen N, Terracciano CM (2016) The fallacy of assigning chamber specificity to iPSC cardiac myocytes from action potential morphology. Biophys J 110(1):281–283

Karakikes I, Stillitano F, Nonnenmacher M, Tzimas C, Sanoudou D, Termglinchan V, Kong CW, Rushing S, Hansen J, Ceholski D, Kolokathis F, Kremastinos D, Katoulis A, Ren L, Cohen N, Gho JM, Tsiapras D, Vink A, Wu JC, Asselbergs FW, Li RA, Hulot JS, Kranias EG, Hajjar RJ (2015) Correction of human phospholamban R14del mutation associated with cardiomyopathy using targeted nucleases and combination therapy. Nat Commun 6:6955

Kattman SJ, Witty AD, Gagliardi M, Dubois NC, Niapour M, Hotta A, Ellis J, Keller G (2011) Stage-specific optimization of activin/nodal and BMP signaling promotes cardiac differentiation of mouse and human pluripotent stem cell lines. Cell Stem Cell 8(2):228–240

Kehat I, Kenyagin-Karsenti D, Snir M, Segev H, Amit M, Gepstein A, Livne E, Binah O, Itskovitz-Eldor J, Gepstein L (2001) Human embryonic stem cells can differentiate into myocytes with structural and functional properties of cardiomyocytes. J Clin Investig 108(3):407–414

Kehat I, Khimovich L, Caspi O, Gepstein A, Shofti R, Arbel G, Huber I, Satin J, Itskovitz-Eldor J, Gepstein L (2004) Electromechanical integration of cardiomyocytes derived from human embryonic stem cells. Nat Biotechnol 22(10):1282–1289

Kelly RG (2012) The second heart field. Curr Top Dev Biol 100:33–65

Kempf H, Olmer R, Kropp C, Ruckert M, Jara-Avaca M, Robles-Diaz D, Franke A, Elliott DA, Wojciechowski D, Fischer M, Roa Lara A, Kensah G, Gruh I, Haverich A, Martin U, Zweigerdt R (2014) Controlling expansion and cardiomyogenic differentiation of human pluripotent stem cells in scalable suspension culture. Stem Cell Reports 3(6):1132–1146

Kempf H, Kropp C, Olmer R, Martin U, Zweigerdt R (2015) Cardiac differentiation of human pluripotent stem cells in scalable suspension culture. Nat Protoc 10(9):1345–1361

Kempf H, Andree B, Zweigerdt R (2016) Large-scale production of human pluripotent stem cell derived cardiomyocytes. Adv Drug Deliv Rev 96:18–30

Khanabdali R, Rosdah AA, Dusting GJ, Lim SY (2016) Harnessing the secretome of cardiac stem cells as therapy for ischemic heart disease. Biochem Pharmacol 113:1–11

Kim C, Wong J, Wen J, Wang S, Wang C, Spiering S, Kan NG, Forcales S, Puri PL, Leone TC, Marine JE, Calkins H, Kelly DP, Judge DP, Chen HS (2013) Studying arrhythmogenic right ventricular dysplasia with patient-specific iPSCs. Nature 494(7435):105–110

Kimelman D (2006) Mesoderm induction: from caps to chips. Nat Rev Genet 7(5):360–372

Kita-Matsuo H, Barcova M, Prigozhina N, Salomonis N, Wei K, Jacot JG, Nelson B, Spiering S, Haverslag R, Kim C, Talantova M, Bajpai R, Calzolari D, Terskikh A, McCulloch AD, Price JH, Conklin BR, Chen HS, Mercola M (2009) Lentiviral vectors and protocols for creation of stable hESC lines for fluorescent tracking and drug resistance selection of cardiomyocytes. PLoS One 4(4):e5046

Kobayashi J, Yoshida M, Tarui S, Hirata M, Nagai Y, Kasahara S, Naruse K, Ito H, Sano S, Oh H (2014) Directed differentiation of patient-specific induced pluripotent stem cells identifies the transcriptional repression and epigenetic modification of NKX2-5, HAND1, and NOTCH1 in hypoplastic left heart syndrome. PLoS One 9(7):e102796

Kosmidis G, Bellin M, Ribeiro MC, van Meer B, Ward-van Oostwaard D, Passier R, Tertoolen LG, Mummery CL, Casini S (2015) Altered calcium handling and increased contraction force in human embryonic stem cell derived cardiomyocytes following short term dexamethasone exposure. Biochem Biophys Res Commun 467(4):998–1005

Kujala K, Paavola J, Lahti A, Larsson K, Pekkanen-Mattila M, Viitasalo M, Lahtinen AM, Toivonen L, Kontula K, Swan H, Laine M, Silvennoinen O, Aalto-Setala K (2012) Cell model of catecholaminergic polymorphic ventricular tachycardia reveals early and delayed afterdepolarizations. PLoS One 7(9):e44660

Kumar N, Richter J, Cutts J, Bush KT, Trujillo C, Nigam SK, Gaasterland T, Brafman D, Willert K (2015) Generation of an expandable intermediate mesoderm restricted progenitor cell line from human pluripotent stem cells. elife 4:e08413

Kwon C, Qian L, Cheng P, Nigam V, Arnold J, Srivastava D (2009) A regulatory pathway involving Notch1/beta-catenin/Isl1 determines cardiac progenitor cell fate. Nat Cell Biol 11(8):951–957

Laflamme MA, Chen KY, Naumova AV, Muskheli V, Fugate JA, Dupras SK, Reinecke H, Xu C, Hassanipour M, Police S, O'Sullivan C, Collins L, Chen Y, Minami E, Gill EA, Ueno S, Yuan C, Gold J, Murry CE (2007) Cardiomyocytes derived from human embryonic stem cells in pro-survival factors enhance function of infarcted rat hearts. Nat Biotechnol 25(9):1015–1024

Lahti AL, Kujala VJ, Chapman H, Koivisto AP, Pekkanen-Mattila M, Kerkela E, Hyttinen J, Kontula K, Swan H, Conklin BR, Yamanaka S, Silvennoinen O, Aalto-Setala K (2012) Model for long QT syndrome type 2 using human iPS cells demonstrates arrhythmogenic characteristics in cell culture. Dis Model Mech 5(2):220–230

Lalit PA, Salick MR, Nelson DO, Squirrell JM, Shafer CM, Patel NG, Saeed I, Schmuck EG, Markandeya YS, Wong R, Lea MR, Eliceiri KW, Hacker TA, Crone WC, Kyba M, Garry DJ, Stewart R, Thomson JA, Downs KM, Lyons GE, Kamp TJ (2016) Lineage reprogramming of fibroblasts into proliferative induced cardiac progenitor cells by defined factors. Cell Stem Cell 18(3):354–367

Lam AT, Chen AK, Li J, Birch WR, Reuveny S, Oh SK (2014) Conjoint propagation and differentiation of human embryonic stem cells to cardiomyocytes in a defined microcarrier spinner culture. Stem Cell Res Ther 5(5):110

Lam AT, Chen AK, Ting SQ, Reuveny S, Oh SK (2016) Integrated processes for expansion and differentiation of human pluripotent stem cells in suspended microcarriers cultures. Biochem Biophys Res Commun 473(3):764–768

Lan F, Lee AS, Liang P, Sanchez-Freire V, Nguyen PK, Wang L, Han L, Yen M, Wang Y, Sun N, Abilez OJ, Hu S, Ebert AD, Navarrete EG, Simmons CS, Wheeler M, Pruitt B, Lewis R, Yamaguchi Y, Ashley EA, Bers DM, Robbins RC, Longaker MT, Wu JC (2013) Abnormal calcium handling properties underlie familial hypertrophic cardiomyopathy pathology in patient-specific induced pluripotent stem cells. Cell Stem Cell 12(1):101–113

Leahy A, Xiong JW, Kuhnert F, Stuhlmann H (1999) Use of developmental marker genes to define temporal and spatial patterns of differentiation during embryoid body formation. J Exp Zool 284(1):67–81

Lecina M, Ting S, Choo A, Reuveny S, Oh S (2010) Scalable platform for human embryonic stem cell differentiation to cardiomyocytes in suspended microcarrier cultures. Tissue Eng Part C Methods 16(6):1609–1619

Lee YK, Ng KM, Lai WH, Chan YC, Lau YM, Lian Q, Tse HF, Siu CW (2011) Calcium homeostasis in human induced pluripotent stem cell-derived cardiomyocytes. Stem Cell Rev 7(4):976–986

Leiker M, Suzuki G, Iyer VS, Canty JM Jr, Lee T (2008) Assessment of a nuclear affinity labeling method for tracking implanted mesenchymal stem cells. Cell Transplant 17(8):911–922

Li Y, Powell S, Brunette E, Lebkowski J, Mandalam R (2005) Expansion of human embryonic stem cells in defined serum-free medium devoid of animal-derived products. Biotechnol Bioeng 91(6):688–698

Lian X, Hsiao C, Wilson G, Zhu K, Hazeltine LB, Azarin SM, Raval KK, Zhang J, Kamp TJ, Palecek SP (2012) Robust cardiomyocyte differentiation from human pluripotent stem cells via temporal modulation of canonical Wnt signaling. Proc Natl Acad Sci U S A 109(27):E1848–E1857

Lian X, Zhang J, Azarin SM, Zhu K, Hazeltine LB, Bao X, Hsiao C, Kamp TJ, Palecek SP (2013) Directed cardiomyocyte differentiation from human pluripotent stem cells by modulating Wnt/beta-catenin signaling under fully defined conditions. Nat Protoc 8(1):162–175

Lian X, Bao X, Zilberter M, Westman M, Fisahn A, Hsiao C, Hazeltine LB, Dunn KK, Kamp TJ, Palecek SP (2015) Chemically defined, albumin-free human cardiomyocyte generation. Nat Methods 12(7):595–596

Liang P, Lan F, Lee AS, Gong T, Sanchez-Freire V, Wang Y, Diecke S, Sallam K, Knowles JW, Wang PJ, Nguyen PK, Bers DM, Robbins RC, Wu JC (2013) Drug screening using a library of human induced pluripotent stem cell-derived cardiomyocytes reveals disease-specific patterns of cardiotoxicity. Circulation 127(16):1677–1691

Lieu DK, Fu JD, Chiamvimonvat N, Tung KC, McNerney GP, Huser T, Keller G, Kong CW, Li RA (2013) Mechanism-based facilitated maturation of human pluripotent stem cell-derived cardiomyocytes. Circ Arrhythm Electrophysiol 6(1):191–201

Lin B, Li Y, Han L, Kaplan AD, Ao Y, Kalra S, Bett GC, Rasmusson RL, Denning C, Yang L (2015) Modeling and study of the mechanism of dilated cardiomyopathy using induced pluripotent stem cells derived from individuals with Duchenne muscular dystrophy. Dis Model Mech 8(5):457–466

Lin Y, Linask KL, Mallon B, Johnson K, Klein M, Beers J, Xie W, Du Y, Liu C, Lai Y, Zou J, Haigney M, Yang H, Rao M, Chen G (2016) Heparin promotes cardiac differentiation of human pluripotent stem cells in chemically defined albumin-free medium, enabling consistent manufacture of cardiomyocytes. Stem Cells Transl Med 6(2):527–538

Ludwig TE, Levenstein ME, Jones JM, Berggren WT, Mitchen ER, Frane JL, Crandall LJ, Daigh CA, Conard KR, Piekarczyk MS, Llanas RA, Thomson JA (2006) Derivation of human embryonic stem cells in defined conditions. Nat Biotechnol 24(2):185–187

Lundy SD, Zhu WZ, Regnier M, Laflamme MA (2013) Structural and functional maturation of cardiomyocytes derived from human pluripotent stem cells. Stem Cells Dev 22(14): 1991–2002

Ma J, Guo L, Fiene SJ, Anson BD, Thomson JA, Kamp TJ, Kolaja KL, Swanson BJ, January CT (2011) High purity human-induced pluripotent stem cell-derived cardiomyocytes: electrophysiological properties of action potentials and ionic currents. Am J Physiol Heart Circ Physiol 301(5):H2006–H2017

Ma D, Wei H, Lu J, Ho S, Zhang G, Sun X, Oh Y, Tan SH, Ng ML, Shim W, Wong P, Liew R (2013a) Generation of patient-specific induced pluripotent stem cell-derived cardiomyocytes as a cellular model of arrhythmogenic right ventricular cardiomyopathy. Eur Heart J 34(15):1122–1133

Ma D, Wei H, Zhao Y, Lu J, Li G, Sahib NB, Tan TH, Wong KY, Shim W, Wong P, Cook SA, Liew R (2013b) Modeling type 3 long QT syndrome with cardiomyocytes derived from patient-specific induced pluripotent stem cells. Int J Cardiol 168(6):5277–5286

Ma D, Wei H, Lu J, Huang D, Liu Z, Loh LJ, Islam O, Liew R, Shim W, Cook SA (2015) Characterization of a novel KCNQ1 mutation for type 1 long QT syndrome and assessment of the therapeutic potential of a novel IKs activator using patient-specific induced pluripotent stem cell-derived cardiomyocytes. Stem Cell Res Ther 6:39

Maherali N, Sridharan R, Xie W, Utikal J, Eminli S, Arnold K, Stadtfeld M, Yachechko R, Tchieu J, Jaenisch R, Plath K, Hochedlinger K (2007) Directly reprogrammed fibroblasts show global epigenetic remodeling and widespread tissue contribution. Cell Stem Cell 1(1):55–70

Malan D, Zhang M, Stallmeyer B, Muller J, Fleischmann BK, Schulze-Bahr E, Sasse P, Greber B (2016) Human iPS cell model of type 3 long QT syndrome recapitulates drug-based phenotype correction. Basic Res Cardiol 111(2):14

Marinho PA, Chailangkarn T, Muotri AR (2015) Systematic optimization of human pluripotent stem cells media using design of experiments. Sci Rep 5:9834

Matsa E, Rajamohan D, Dick E, Young L, Mellor I, Staniforth A, Denning C (2011) Drug evaluation in cardiomyocytes derived from human induced pluripotent stem cells carrying a long QT syndrome type 2 mutation. Eur Heart J 32(8):952–962

Matsa E, Dixon JE, Medway C, Georgiou O, Patel MJ, Morgan K, Kemp PJ, Staniforth A, Mellor I, Denning C (2014) Allele-specific RNA interference rescues the long-QT syndrome phenotype in human-induced pluripotency stem cell cardiomyocytes. Eur Heart J 35(16):1078–1087

Mehta A, Sequiera GL, Ramachandra CJ, Sudibyo Y, Chung Y, Sheng J, Wong KY, Tan TH, Wong P, Liew R, Shim W (2014) Re-trafficking of hERG reverses long QT syndrome 2 phenotype in human iPS-derived cardiomyocytes. Cardiovasc Res 102(3):497–506

Minami I, Yamada K, Otsuji TG, Yamamoto T, Shen Y, Otsuka S, Kadota S, Morone N, Barve M, Asai Y, Tenkova-Heuser T, Heuser JE, Uesugi M, Aiba K, Nakatsuji N (2012) A small molecule that promotes cardiac differentiation of human pluripotent stem cells under defined, cytokine- and xeno-free conditions. Cell Rep 2(5):1448–1460

Montserrat N, Nivet E, Sancho-Martinez I, Hishida T, Kumar S, Miquel L, Cortina C, Hishida Y, Xia Y, Esteban CR, Izpisua Belmonte JC (2013) Reprogramming of human fibroblasts to pluripotency with lineage specifiers. Cell Stem Cell 13(3):341–350

Mordwinkin NM, Burridge PW, Wu JC (2013) A review of human pluripotent stem cell-derived cardiomyocytes for high-throughput drug discovery, cardiotoxicity screening, and publication standards. J Cardiovasc Transl Res 6(1):22–30

Moretti A, Bellin M, Welling A, Jung CB, Lam JT, Bott-Flugel L, Dorn T, Goedel A, Hohnke C, Hofmann F, Seyfarth M, Sinnecker D, Schomig A, Laugwitz KL (2010) Patient-specific induced pluripotent stem-cell models for long-QT syndrome. N Engl J Med 363(15): 1397–1409

Mozaffarian D, Benjamin EJ, Go AS, Arnett DK, Blaha MJ, Cushman M, Das SR, de Ferranti S, Després J-P, Fullerton HJ, Howard VJ, Huffman MD, Isasi CR, Jiménez MC, Judd SE, Kissela BM, Lichtman JH, Lisabeth LD, Liu S, Mackey RH, Magid DJ, McGuire DK, Mohler ER, Moy CS, Muntner P, Mussolino ME, Nasir K, Neumar RW, Nichol G, Palaniappan L, Pandey DK, Reeves MJ, Rodriguez CJ, Rosamond W, Sorlie PD, Stein J, Towfighi A, Turan TN, Virani SS, Woo D, Yeh RW, Turner MB (2016) Heart disease and stroke statistics—2016 update: a report from the American Heart Association. Circulation 133(4):e38–e360

Mummery C, Ward-van Oostwaard D, Doevendans P, Spijker R, van den Brink S, Hassink R, van der Heyden M, Opthof T, Pera M, de la Riviere AB, Passier R, Tertoolen L (2003) Differentiation of human embryonic stem cells to cardiomyocytes: role of coculture with visceral endoderm-like cells. Circulation 107(21):2733–2740

Murry CE, Reinecke H, Pabon LM (2006) Regeneration gaps: observations on stem cells and cardiac repair. J Am Coll Cardiol 47(9):1777–1785

Ng ES, Davis RP, Azzola L, Stanley EG, Elefanty AG (2005) Forced aggregation of defined numbers of human embryonic stem cells into embryoid bodies fosters robust, reproducible hematopoietic differentiation. Blood 106(5):1601–1603

Nichols J, Smith A (2009) Naive and primed pluripotent states. Cell Stem Cell 4(6):487–492

Niebruegge S, Bauwens CL, Peerani R, Thavandiran N, Masse S, Sevaptisidis E, Nanthakumar K, Woodhouse K, Husain M, Kumacheva E, Zandstra PW (2009) Generation of human embryonic stem cell-derived mesoderm and cardiac cells using size-specified aggregates in an oxygen-controlled bioreactor. Biotechnol Bioeng 102(2):493–507

Noseda M, Peterkin T, Simoes FC, Patient R, Schneider MD (2011) Cardiopoietic factors: extracellular signals for cardiac lineage commitment. Circ Res 108(1):129–152

Novak A, Barad L, Zeevi-Levin N, Shick R, Shtrichman R, Lorber A, Itskovitz-Eldor J, Binah O (2012) Cardiomyocytes generated from CPVTD307H patients are arrhythmogenic in response to beta-adrenergic stimulation. J Cell Mol Med 16(3):468–482

Novak A, Barad L, Lorber A, Gherghiceanu M, Reiter I, Eisen B, Eldor L, Itskovitz-Eldor J, Eldar M, Arad M, Binah O (2015) Functional abnormalities in iPSC-derived cardiomyocytes generated from CPVT1 and CPVT2 patients carrying ryanodine or calsequestrin mutations. J Cell Mol Med 19(8):2006–2018

Ojala M, Prajapati C, Polonen RP, Rajala K, Pekkanen-Mattila M, Rasku J, Larsson K, Aalto-Setala K (2016) Mutation-specific phenotypes in hiPSC-derived cardiomyocytes carrying either myosin-binding protein C or alpha-tropomyosin mutation for hypertrophic cardiomyopathy. Stem Cells Int 2016:1684792

Okawa S, del Sol A (2015) A computational strategy for predicting lineage specifiers in stem cell subpopulations. Stem Cell Res 15(2):427–434

Okita K, Ichisaka T, Yamanaka S (2007) Generation of germline-competent induced pluripotent stem cells. Nature 448(7151):313–317

Ong SG, Huber BC, Lee WH, Kodo K, Ebert AD, Ma Y, Nguyen PK, Diecke S, Chen WY, Wu JC (2015) Microfluidic single-cell analysis of transplanted human induced pluripotent stem cell-derived cardiomyocytes after acute myocardial infarction. Circulation 132(8):762–771

Osafune K, Caron L, Borowiak M, Martinez RJ, Fitz-Gerald CS, Sato Y, Cowan CA, Chien KR, Melton DA (2008) Marked differences in differentiation propensity among human embryonic stem cell lines. Nat Biotechnol 26(3):313–315

Park IH, Zhao R, West JA, Yabuuchi A, Huo H, Ince TA, Lerou PH, Lensch MW, Daley GQ (2008) Reprogramming of human somatic cells to pluripotency with defined factors. Nature 451(7175):141–146

Penttinen K, Swan H, Vanninen S, Paavola J, Lahtinen AM, Kontula K, Aalto-Setala K (2015) Antiarrhythmic effects of dantrolene in patients with catecholaminergic polymorphic ventricular tachycardia and replication of the responses using iPSC models. PLoS One 10(5):e0125366

Rana P, Anson B, Engle S, Will Y (2012) Characterization of human-induced pluripotent stem cell-derived cardiomyocytes: bioenergetics and utilization in safety screening. Toxicol Sci 130(1):117–131

Rao J, Pfeiffer MJ, Frank S, Adachi K, Piccini I, Quaranta R, Arauzo-Bravo M, Schwarz J, Schade D, Leidel S, Scholer HR, Seebohm G, Greber B (2016) Stepwise clearance of repressive roadblocks drives cardiac induction in human ESCs. Cell Stem Cell 18(3):341–353

Raval KK, Tao R, White BE, De Lange WJ, Koonce CH, Yu J, Kishnani PS, Thomson JA, Mosher DF, Ralphe JC, Kamp TJ (2015) Pompe disease results in a Golgi-based glycosylation deficit in human induced pluripotent stem cell-derived cardiomyocytes. J Biol Chem 290(5):3121–3136

Ren Y, Lee MY, Schliffke S, Paavola J, Amos PJ, Ge X, Ye M, Zhu S, Senyei G, Lum L, Ehrlich BE, Qyang Y (2011) Small molecule Wnt inhibitors enhance the efficiency of BMP-4-directed cardiac differentiation of human pluripotent stem cells. J Mol Cell Cardiol 51(3): 280–287

Rosen MR, Robinson RB, Brink PR, Cohen IS (2011) The road to biological pacing. Nat Rev Cardiol 8(11):656–666

Saito Y, Nakamura K, Yoshida M, Sugiyama H, Ohe T, Kurokawa J, Furukawa T, Takano M, Nagase S, Morita H, Kusano KF, Ito H (2015) Enhancement of spontaneous activity by HCN4 overexpression in mouse embryonic stem cell-derived cardiomyocytes – a possible biological pacemaker. PLoS One 10(9):e0138193

Sato Y, Kobayashi H, Higuchi T, Shimada Y, Era T, Kimura S, Eto Y, Ida H, Ohashi T (2015) Disease modeling and lentiviral gene transfer in patient-specific induced pluripotent stem cells from late-onset Pompe disease patient. Mol Ther Methods Clin Dev 2:15023

Schleich JM, Abdulla T, Summers R, Houyel L (2013) An overview of cardiac morphogenesis. Arch Cardiovasc Dis 106(11):612–623

Shu J, Wu C, Wu Y, Li Z, Shao S, Zhao W, Tang X, Yang H, Shen L, Zuo X, Yang W, Shi Y, Chi X, Zhang H, Gao G, Shu Y, Yuan K, He W, Tang C, Zhao Y, Deng H (2013) Induction of pluripotency in mouse somatic cells with lineage specifiers. Cell 153(5):963–975

Siu CW, Lee YK, Ho JC, Lai WH, Chan YC, Ng KM, Wong LY, Au KW, Lau YM, Zhang J, Lay KW, Colman A, Tse HF (2012) Modeling of lamin a/C mutation premature cardiac aging using patient-specific induced pluripotent stem cells. Aging (Albany NY) 4(11):803–822

Skrzynia C, Berg JS, Willis MS, Jensen BC (2015) Genetics and heart failure: a concise guide for the clinician. Curr Cardiol Rev 11(1):10–17

Smith AG, Heath JK, Donaldson DD, Wong GG, Moreau J, Stahl M, Rogers D (1988) Inhibition of pluripotential embryonic stem cell differentiation by purified polypeptides. Nature 336(6200):688–690

Sun N, Yazawa M, Liu J, Han L, Sanchez-Freire V, Abilez OJ, Navarrete EG, Hu S, Wang L, Lee A, Pavlovic A, Lin S, Chen R, Hajjar RJ, Snyder MP, Dolmetsch RE, Butte MJ, Ashley EA, Longaker MT, Robbins RC, Wu JC (2012) Patient-specific induced pluripotent stem cells as a model for familial dilated cardiomyopathy. Sci Transl Med 4 (130):130ra147.

Takahashi K, Yamanaka S (2006) Induction of pluripotent stem cells from mouse embryonic and adult fibroblast cultures by defined factors. Cell 126(4):663–676

Takahashi K, Tanabe K, Ohnuki M, Narita M, Ichisaka T, Tomoda K, Yamanaka S (2007) Induction of pluripotent stem cells from adult human fibroblasts by defined factors. Cell 131(5):861–872

Takei S, Ichikawa H, Johkura K, Mogi A, No H, Yoshie S, Tomotsune D, Sasaki K (2009) Bone morphogenetic protein-4 promotes induction of cardiomyocytes from human embryonic stem cells in serum-based embryoid body development. Am J Physiol Heart Circ Physiol 296(6):H1793–H1803

Tanaka A, Yuasa S, Mearini G, Egashira T, Seki T, Kodaira M, Kusumoto D, Kuroda Y, Okata S, Suzuki T, Inohara T, Arimura T, Makino S, Kimura K, Kimura A, Furukawa T, Carrier L, Node K, Fukuda K (2014) Endothelin-1 induces myofibrillar disarray and contractile vector variability in hypertrophic cardiomyopathy-induced pluripotent stem cell-derived cardiomyocytes. J Am Heart Assoc 3(6):e001263

Tang XL, Rokosh G, Sanganalmath SK, Yuan F, Sato H, Mu J, Dai S, Li C, Chen N, Peng Y, Dawn B, Hunt G, Leri A, Kajstura J, Tiwari S, Shirk G, Anversa P, Bolli R (2010) Intracoronary administration of cardiac progenitor cells alleviates left ventricular dysfunction in rats with a 30-day-old infarction. Circulation 121(2):293–305

Terrenoire C, Wang K, Tung KW, Chung WK, Pass RH, Lu JT, Jean JC, Omari A, Sampson KJ, Kotton DN, Keller G, Kass RS (2013) Induced pluripotent stem cells used to reveal drug actions in a long QT syndrome family with complex genetics. J Gen Physiol 141(1):61–72

Tesar PJ, Chenoweth JG, Brook FA, Davies TJ, Evans EP, Mack DL, Gardner RL, McKay RD (2007) New cell lines from mouse epiblast share defining features with human embryonic stem cells. Nature 448(7150):196–199

Theis JL, Hrstka SC, Evans JM, O'Byrne MM, de Andrade M, O'Leary PW, Nelson TJ, Olson TM (2015) Compound heterozygous NOTCH1 mutations underlie impaired cardiogenesis in a patient with hypoplastic left heart syndrome. Hum Genet 134(9):1003–1011

Theunissen TW, Powell BE, Wang H, Mitalipova M, Faddah DA, Reddy J, Fan ZP, Maetzel D, Ganz K, Shi L, Lungjangwa T, Imsoonthornruksa S, Stelzer Y, Rangarajan S, D'Alessio A, Zhang J, Gao Q, Dawlaty MM, Young RA, Gray NS, Jaenisch R (2014) Systematic identification of culture conditions for induction and maintenance of naive human pluripotency. Cell Stem Cell 15(4):471–487

Thomson JA, Itskovitz-Eldor J, Shapiro SS, Waknitz MA, Swiergiel JJ, Marshall VS, Jones JM (1998) Embryonic stem cell lines derived from human blastocysts. Science 282(5391):1145–1147

Thomson M, Liu SJ, Zou LN, Smith Z, Meissner A, Ramanathan S (2011) Pluripotency factors in embryonic stem cells regulate differentiation into germ layers. Cell 145(6):875–889

Tohyama S, Hattori F, Sano M, Hishiki T, Nagahata Y, Matsuura T, Hashimoto H, Suzuki T, Yamashita H, Satoh Y, Egashira T, Seki T, Muraoka N, Yamakawa H, Ohgino Y, Tanaka T, Yoichi M, Yuasa S, Murata M, Suematsu M, Fukuda K (2013) Distinct metabolic flow enables large-scale purification of mouse and human pluripotent stem cell-derived cardiomyocytes. Cell Stem Cell 12(1):127–137

Tompkins JD, Jung M, Chen CY, Lin Z, Ye J, Godatha S, Lizhar E, Wu X, Hsu D, Couture LA, Riggs AD (2016) Mapping human pluripotent-to-cardiomyocyte differentiation: methylomes, transcriptomes, and exon DNA methylation "memories". EBioMedicine 4:74–85

Tran TH, Wang X, Browne C, Zhang Y, Schinke M, Izumo S, Burcin M (2009) Wnt3a-induced mesoderm formation and cardiomyogenesis in human embryonic stem cells. Stem Cells 27(8):1869–1878

Tse HF, Ho JC, Choi SW, Lee YK, Butler AW, Ng KM, Siu CW, Simpson MA, Lai WH, Chan YC, Au KW, Zhang J, Lay KW, Esteban MA, Nicholls JM, Colman A, Sham PC (2013) Patient-specific induced-pluripotent stem cells-derived cardiomyocytes recapitulate the pathogenic phenotypes of dilated cardiomyopathy due to a novel DES mutation identified by whole exome sequencing. Hum Mol Genet 22(7):1395–1403

Uosaki H, Fukushima H, Takeuchi A, Matsuoka S, Nakatsuji N, Yamanaka S, Yamashita JK (2011) Efficient and scalable purification of cardiomyocytes from human embryonic and induced pluripotent stem cells by VCAM1 surface expression. PLoS One 6(8):e23657

Vallier L, Alexander M, Pedersen RA (2005) Activin/nodal and FGF pathways cooperate to maintain pluripotency of human embryonic stem cells. J Cell Sci 118(Pt 19):4495–4509

Wang L, Schulz TC, Sherrer ES, Dauphin DS, Shin S, Nelson AM, Ware CB, Zhan M, Song CZ, Chen X, Brimble SN, McLean A, Galeano MJ, Uhl EW, D'Amour KA, Chesnut JD, Rao MS, Blau CA, Robins AJ (2007) Self-renewal of human embryonic stem cells requires insulin-like growth factor-1 receptor and ERBB2 receptor signaling. Blood 110(12):4111–4119

Wang G, McCain ML, Yang L, He A, Pasqualini FS, Agarwal A, Yuan H, Jiang D, Zhang D, Zangi L, Geva J, Roberts AE, Ma Q, Ding J, Chen J, Wang DZ, Li K, Wang J, Wanders RJ, Kulik W, Vaz FM, Laflamme MA, Murry CE, Chien KR, Kelley RI, Church GM, Parker KK, Pu WT (2014) Modeling the mitochondrial cardiomyopathy of Barth syndrome with induced pluripotent stem cell and heart-on-chip technologies. Nat Med 20(6):616–623

Ware CB, Nelson AM, Mecham B, Hesson J, Zhou W, Jonlin EC, Jimenez-Caliani AJ, Deng X, Cavanaugh C, Cook S, Tesar PJ, Okada J, Margaretha L, Sperber H, Choi M, Blau CA, Treuting PM, Hawkins RD, Cirulli V, Ruohola-Baker H (2014) Derivation of naive human embryonic stem cells. Proc Natl Acad Sci U S A 111(12):4484–4489

Wen JY, Wei CY, Shah K, Wong J, Wang C, Chen HS (2015) Maturation-based model of arrhythmogenic right ventricular dysplasia using patient-specific induced pluripotent stem cells. Circ J 79(7):1402–1408

Wernig M, Meissner A, Foreman R, Brambrink T, Ku M, Hochedlinger K, Bernstein BE, Jaenisch R (2007) In vitro reprogramming of fibroblasts into a pluripotent ES-cell-like state. Nature 448(7151):318–324

Willems E, Spiering S, Davidovics H, Lanier M, Xia Z, Dawson M, Cashman J, Mercola M (2011) Small-molecule inhibitors of the Wnt pathway potently promote cardiomyocytes from human embryonic stem cell-derived mesoderm. Circ Res 109(4):360–364

Willems E, Cabral-Teixeira J, Schade D, Cai W, Reeves P, Bushway PJ, Lanier M, Walsh C, Kirchhausen T, Izpisua Belmonte JC, Cashman J, Mercola M (2012) Small molecule-mediated TGF-beta type II receptor degradation promotes cardiomyogenesis in embryonic stem cells. Cell Stem Cell 11(2):242–252

Wu H, Lee J, Vincent LG, Wang Q, Gu M, Lan F, Churko JM, Sallam KI, Matsa E, Sharma A, Gold JD, Engler AJ, Xiang YK, Bers DM, Wu JC (2015) Epigenetic regulation of phosphodiesterases 2A and 3A underlies compromised beta-adrenergic signaling in an iPSC model of dilated cardiomyopathy. Cell Stem Cell 17(1):89–100

Xiong Q, Ye L, Zhang P, Lepley M, Tian J, Li J, Zhang L, Swingen C, Vaughan JT, Kaufman DS, Zhang J (2013) Functional consequences of human induced pluripotent stem cell therapy: myocardial ATP turnover rate in the in vivo swine heart with postinfarction remodeling. Circulation 127(9):997–1008

Xu C, Inokuma MS, Denham J, Golds K, Kundu P, Gold JD, Carpenter MK (2001) Feeder-free growth of undifferentiated human embryonic stem cells. Nat Biotechnol 19(10):971–974

Xu RH, Chen X, Li DS, Li R, Addicks GC, Glennon C, Zwaka TP, Thomson JA (2002) BMP4 initiates human embryonic stem cell differentiation to trophoblast. Nat Biotechnol 20(12):1261–1264

Xu C, Rosler E, Jiang J, Lebkowski JS, Gold JD, O'Sullivan C, Delavan-Boorsma K, Mok M, Bronstein A, Carpenter MK (2005a) Basic fibroblast growth factor supports undifferentiated human embryonic stem cell growth without conditioned medium. Stem Cells 23(3):315–323

Xu RH, Peck RM, Li DS, Feng X, Ludwig T, Thomson JA (2005b) Basic FGF and suppression of BMP signaling sustain undifferentiated proliferation of human ES cells. Nat Methods 2(3):185–190

Xu XQ, Graichen R, Soo SY, Balakrishnan T, Rahmat SN, Sieh S, Tham SC, Freund C, Moore J, Mummery C, Colman A, Zweigerdt R, Davidson BP (2008) Chemically defined medium supporting cardiomyocyte differentiation of human embryonic stem cells. Differentiation 76(9):958–970

Yang L, Soonpaa MH, Adler ED, Roepke TK, Kattman SJ, Kennedy M, Henckaerts E, Bonham K, Abbott GW, Linden RM, Field LJ, Keller GM (2008) Human cardiovascular progenitor cells develop from a KDR+ embryonic-stem-cell-derived population. Nature 453(7194):524–528

Yang X, Pabon L, Murry CE (2014a) Engineering adolescence: maturation of human pluripotent stem cell-derived cardiomyocytes. Circ Res 114(3):511–523

Yang X, Rodriguez M, Pabon L, Fischer KA, Reinecke H, Regnier M, Sniadecki NJ, Ruohola-Baker H, Murry CE (2014b) Tri-iodo-l-thyronine promotes the maturation of human cardiomyocytes-derived from induced pluripotent stem cells. J Mol Cell Cardiol 72: 296–304

Yao S, Chen S, Clark J, Hao E, Beattie GM, Hayek A, Ding S (2006) Long-term self-renewal and directed differentiation of human embryonic stem cells in chemically defined conditions. Proc Natl Acad Sci U S A 103(18):6907–6912

Yazawa M, Hsueh B, Jia X, Pasca AM, Bernstein JA, Hallmayer J, Dolmetsch RE (2011) Using induced pluripotent stem cells to investigate cardiac phenotypes in Timothy syndrome. Nature 471(7337):230–234

Ye L, Chang YH, Xiong Q, Zhang P, Zhang L, Somasundaram P, Lepley M, Swingen C, Su L, Wendel JS, Guo J, Jang A, Rosenbush D, Greder L, Dutton JR, Zhang J, Kamp TJ, Kaufman DS, Ge Y, Zhang J (2014) Cardiac repair in a porcine model of acute myocardial infarction with human induced pluripotent stem cell-derived cardiovascular cells. Cell Stem Cell 15(6):750–761

Ying QL, Nichols J, Chambers I, Smith A (2003) BMP induction of id proteins suppresses differentiation and sustains embryonic stem cell self-renewal in collaboration with STAT3. Cell 115(3):281–292

Yu P, Pan G, Yu J, Thomson JA (2011) FGF2 sustains NANOG and switches the outcome of BMP4-induced human embryonic stem cell differentiation. Cell Stem Cell 8(3):326–334

Zhang J, Wilson GF, Soerens AG, Koonce CH, Yu J, Palecek SP, Thomson JA, Kamp TJ (2009) Functional cardiomyocytes derived from human induced pluripotent stem cells. Circ Res 104(4):e30–e41

Zhang Q, Jiang J, Han P, Yuan Q, Zhang J, Zhang X, Xu Y, Cao H, Meng Q, Chen L, Tian T, Wang X, Li P, Hescheler J, Ji G, Ma Y (2011) Direct differentiation of atrial and ventricular myocytes from human embryonic stem cells by alternating retinoid signals. Cell Res 21(4):579–587

Zhang J, Klos M, Wilson GF, Herman AM, Lian X, Raval KK, Barron MR, Hou L, Soerens AG, Yu J, Palecek SP, Lyons GE, Thomson JA, Herron TJ, Jalife J, Kamp TJ (2012) Extracellular matrix promotes highly efficient cardiac differentiation of human pluripotent stem cells: the matrix sandwich method. Circ Res 111(9):1125–1136

Zhang XH, Haviland S, Wei H, Saric T, Fatima A, Hescheler J, Cleemann L, Morad M (2013) Ca2+ signaling in human induced pluripotent stem cell-derived cardiomyocytes (iPS-CM) from normal and catecholaminergic polymorphic ventricular tachycardia (CPVT)-afflicted subjects. Cell Calcium 54(2):57–70

Zhang M, Schulte JS, Heinick A, Piccini I, Rao J, Quaranta R, Zeuschner D, Malan D, Kim KP, Ropke A, Sasse P, Arauzo-Bravo M, Seebohm G, Scholer H, Fabritz L, Kirchhof P, Muller FU, Greber B (2015) Universal cardiac induction of human pluripotent stem cells in two and three-dimensional formats: implications for in vitro maturation. Stem Cells 33(5):1456–1469

Zhang Y, Cao N, Huang Y, Spencer CI, Fu JD, Yu C, Liu K, Nie B, Xu T, Li K, Xu S, Bruneau BG, Srivastava D, Ding S (2016) Expandable cardiovascular progenitor cells reprogrammed from fibroblasts. Cell Stem Cell 18(3):368–381

Zhu WZ, Xie Y, Moyes KW, Gold JD, Askari B, Laflamme MA (2010) Neuregulin/ErbB signaling regulates cardiac subtype specification in differentiating human embryonic stem cells. Circ Res 107(6):776–786

Zhu R, Blazeski A, Poon E, Costa KD, Tung L, Boheler KR (2014) Physical developmental cues for the maturation of human pluripotent stem cell-derived cardiomyocytes. Stem Cell Res Ther 5(5):117

Differentiation and Use of Induced Pluripotent Stem Cells for Cardiovascular Therapy and Tissue Engineering

5

Saidulu Mattapally, W. Kevin Cukier-Meisner, and Jianyi Zhang

Abstract
Human induced pluripotent stem cells (hiPSCs) are capable of differentiating into an unlimited number of cells of any lineage. Collectively, these properties suggest that hiPSC products may be an ideal source of cells for medical science and therapeutic use. Here, we discuss recent advancements in techniques for differentiating hiPSCs into cardiomyocytes, endothelial cells, and smooth muscle cells and in the use of hiPSC-derived cells for cardiovascular therapy and tissue engineering.

5.1 Introduction

The optimal cell lineage, or combination of cell lineages, for use in tissue engineering is likely to remain a topic of debate for some time. Cells of human origin, particularly autologous human origin, are most desirable for therapeutic applications, because they minimize concerns about immunogenicity and other risks that may be associated with xenogeneic and allogeneic cells, but adequate numbers of cells may not be obtainable from a patient's somatic tissues, or the tissues may be compromised by the underlying disease. Thus, patient-specific induced pluripotent stem cells (iPSCs) are among the most promising sources of cells for therapeutic use, because they can, at least in principle, be used to generate an unlimited number of cells of any lineage. However, several of the reprogramming factors that have been used to generate human iPSCs (hiPSCs) may be oncogenic (Hochedlinger and Plath 2009; Kaji et al. 2009; Loh et al. 2009; Yu et al. 2009), and, consequently, hiPSCs must be differentiated into specific cell types before administration to patients

S. Mattapally • W.K. Cukier-Meisner • J. Zhang (✉)
Department of Biomedical Engineering, School of Medicine, School of Engineering, The University of Alabama at Birmingham, Birmingham, AL 35294, USA
e-mail: jayzhang@uab.edu

M. Ieda, W.-H. Zimmermann (eds.), *Cardiac Regeneration*, Cardiac and Vascular Biology 4, DOI 10.1007/978-3-319-56106-6_5

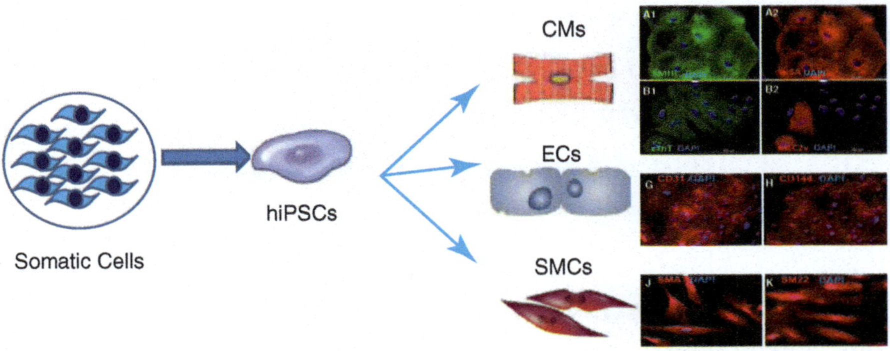

Fig. 5.1 Generation and differentiation of human induced pluripotent stem cells (hiPSCs) for cardiovascular therapy. Patient-specific hiPSCs are generated by transfecting the patient's own somatic cells with vectors coding for reprogramming factors; then, the hiPSCs are differentiated into cardiomyocytes (CMs), endothelial cells (ECs), and/or smooth muscle cells (SMCs) for use in cardiovascular tissue engineering or cell therapy (scale bar = 100 μm). The identity of the differentiated cells can be confirmed via immunofluorescence analyses of marker proteins (CMs: cTnT, SMHC, α-SA, and MLC2v; ECs: CD31 and CD144; SMCs: SMA and SM22)

(Fig. 5.1). Here, we summarize some of the challenges and recent developments in techniques for differentiating hiPSCs into cardiomyocytes (CMs), endothelial cells (ECs), and smooth muscle cells (SMCs), as well as observations from investigations of hiPSC-derived vascular cells in preclinical models of cardiovascular disease (Yang et al. 2016; Ye et al. 2013, 2014; Zhang et al. 2014, 2015; Zhu et al. 2017; Xiong et al. 2012, 2013).

5.2 Differentiating hiPSCs into Cardiomyocytes

Strategies for differentiating hiPSCs into cardiomyocytes can be distributed into three general categories: (1) embryoid body (EB)-mediated differentiation, (2) co-culturing with inductive stromal cells, or (3) coordinated monolayer differentiation. EB-mediated protocols mimic the spontaneous differentiation that occurs during normal embryonic development and are the most well-established methods. Co-culture-based protocols typically use inductive endodermal-like cells to trigger the differentiation of cardiogenic antecedents; OP9 stroma cells, for example, can induce the differentiation of murine iPSCs into Flk1$^+$ cells, which are early cardiovascular progenitor cells. However, the utility of this approach has been limited by concerns about potential immunogenic or other complications that may be caused by proteins or other factors that are produced by the inductive cells. In the monolayer differentiation method, iPSCs are scattered over a Matrigel-coated surface and then cultured with cytokines such as activin A, bone-morphogenic protein 4 (BMP4), basic fibroblast growth factor (bFGF), and vascular endothelial growth factor (VEGF). The cells can also be sandwiched between two layers of Matrigel, which can produce populations of hiPSC-derived cardiomyocytes (hiPSC-CMs) that are up to 98% pure (Fig. 5.2a), or cultured on a gelatin-coated, three-dimensional, poly-ε-caprolactone nanofiber scaffold, which induces differentiation spontaneously and does not require the addition of cytokines.

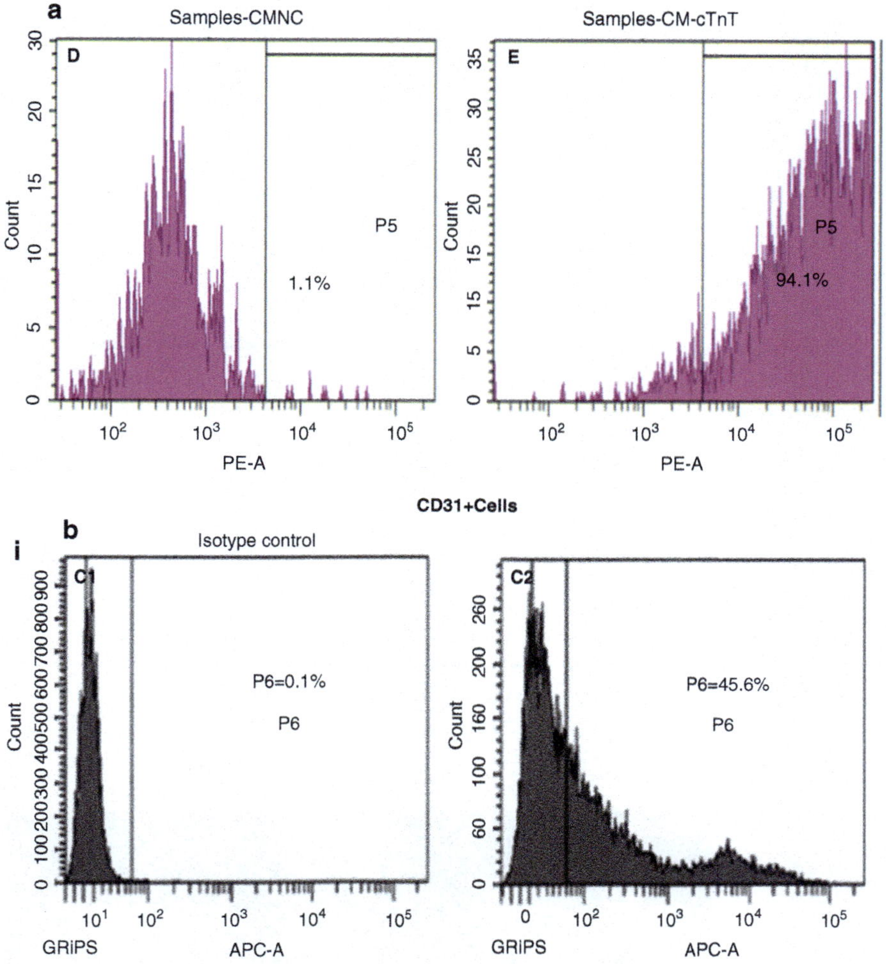

Fig. 5.2 Characteristics of hiPSC-derived cell populations. (**a**) Flow cytometry analyses of cardiac troponin T (cTnT) expression suggest that coordinated monolayer differentiation protocols can produce hiPSC-derived CM populations that are more than 90% pure. (**b**) (i) The efficiency of hiPSC-EC differentiation (assessed via flow cytometry analyses of CD31 expression) can vary depending on which hiPSC line is used (*top panels*, GRiPS cells; *bottom panels*, PCBC16iPS cells); (ii) after purification, the EC phenotype, as defined by the presence of CD31 and CD144 expression, can remain stable for 4 weeks. (**c**) Current hiPSC-SMC differentiation protocols are capable of producing cells with predominantly synthetic or contractile phenotypes. The two phenotypes can be characterized by substantial differences in (i) marker expression (smooth muscle actin [SMA], collagen 1 [Col 1], connexin 43 [Cx 43], and vimentin [Vim]; scale bar =200 μm), (ii) migration, (iii) proliferation, and (iv) contractile response. Data are presented as mean ± SEM. Panel (**a**) is from Ye Y et al. Cell Stem Cells, 2014;15:750–761. Panel (**b**) is from Zhang L et al. Biomaterials, 2014; 35:3786–3793. Panel (**c**) is modified from Yang L et al. PLoS ONE, 2016; 11:e0147155

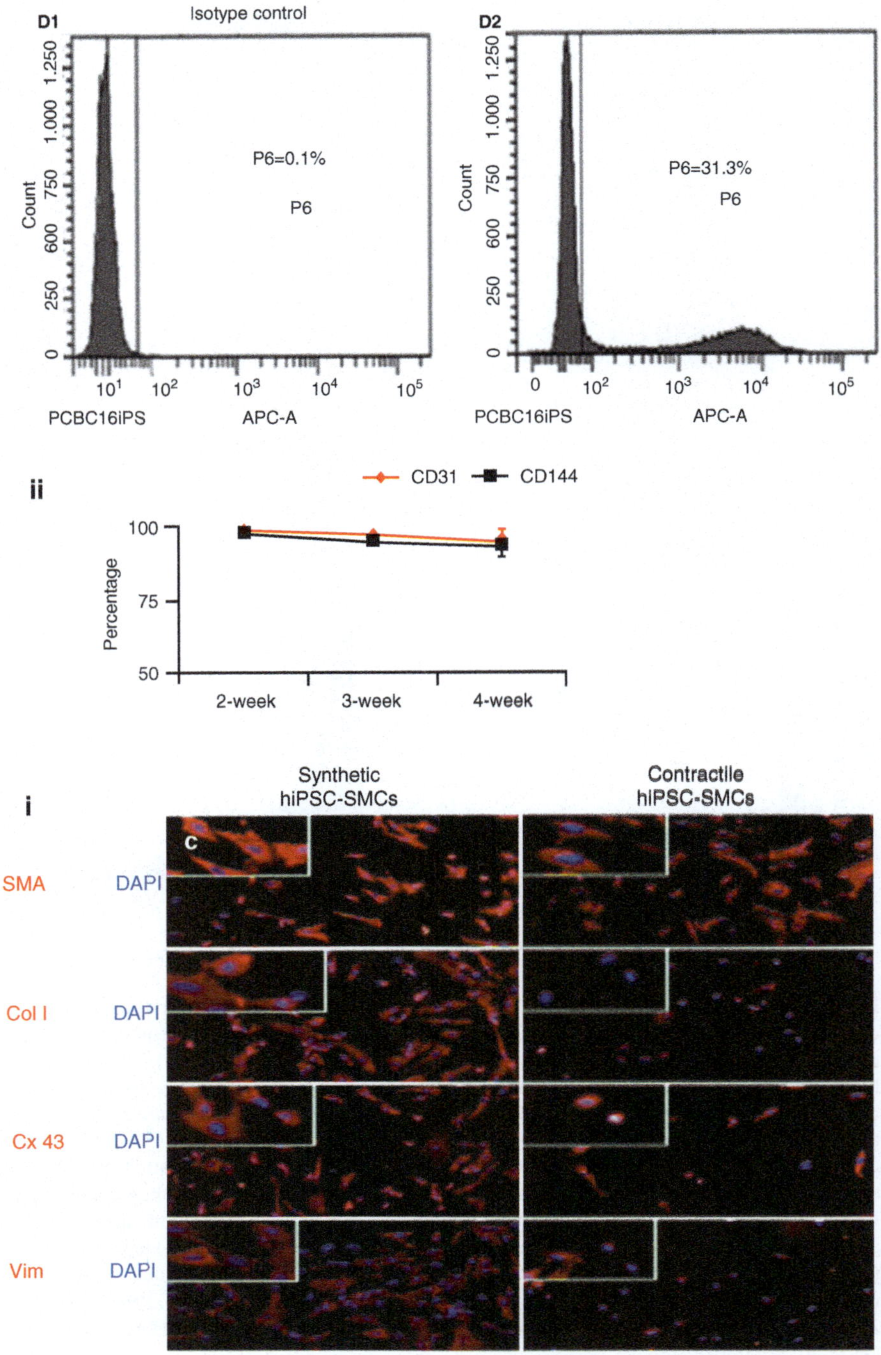

Fig. 5.2 (continued)

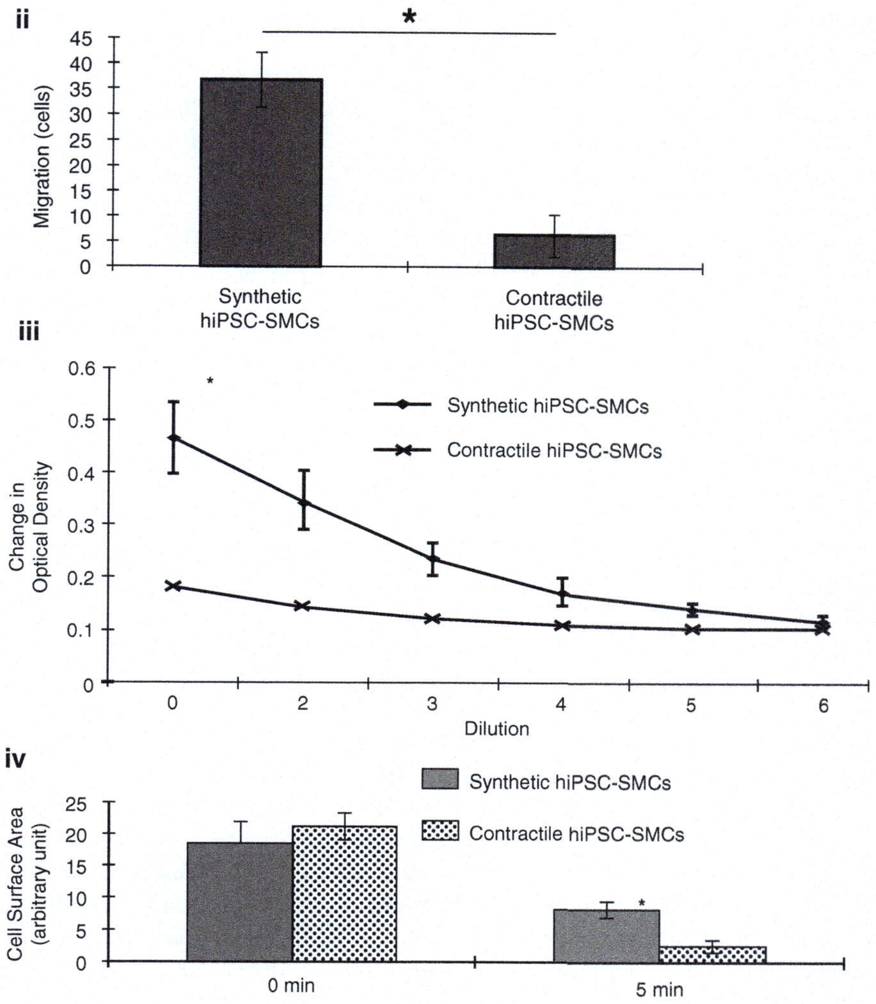

Fig. 5.2 (continued)

Protocols in all three categories are limited by concerns about efficiency, the purity and maturation of the differentiated cells, and cost, but monolayer-based protocols are gaining in popularity as our knowledge of the differentiation medium becomes more refined. Recent studies have also begun to investigate whether small chemical molecules can promote hiPSC-CM differentiation. Quattrocelli et al. (2011) have shown that sulfonyl-hydrazone-1 increases the number of beating foci and upregulates the expression of cardiomyogenic microRNAs such as miR-1, miR-133a, and miR-208a in differentiating hiPSC-CMs. Liu et al. (2016) found that resveratrol, a common polyphenol compound, enhances the cardiac differentiation of hiPSCs by suppressing the canonical Wnt signaling pathway and by activating the SRF/miR-1 axis, which suggests that other chemically synthesized Wnt signaling inhibitors, such as the small molecule KY02111, could also improve hiPSC-CM differentiation. Ascorbic acid, cyclosporin A, and trichostatin A also promote hiPSC-CM differentiation, while nitrous oxide and hydrogen sulfide are likely to do so, because they are known to participate in cardiac development. Furthermore, recent studies indicate that the maturation of murine iPSC-CMs can be promoted by the proteins cardiotrophin-1 (Liu et al. 2015), which regulates the JAK2/STAT3/Pim-1 signaling pathway, and neuregulin-1β (Iglesias-Garcia et al. 2015), which also appears to promote ventricular specificity. Cardiomyocyte differentiation and maturation may also be promoted by reproducing the electrical fields present in the developing myocardium; however, the effectiveness of any of these approaches, and whether that effectiveness may vary depending on which hiPSC line is used, has yet to be determined.

5.3 Differentiating hiPSCs into ECs

Traditionally, the differentiation of hiPSCs into ECs (hiPSC-ECs) has been directed through the formation of EBs or co-culture with murine stromal cells (Zhu et al. 2017; Choi et al. 2009; Woll et al. 2008). However, the efficiency of the EB method is low—less than 15% of the differentiated cells express endothelial markers such as CD31—and co-cultured cells may be contaminated with murine cells or proteins. Differentiation is also typically performed on a two-dimensional surface (Choi et al. 2009; Li et al. 2011; Rufaihah et al. 2011), perhaps because the endothelium is a two-dimensional organ, whereas the microenvironment of pluripotent cells has three dimensions during development. More recently, hiPSCs have been seeded into a three-dimensional fibrin scaffold and differentiated via a two-step process that avoids the use of feeder cells (Fig. 5.2b) (Zhang et al. 2014). First, the cells were committed to the mesodermal lineage by culturing them with activin A and BMP4 (Ye et al. 2013); then, the EC phenotype was induced with VEGF, transforming growth factor (TGF) β1, and erythropoietin, which have been used to promote the differentiation of embryonic stem cells (ESCs) and/or endothelial progenitor cells into ECs (Bahlmann et al. 2003; Gualandris et al. 2000; Yamashita et al. 2000). Up to 45% of the hiPSC-ECs generated via this three-dimensional differentiation protocol expressed CD31, compared to just ~5% when the cells were differentiated in monolayers, and the protocol could be made even more efficient by modifying the scaffold to incorporate factors that regulate EC differentiation, proliferation, and maturation. After purification, the cells were maintained with VEGF and a TGFβ1 inhibitor (SB-431542), because TGFβ1 limits the proliferation of cultured ECs (Beck and D'Amore 1997; Watabe et al. 2003), and the EC phenotype remained stable for at least 4 weeks. Even in the absence of exogenous growth factors, 50% pure populations of hiPSC-ECs can be generated in just 5 days by inhibiting glycogen synthase kinase 3 (GSK3), and the purity increased to >99% after a single round of magnetic affinity selection.

5.4 Differentiating hiPSCs into SMCs

The most prominent characteristic of vascular SMCs is their ability to influence blood pressure and tissue perfusion by contracting or relaxing to change the diameter of the vessel lumen. However, vascular SMCs also proliferate, migrate, and produce extracellular matrix components, thereby contributing to long-term changes in the vasculature that occur in response to injury, pregnancy, and other physiological demands (Owens et al. 2004). The range of functions that SMCs possess is accompanied by a diverse set of phenotypes with predominantly contractile or synthetic characteristics (Yang et al. 2016; Zhu et al. 2017). This diversity is manifested morphologically, in the cells' pattern of gene expression, and in measurements of cellular activity. Compared to contractile SMCs, synthetic SMCs contract less extensively and generally express lower levels of conventional SMC markers (such as collagen 1, connexin 43, and vimentin) but are more proliferative and migratory (Fig. 5.2c). This variability can likely be attributed, in part, to the developmental origins of SMCs, which are descended from a number of different embryonic regions (Gittenberger-de Groot et al. 1999), including the neural crest (Jiang et al. 2000), the paraxial/somatic mesoderm (Wasteson et al. 2008), the lateral plate mesoderm (Mikawa and Gourdie 1996), and the secondary heart field (Waldo et al. 2005). Furthermore, the phenotypes of individual SMCs are also believed to fluctuate (Hao et al. 2003), and this phenotypic modulation may have an important role in vascular disease and in the recovery from vascular injury. Thus, the predominant phenotype of a particular population of SMCs may determine how suitable the cells are for a given application: contractile SMCs may be most appropriate for in vitro studies of vasoconstriction, while a heterogeneous combination of synthetic and contractile cells may be needed for cell therapy and tissue engineering (Fernandez et al. 2014; Franck et al. 2013; Sundaram et al. 2014; Wang et al. 2014; Watson et al. 2014; Wong et al. 2015; Ye et al. 2015).

Effective methods for generating SMCs from hiPSCs (hiPSC-SMCs) have been available for a number of years (Cheung et al. 2012; Hill et al. 2010; Park et al. 2010), but protocols for inducing a predominantly contractile or synthetic phenotype in the differentiated cells have only recently been developed (Yang et al. 2016). As with conventional methods, these phenotype-specific protocols begin by directing the undifferentiated hiPSCs toward an intermediate (mesodermal) lineage before inducing the terminal SMC phenotype, and the entire procedure is performed without feeder cells, so the final, purified cell populations are unlikely to be contaminated with nonhuman cells or proteins. The mesodermal specification is produced by culturing the cells with a GSK inhibitor and BMP4 for 3 days; then, the synthetic SMC phenotype is induced with VEGF and fibroblast growth factor (FGF), or contractile hiPSC-SMCs are produced by culturing the cells with varying combinations of platelet-derived growth factor (PDGF), TGF, and FGF. The differentiated cells can be purified to ~95% via metabolic selection, and the SMC phenotype remains stable for at least 20 population doublings. Cells obtained via the two protocols differ substantially in marker expression: levels of smooth muscle actin and myosin heavy chain 11 (SMHC) were higher in contractile hiPSC-SMCs, while the synthetic hiPSC-SMCs produced less calponin, more collagen 1 and more connexin 43, which is consistent with observations in primary swine coronary artery SMCs (Beamish et al. 2010; Christen et al. 1999). Contractile hiPSC-SMCs also responded more strongly to treatment with carbachol—contracting to 12% of their original size, compared to 44% for synthetic hiPSC-SMCs—but measurements of proliferation and migration were greater in synthetic hiPSC-SMCs.

5.5 Epigenetic Effects

Induced pluripotent stem cells appear to retain some of the epigenetic characteristics associated with their tissues of origin, and, consequently, the properties of hiPSC-derived cells may depend, to some extent, on which cells were used to generate the hiPSCs (Bar-Nur et al. 2011; Kim et al. 2010, 2011). For example, when hiPSCs generated from cardiac fibroblasts (hciPSCs) were differentiated into cardiomyocytes and compared with cardiomyocytes derived from established hiPSC lines (which had been reprogrammed from dermal fibroblasts [hdiPSCs] or umbilical cord blood cells [hUBCiPSCs]), beating was observed in more than 90% of the clusters derived from hciPSCs, compared to <70% for hdiPSC-derived clusters and 50% of clusters from hUBCiPSCs. The hciPSC-derived cells were also much more likely to co-express the cardiac-specific isoforms of troponin T and alpha sarcomeric actin, and when sheets of the hciPSC-derived cardiomyocytes were transplanted into the infarcted hearts of mice, the proportion of cells that were retained and continued to survive for at least 4 weeks after administration exceeded 30% (Zhang et al. 2015), which is several-fold greater than the retention/survival rate observed in other studies of myocardial cell therapy, regardless of the type of cell administered. Collectively, these observations suggest that the effectiveness of hiPSC-derived cell therapy will be greatest if (whenever possible) the hiPSCs are generated from cells that closely resemble those in the tissue or organ to be treated.

5.6 Tumorigenesis

The earliest studies to investigate the use of iPSCs for the treatment of myocardial infarction were performed with undifferentiated cells and yielded promising results (Nelson et al. 2009; Singla et al. 2011). When iPSCs were generated from murine fibroblasts (Nelson et al. 2009) or vascular SMCs (Singla et al. 2011) and injected into infarcted mouse hearts, the transplanted cells appeared to differentiate into cardiomyocytes, SMCs, and ECs (Nelson et al. 2009), and the treatments led to improvements in both myocardial performance and remodeling without inducing the formation of tumors in animals with uncompromised immune systems. However, subsequent studies found evidence of tumor formation in the hearts of 37.5% of immunocompetent mice that had been treated with murine iPSCs (Ahmed et al. 2011), and observations in a rat myocardial infarction model (Zhang et al. 2011) indicated that rat iPSC-induced tumors had spread to regions outside of the heart.

Thus, despite promising evidence of potential efficacy, these investigations convincingly demonstrated that the tumorigenic potential of iPSCs could not be adequately suppressed by the host animal's immune system and, consequently, that hiPSCs must undergo some degree of differentiation before they can be safely administered to patients. However, a number of studies found evidence of tumor formation even when embryonic stem cells (ESCs) were differentiated into neuronal lineage cells before transplantation into the eyes of rhodopsin-knockout mice (Arnhold et al. 2004) or into the brains of fetal (Wernig et al. 2004) and adult Parkinsonian mice (Roy et al. 2006). Furthermore, the global pattern of gene expression in cancer cells, as well as the expression of a subset of oncogenic genes, more closely resembles the expression patterns in ESC- or iPSC-derived ECs than in somatic ECs (Ghosh et al. 2011). Thus, concerns about the potential tumorigenicity of ESC- and iPSC-derived cells continue to exist and must be addressed before large-scale clinical trials are initiated.

The tumorigenicity of iPSC-derived cells likely evolves from at least two sources: (1) any undifferentiated or incompletely differentiated cells that remain in the population of iPSC-derived cells and (2) residual oncogenic activity from the reprogramming factors used to generate the iPSCs. The former cause can be addressed by continuing to identify more efficient differentiation and selection protocols, while attempts to address the latter concern have focused primarily on the development of new methods for iPSC generation that do not rely on viral integration of exogenous reprogramming factors, such as the delivery of episomal vectors (Yu et al. 2009) and mRNA (Warren et al. 2010) or the direct delivery of reprogramming proteins (Kim et al. 2009). Quantitative RT-PCR analyses suggested that the expression of several oncogenic factors can be reduced by reprogramming the iPSCs with "minicircle" DNA, which contains no bacterial DNA and is not integrated into the host genome, but expression levels of cancer-related genes remained somewhat higher than the levels observed in somatic ECs (Ghosh et al. 2011).

5.6.1 Myocardial Tissue Equivalent Patches

Cell-based regenerative myocardial therapy can be administered as a solution of individual cells or as a tissue-engineered myocardial patch. Cell solutions are more amenable to delivery via pericardial endoscopy (Kimura et al. 2012, 2011) and, consequently, avoid the invasiveness of a thoracotomy or open-chest surgery; however, the proportion of cells that are engrafted by the native myocardium after intracoronary infusion (Barbash et al. 2003) or direct myocardial injection (Hattan et al. 2005; Ptaszek et al. 2012; van Laake et al. 2007) is rarely much greater than 1%, in part, because the cells are washed away by the circulation or squeezed out through the needle track by the pressures produced during left ventricular systole. The engraftment of injected cells can be improved by sealing the needle track with a patch of biomaterial, and the patch can also be modified to release paracrine factors, such as insulin-like growth factor (IGF), that may enhance the survival of the injected cells. This method has been used to improve the engraftment of hiPSC-ECs, -SMCs, and -CMs after injection into the hearts of pigs with experimentally induced ischemia-reperfusion injury (Ye et al. 2014); ~9% of the hiPSC-derived cardiac cells remained engrafted for at least 4 weeks after administration, and the treatment was associated with significant improvements in infarct size, perfusion, metabolism, and cardiac function.

The engraftment and survival rate of transplanted cells is generally higher when the cells are administered as an engineered myocardial patch than when injected directly into the myocardium, and the survival of a transplanted patch is critically dependent on vascular growth. Patches composed of unselected cardiac cells appear to be intrinsically vasculogenic, because they have been shown to express VEGF, cyclooxygenase-2, Tie-2, and angiopoietin-1 and to form networks of ECs in culture that migrate and connect to the host animal's vasculature after implantation (Sekiya et al. 2006). Furthermore, the extent of vascular growth was positively correlated with the number of ECs that were present in the population of cardiac cells used to create the patch, and when patches that contained varying proportions of isolated neonatal rat cardiomyocytes and ECs (12:1, 6:1, and 3:1 cardiomyocytes:ECs) (Sekine et al. 2008) were tested in a rat MI model, measurements of cardiac function were most improved in rats that received patches with the greatest proportion of ECs. ECs are also known to release a variety of autocrine and paracrine agents that regulate myocardial metabolism and perfusion (Cines et al. 1998; Masoli et al. 2000; Ramaciotti et al. 1992; Versari et al. 2009), as well as the contractile state and rhythmicity of adjacent cardiomyocytes (Brutsaert 2003), and when cardiac tissue patches were created by seeding hiPSC-ECs, -SMCs, and -CMs into a fibrinogen scaffold, the patches began to contract within 3 days of manufacture, compared to 7 days when the hiPSC-ECs were omitted (Zhang et al. 2014). Collectively, these observations suggest that ECs are likely to improve the therapeutic potential of engineered cardiac tissue not only by increasing perfusion but also through direct effects on the contractile activity of cardiomyocytes.

To our knowledge, treatment with hiPSC-SMCs alone has never been investigated in a myocardial injury model; however, several studies have been performed with patches that contained combinations of hiPSC-SMCs and hiPSC-ECs (referred to collectively as hiPSC-derived vascular cells [hiPSC-VCs]), including a large-animal investigation by members of the Zhang group (Xiong et al. 2013). The cells were suspended in a fibrinogen solution and then co-injected with a solution of thrombin into a ring positioned over the site of injury. The solutions solidified to form a semisolid patch within 1 min of mixing, and 4 weeks later, the treatment was associated with significant improvements in cardiac function and infarct size. The hiPSC-VCs also released paracrine factors that appeared to mobilize endogenous progenitor cells, attenuate regional wall stress, stimulate neovascularization, and improve perfusion, while the structural support provided by the patch may help preserve the integrity of the adjacent, healthy myocardium by preventing the thin, fibrous scar tissue from bulging during systole. The overstretched cardiomyocytes surrounding the bulge are believed to undergo a series of metabolic changes that lead to declines in the ratio of phosphocreatine to ATP (i.e., the PCr/ATP ratio) and in the rate of ATP turnover (Xiong et al. 2013; Bolognese et al. 2002; Feygin et al. 2007; Hu et al. 2006), but the severity of these bioenergetic abnormalities was significantly reduced in pigs with ischemia-reperfusion injury after treatment with a fibrin patch containing hESC-derived vascular cells (Xiong et al. 2012).

5.6.2 Tissue-Engineered Blood Vessels

Patients undergoing coronary artery bypass graft (CABG) surgery, hemodialysis, or treatment for peripheral vascular disease often have few small-diameter vessels that are suitable for autologous transplantation (Niklason et al. 1999; McAllister et al. 2009). Tissue-engineered blood vessels (TEBVs) could relieve this scarcity, and early-stage clinical trials have been conducted with TEBVs generated from autologous bone marrow cells (Shin'oka et al. 2005) or skin fibroblasts and ECs (McAllister et al. 2009), but long wait times may limit the utility of this approach (Dash et al. 2015). TEBVs generated from iPSC-derived cells have been investigated in only a few preclinical studies. Hibino et al. (2012) created TEBVs from murine iPSC-derived SMCs and ECs by wrapping a sheet of the differentiated cells around a 0.8-mm-diameter biodegradable scaffold and then implanted the TEBVs in the inferior vena cava of severe combined immunodeficiency (SCID) mice. All of the mice survived with no evidence of thrombosis, aneurysm formation, graft rupture, or calcification, and histological assessments performed 10 weeks after implantation indicated that the TEBVs were endothelialized and contained a layer of cells that expressed smooth muscle actin and calponin. However, just 10% of the iPSC-derived cells remained in the TEBV for at least 4 weeks after transplantation; thus, the primary structural and functional components of the implanted graft were generated via the growth of native tissue, which may have been stimulated by the paracrine activity of the iPSC-derived cells. TEBVs have also been produced by seeding decellularized vessels with SMCs and ECs derived from human partially induced pluripotent stem cells (PiPSCs), which may be less tumorigenic than fully induced iPSCs (Margariti et al. 2012); when implanted into the carotid artery of SCID mice, 60% of the animals survived, and the grafts appeared to contain substantial numbers of PiPSC-derived cells for 3 weeks after transplantation (Karamariti et al. 2013). Collectively, the results from this very limited number of studies suggest that hiPSC-VCs may be useful for generating TEBVs and support the continued development of this technology.

5.7 Conclusion and Future Directions

Recent reports have demonstrated that iPSC technology is very useful for providing tools for basic science as well as for future clinical applications. Small-molecule protocols for cardiomyocyte, EC, and SMC differentiation are available today and result in considerable cell yields. All direct progression through a mesodermal progenitor cell stage via GSK3 inhibition; then, the mesodermal progenitors can be directed into spontaneously contracting cardiomyocytes via WNT inhibition, while permitting endogenous canonical WNT signaling yields a population enriched for vascular progenitors. VEGF, PDGF, TGF, and erythropoietin are all implicated in EC and SMC development from mesodermal progenitors. Together, these results support the paradigm of guided differentiation to distinct cell types via modulation of master regulators of cell fate and represent a promising approach to efficiently produce all cell types needed for efficient tissue repair. Nevertheless, unwanted proliferation has to be considered, and techniques for purification, particularly with regard to reducing the risk of tumorigenesis, are required for clinical translation. Future studies using metabolic and mechanical matrix/scaffold, electrical interventions, and gene editing are warranted to further promote the maturation of hiPSC-derived cells and enhance their biological activity.

Compliance with Ethical Standards

Conflict of Interest The authors declare that they have no conflict of interest.

Ethical approval This article does not contain any studies with human participants performed by any of the authors.

References

Ahmed RP, Ashraf M, Buccini S, Shujia J, Haider H (2011) Cardiac tumorigenic potential of induced pluripotent stem cells in an immunocompetent host with myocardial infarction. Regen Med 6:171–178

Arnhold S, Klein H, Semkova I, Addicks K, Schraermeyer U (2004) Neurally selected embryonic stem cells induce tumor formation after long-term survival following engraftment into the subretinal space. Invest Ophthalmol Vis Sci 45:4251–4255

Bahlmann FH, DeGroot K, Duckert T, Niemczyk E, Bahlmann E, Boehm SM, Haller H, Fliser D (2003) Endothelial progenitor cell proliferation and differentiation is regulated by erythropoietin. Kidney Int 64:1648–1652

Barbash IM, Chouraqui P, Baron J, Feinberg MS, Etzion S, Tessone A, Miller L, Guetta E, Zipori D, Kedes LH, Kloner RA, Leor J (2003) Systemic delivery of bone marrow-derived mesenchymal stem cells to the infarcted myocardium: feasibility, cell migration, and body distribution. Circulation 108:863–868

Bar-Nur O, Russ HA, Efrat S, Benvenisty N (2011) Epigenetic memory and preferential lineage-specific differentiation in induced pluripotent stem cells derived from human pancreatic islet beta cells. Cell Stem Cell 9:17–23

Beamish JA, He P, Kottke-Marchant K, Marchant RE (2010) Molecular regulation of contractile smooth muscle cell phenotype: implications for vascular tissue engineering. Tissue Eng Part B Rev 16:467–491

Beck L Jr, D'Amore PA (1997) Vascular development: cellular and molecular regulation. FASEB J 11:365–373

Bolognese L, Neskovic AN, Parodi G, Cerisano G, Buonamici P, Santoro GM, Antoniucci D (2002) Left ventricular remodeling after primary coronary angioplasty: patterns of left ventricular dilation and long-term prognostic implications. Circulation 106:2351–2357

Brutsaert DL (2003) Cardiac endothelial-myocardial signaling: its role in cardiac growth, contractile performance, and rhythmicity. Physiol Rev 83:59–115

Cheung C, Bernardo AS, Trotter MW, Pedersen RA, Sinha S (2012) Generation of human vascular smooth muscle subtypes provides insight into embryological origin-dependent disease susceptibility. Nat Biotechnol 30:165–173

Choi KD, Yu J, Smuga-Otto K, Salvagiotto G, Rehrauer W, Vodyanik M, Thomson J, Slukvin I (2009) Hematopoietic and endothelial differentiation of human induced pluripotent stem cells. Stem Cells 27:559–567

Christen T, Bochaton-Piallat ML, Neuville P, Rensen S, Redard M, van Eys G, Gabbiani G (1999) Cultured porcine coronary artery smooth muscle cells. A new model with advanced differentiation. Circ Res 85:99–107

Cines DB, Pollak ES, Buck CA, Loscalzo J, Zimmerman GA, McEver RP, Pober JS, Wick TM, Konkle BA, Schwartz BS, Barnathan ES, McCrae KR, Hug BA, Schmidt AM, Stern DM (1998) Endothelial cells in physiology and in the pathophysiology of vascular disorders. Blood 91:3527–3561

Dash BC, Jiang Z, Suh C, Qyang Y (2015) Induced pluripotent stem cell-derived vascular smooth muscle cells: methods and application. Biochem J 465:185–194

Fernandez CE, Achneck HE, Reichert WM, Truskey GA (2014) Biological and engineering design considerations for vascular tissue engineered blood vessels (TEBVs). Curr Opin Chem Eng 3:83–90

Feygin J, Mansoor A, Eckman P, Swingen C, Zhang J (2007) Functional and bioenergetic modulations in the infarct border zone following autologous mesenchymal stem cell transplantation. Am J Physiol Heart Circ Physiol 293:H1772–H1780

Franck D, Gil ES, Adam RM, Kaplan DL, Chung YG, Estrada CR Jr, Mauney JR (2013) Evaluation of silk biomaterials in combination with extracellular matrix coatings for bladder tissue engineering with primary and pluripotent cells. PLoS One 8:e56237

Ghosh Z, Huang M, Hu S, Wilson KD, Dey D, Wu JC (2011) Dissecting the oncogenic and tumorigenic potential of differentiated human induced pluripotent stem cells and human embryonic stem cells. Cancer Res 71:5030–5039

Gittenberger-de Groot AC, DeRuiter MC, Bergwerff M, Poelmann RE (1999) Smooth muscle cell origin and its relation to heterogeneity in development and disease. Arterioscler Thromb Vasc Biol 19:1589–1594

Gualandris A, Annes JP, Arese M, Noguera I, Jurukovski V, Rifkin DB (2000) The latent transforming growth factor-beta-binding protein-1 promotes in vitro differentiation of embryonic stem cells into endothelium. Mol Biol Cell 11:4295–4308

Hao H, Gabbiani G, Bochaton-Piallat ML (2003) Arterial smooth muscle cell heterogeneity: implications for atherosclerosis and restenosis development. Arterioscler Thromb Vasc Biol 23:1510–1520

Hattan N, Kawaguchi H, Ando K, Kuwabara E, Fujita J, Murata M, Suematsu M, Mori H, Fukuda K (2005) Purified cardiomyocytes from bone marrow mesenchymal stem cells produce stable intracardiac grafts in mice. Cardiovasc Res 65:334–344

Hibino N, Duncan DR, Nalbandian A, Yi T, Qyang Y, Shinoka T, Breuer CK (2012) Evaluation of the use of an induced puripotent stem cell sheet for the construction of tissue-engineered vascular grafts. J Thorac Cardiovasc Surg 143:696–703

Hill KL, Obrtlikova P, Alvarez DF, King JA, Keirstead SA, Allred JR, Kaufman DS (2010) Human embryonic stem cell-derived vascular progenitor cells capable of endothelial and smooth muscle cell function. Exp Hematol 38:246.e1–257.e1

Hochedlinger K, Plath K (2009) Epigenetic reprogramming and induced pluripotency. Development 136:509–523

Hu Q, Wang X, Lee J, Mansoor A, Liu J, Zeng L, Swingen C, Zhang G, Feygin J, Ochiai K, Bransford TL, From AH, Bache RJ, Zhang J (2006) Profound bioenergetic abnormalities in peri-infarct myocardial regions. Am J Physiol Heart Circ Physiol 291:H648–H657

Iglesias-Garcia O, Baumgartner S, Macri-Pellizzeri L, Rodriguez-Madoz JR, Abizanda G, Guruceaga E, Albiasu E, Corbacho D, Benavides-Vallve C, Soriano-Navarro M, Gonzalez-Granero S, Gavira JJ, Krausgrill B, Rodriguez-Manero M, Garcia-Verdugo JM, Ortiz-de-Solorzano C, Halbach M, Hescheler J, Pelacho B, Prosper F (2015) Neuregulin-1beta induces mature ventricular cardiac differentiation from induced pluripotent stem cells contributing to cardiac tissue repair. Stem Cells Dev 24:484–496

Jiang X, Rowitch DH, Soriano P, McMahon AP, Sucov HM (2000) Fate of the mammalian cardiac neural crest. Development 127:1607–1616

Kaji K, Norrby K, Paca A, Mileikovsky M, Mohseni P, Woltjen K (2009) Virus-free induction of pluripotency and subsequent excision of reprogramming factors. Nature 458:771–775

Karamariti E, Margariti A, Winkler B, Wang X, Hong X, Baban D, Ragoussis J, Huang Y, Han JD, Wong MM, Sag CM, Shah AM, Hu Y, Xu Q (2013) Smooth muscle cells differentiated from reprogrammed embryonic lung fibroblasts through DKK3 signaling are potent for tissue engineering of vascular grafts. Circ Res 112:1433–1443

Kim D, Kim CH, Moon JI, Chung YG, Chang MY, Han BS, Ko S, Yang E, Cha KY, Lanza R, Kim KS (2009) Generation of human induced pluripotent stem cells by direct delivery of reprogramming proteins. Cell Stem Cell 4:472–476

Kim K, Doi A, Wen B, Ng K, Zhao R, Cahan P, Kim J, Aryee MJ, Ji H, Ehrlich LI, Yabuuchi A, Takeuchi A, Cunniff KC, Hongguang H, McKinney-Freeman S, Naveiras O, Yoon TJ, Irizarry RA, Jung N, Seita J, Hanna J, Murakami P, Jaenisch R, Weissleder R, Orkin SH, Weissman IL, Feinberg AP, Daley GQ (2010) Epigenetic memory in induced pluripotent stem cells. Nature 467:285–290

Kim K, Zhao R, Doi A, Ng K, Unternaehrer J, Cahan P, Huo H, Loh YH, Aryee MJ, Lensch MW, Li H, Collins JJ, Feinberg AP, Daley GQ (2011) Donor cell type can influence the epigenome and differentiation potential of human induced pluripotent stem cells. Nat Biotechnol 29:1117–1119

Kimura T, Miyoshi S, Takatsuki S, Tanimoto K, Fukumoto K, Soejima K, Fukuda K (2011) Safety and efficacy of pericardial endoscopy by percutaneous subxyphoid approach in swine heart in vivo. J Thorac Cardiovasc Surg 142:181–190

Kimura T, Miyoshi S, Okamoto K, Fukumoto K, Tanimoto K, Soejima K, Takatsuki S, Fukuda K (2012) The effectiveness of rigid pericardial endoscopy for minimally invasive minor surgeries: cell transplantation, epicardial pacemaker lead implantation, and epicardial ablation. J Cardiothorac Surg 7:117

van Laake LW, Passier R, Monshouwer-Kloots J, Verkleij AJ, Lips DJ, Freund C, den Ouden K, Ward-van Oostwaard D, Korving J, Tertoolen LG, van Echteld CJ, Doevendans PA, Mummery CL (2007) Human embryonic stem cell-derived cardiomyocytes survive and mature in the mouse heart and transiently improve function after myocardial infarction. Stem Cell Res 1:9–24

Li Z, Hu S, Ghosh Z, Han Z, Wu JC (2011) Functional characterization and expression profiling of human induced pluripotent stem cell- and embryonic stem cell-derived endothelial cells. Stem Cells Dev 20:1701–1710

Liu T, Zhang R, Guo T, Ma S, Han D, Li XJ, Jin Y, Fan MM, Wang YB, Chen YD, Cao F (2015) Cardiotrophin-1 promotes cardiomyocyte differentiation from mouse induced pluripotent stem cells via JAK2/STAT3/Pim-1 signaling pathway. J Geriatr Cardiol 12:591–599

Liu H, Zhang S, Zhao L, Zhang Y, Li Q, Chai X, Zhang Y (2016) Resveratrol enhances cardiomyocyte differentiation of human induced pluripotent stem cells through inhibiting canonical WNT signal pathway and enhancing serum response factor-miR-1 axis. Stem Cells Int 2016:2524092

Loh YH, Agarwal S, Park IH, Urbach A, Huo H, Heffner GC, Kim K, Miller JD, Ng K, Daley GQ (2009) Generation of induced pluripotent stem cells from human blood. Blood 113:5476–5479

Margariti A, Winkler B, Karamariti E, Zampetaki A, Tsai TN, Baban D, Ragoussis J, Huang Y, Han JD, Zeng L, Hu Y, Xu Q (2012) Direct reprogramming of fibroblasts into endothelial cells capable of angiogenesis and reendothelialization in tissue-engineered vessels. Proc Natl Acad Sci U S A 109:13793–13798

Masoli O, Balino NP, Sabate D, Jalon J, Meretta A, Cragnolino D, Sarmiento R, DiCarli MF (2000) Effect of endothelial dysfunction on regional perfusion in myocardial territories supplied by normal and diseased vessels in patients with coronary artery disease. J Nucl Cardiol 7:199–204

McAllister TN, Maruszewski M, Garrido SA, Wystrychowski W, Dusserre N, Marini A, Zagalski K, Fiorillo A, Avila H, Manglano X, Antonelli J, Kocher A, Zembala M, Cierpka L, de la Fuente LM, L'Heureux N (2009) Effectiveness of haemodialysis access with an autologous tissue-engineered vascular graft: a multicentre cohort study. Lancet 373:1440–1446

Mikawa T, Gourdie RG (1996) Pericardial mesoderm generates a population of coronary smooth muscle cells migrating into the heart along with ingrowth of the epicardial organ. Dev Biol 174:221–232

Nelson TJ, Martinez-Fernandez A, Yamada S, Perez-Terzic C, Ikeda Y, Terzic A (2009) Repair of acute myocardial infarction by human stemness factors induced pluripotent stem cells. Circulation 120:408–416

Niklason LE, Gao J, Abbott WM, Hirschi KK, Houser S, Marini R, Langer R (1999) Functional arteries grown in vitro. Science 284:489–493

Owens GK, Kumar MS, Wamhoff BR (2004) Molecular regulation of vascular smooth muscle cell differentiation in development and disease. Physiol Rev 84:767–801

Park SW, Jun Koh Y, Jeon J, Cho YH, Jang MJ, Kang Y, Kim MJ, Choi C, Sook Cho Y, Chung HM, Koh GY, Han YM (2010) Efficient differentiation of human pluripotent stem cells into functional CD34+ progenitor cells by combined modulation of the MEK/ERK and BMP4 signaling pathways. Blood 116:5762–5772

Ptaszek LM, Mansour M, Ruskin JN, Chien KR (2012) Towards regenerative therapy for cardiac disease. Lancet 379:933–942

Quattrocelli M, Palazzolo G, Agnolin I, Martino S, Bouche M, Anastasia L, Sampaolesi M (2011) Synthetic sulfonyl-hydrazone-1 positively regulates cardiomyogenic microRNA expression and cardiomyocyte differentiation of induced pluripotent stem cells. J Cell Biochem 112:2006–2014

Ramaciotti C, Sharkey A, McClellan G, Winegrad S (1992) Endothelial cells regulate cardiac contractility. Proc Natl Acad Sci U S A 89:4033–4036

Roy NS, Cleren C, Singh SK, Yang L, Beal MF, Goldman SA (2006) Functional engraftment of human ES cell-derived dopaminergic neurons enriched by coculture with telomerase-immortalized midbrain astrocytes. Nat Med 12:1259–1268

Rufaihah AJ, Huang NF, Jame S, Lee JC, Nguyen HN, Byers B, De A, Okogbaa J, Rollins M, Reijo-Pera R, Gambhir SS, Cooke JP (2011) Endothelial cells derived from human iPSCS increase capillary density and improve perfusion in a mouse model of peripheral arterial disease. Arterioscler Thromb Vasc Biol 31:e72–e79

Sekine H, Shimizu T, Hobo K, Sekiya S, Yang J, Yamato M, Kurosawa H, Kobayashi E, Okano T (2008) Endothelial cell coculture within tissue-engineered cardiomyocyte sheets enhances neovascularization and improves cardiac function of ischemic hearts. Circulation 118:S145–S152

Sekiya S, Shimizu T, Yamato M, Kikuchi A, Okano T (2006) Bioengineered cardiac cell sheet grafts have intrinsic angiogenic potential. Biochem Biophys Res Commun 341:573–582

Shin'oka T, Matsumura G, Hibino N, Naito Y, Watanabe M, Konuma T, Sakamoto T, Nagatsu M, Kurosawa H (2005) Midterm clinical result of tissue-engineered vascular autografts seeded with autologous bone marrow cells. J Thorac Cardiovasc Surg 129:1330–1338

Singla DK, Long X, Glass C, Singla RD, Yan B (2011) Induced pluripotent stem (iPS) cells repair and regenerate infarcted myocardium. Mol Pharm 8:1573–1581

Sundaram S, One J, Siewert J, Teodosescu S, Zhao L, Dimitrievska S, Qian H, Huang AH, Niklason L (2014) Tissue-engineered vascular grafts created from human induced pluripotent stem cells. Stem Cells Transl Med 3:1535–1543

Versari D, Daghini E, Virdis A, Ghiadoni L, Taddei S (2009) Endothelial dysfunction as a target for prevention of cardiovascular disease. Diabetes Care 32(Suppl 2):S314–S321

Waldo KL, Hutson MR, Ward CC, Zdanowicz M, Stadt HA, Kumiski D, Abu-Issa R, Kirby ML (2005) Secondary heart field contributes myocardium and smooth muscle to the arterial pole of the developing heart. Dev Biol 281:78–90

Wang Y, Hu J, Jiao J, Liu Z, Zhou Z, Zhao C, Chang LJ, Chen YE, Ma PX, Yang B (2014) Engineering vascular tissue with functional smooth muscle cells derived from human iPS cells and nanofibrous scaffolds. Biomaterials 35:8960–8969

Warren L, Manos PD, Ahfeldt T, Loh YH, Li H, Lau F, Ebina W, Mandal PK, Smith ZD, Meissner A, Daley GQ, Brack AS, Collins JJ, Cowan C, Schlaeger TM, Rossi DJ (2010) Highly efficient reprogramming to pluripotency and directed differentiation of human cells with synthetic modified mRNA. Cell Stem Cell 7:618–630

Wasteson P, Johansson BR, Jukkola T, Breuer S, Akyurek LM, Partanen J, Lindahl P (2008) Developmental origin of smooth muscle cells in the descending aorta in mice. Development 135:1823–1832

Watabe T, Nishihara A, Mishima K, Yamashita J, Shimizu K, Miyazawa K, Nishikawa S, Miyazono K (2003) TGF-beta receptor kinase inhibitor enhances growth and integrity of embryonic stem cell-derived endothelial cells. J Cell Biol 163:1303–1311

Watson CL, Mahe MM, Munera J, Howell JC, Sundaram N, Poling HM, Schweitzer JI, Vallance JE, Mayhew CN, Sun Y, Grabowski G, Finkbeiner SR, Spence JR, Shroyer NF, Wells JM, Helmrath MA (2014) An in vivo model of human small intestine using pluripotent stem cells. Nat Med 20:1310–1314

Wernig M, Benninger F, Schmandt T, Rade M, Tucker KL, Bussow H, Beck H, Brustle O (2004) Functional integration of embryonic stem cell-derived neurons in vivo. J Neurosci 24:5258–5268

Woll PS, Morris JK, Painschab MS, Marcus RK, Kohn AD, Biechele TL, Moon RT, Kaufman DS (2008) Wnt signaling promotes hematoendothelial cell development from human embryonic stem cells. Blood 111:122–131

Wong MM, Hong X, Karamariti E, Hu Y, Xu Q (2015) Generation and grafting of tissue-engineered vessels in a mouse model. J Vis Exp 97. doi:10.3791/52565

Xiong Q, Ye L, Zhang P, Lepley M, Swingen C, Zhang L, Kaufman DS, Zhang J (2012) Bioenergetic and functional consequences of cellular therapy: activation of endogenous cardiovascular progenitor cells. Circ Res 111:455–468

Xiong Q, Ye L, Zhang P, Lepley M, Tian J, Li J, Zhang L, Swingen C, Vaughan JT, Kaufman DS, Zhang J (2013) Functional consequences of human induced pluripotent stem cell therapy: myocardial ATP turnover rate in the in vivo swine heart with postinfarction remodeling. Circulation 127:997–1008

Yamashita J, Itoh H, Hirashima M, Ogawa M, Nishikawa S, Yurugi T, Naito M, Nakao K, Nishikawa S (2000) Flk1-positive cells derived from embryonic stem cells serve as vascular progenitors. Nature 408:92–96

Yang L, Geng Z, Nickel T, Johnson C, Gao L, Dutton J, Hou C, Zhang J (2016) Differentiation of human induced-pluripotent stem cells into smooth-muscle cells: two novel protocols. PLoS One 11:e0147155

Ye L, Zhang S, Greder L, Dutton J, Keirstead SA, Lepley M, Zhang L, Kaufman D, Zhang J (2013) Effective cardiac myocyte differentiation of human induced pluripotent stem cells requires VEGF. PLoS One 8:e53764

Ye L, Chang YH, Xiong Q, Zhang P, Zhang L, Somasundaram P, Lepley M, Swingen C, Su L, Wendel JS, Guo J, Jang A, Rosenbush D, Greder L, Dutton JR, Zhang J, Kamp TJ, Kaufman DS, Ge Y (2014) Cardiac repair in a porcine model of acute myocardial infarction with human induced pluripotent stem cell-derived cardiovascular cells. Cell Stem Cell 15:750–761

Ye L, Basu J, Zhang J (2015) Fabrication of a myocardial patch with cells differentiated from human-induced pluripotent stem cells. Methods Mol Biol 1299:103–114

Yu J, Hu K, Smuga-Otto K, Tian S, Stewart R, Slukvin II, Thomson JA (2009) Human induced pluripotent stem cells free of vector and transgene sequences. Science 324:797–801

Zhang Y, Wang D, Chen M, Yang B, Zhang F, Cao K (2011) Intramyocardial transplantation of undifferentiated rat induced pluripotent stem cells causes tumorigenesis in the heart. PLoS One 6:e19012

Zhang S, Dutton JR, Su L, Zhang J, Ye L (2014) The influence of a spatiotemporal 3D environment on endothelial cell differentiation of human induced pluripotent stem cells. Biomaterials 35:3786–3793

Zhang L, Guo J, Zhang P, Xiong Q, Wu SC, Xia L, Roy SS, Tolar J, O'Connell TD, Kyba M, Liao K, Zhang J (2015) Derivation and high engraftment of patient-specific cardiomyocyte sheet using induced pluripotent stem cells generated from adult cardiac fibroblast. Circ Heart Fail 8:156–166

Zhu W, Gao L, Zhang J (2017) Pluripotent stem cell derived cardiac cells for myocardial repair. J Vis Exp 120. doi:10.3791/55142

Direct Cardiac Reprogramming

6

Sho Haginiwa and Masaki Ieda

Abstract

Recent advances in medical treatment and the development of new mechanical devices have greatly improved the prognosis for heart disease patients. However, heart disease, particularly heart failure, is still a major health issue with continuously increasing numbers of affected patients. Because adult heart muscle has a low regenerative capacity, cardiac function declines with age after cardiac injury. A potential approach to solve this problem is regenerative medicine, aiming at the remuscularization of damaged hearts. Studies conducted in small animals and humans revealed that transplanting various types of cells into failing hearts resulted in the repair of injured hearts and improved cardiac function, but the effects were modest, and further improvement is needed before the method can be widely applied in the clinic. Moreover, true muscle regeneration or cardiac differentiation from so far clinically tested adult stem cells seems to be a rare event, and the beneficial effects of these cell-based therapies are likely due to paracrine factors secreted by the transplanted cells. To regenerate cardiac muscle, it is important to first understand the mechanism of cardiac cell fate determination. Several groups including ours recently found that somatic cells can be directly reprogrammed into cardiomyocyte-like cells using combinations of master regulators. The cardiac reprogramming approach is applicable not only in vitro but also in vivo. It can repair injured hearts and improve cardiac function. Thus, this new technology may open an avenue for regenerative therapy for heart disease.

S. Haginiwa
Department of Cardiology, Keio University School of Medicine,
35 Shinanomachi, Shinjuku-ku, Tokyo 160-8582, Japan

M. Ieda (✉)
Department of Cardiology, Keio University School of Medicine,
35 Shinanomachi, Shinjuku-ku, Tokyo 160-8582, Japan

AMED, PRIME, 35 Shinanomachi, Shinjuku-ku, Tokyo 160-8582, Japan
e-mail: mieda@z8.keio.jp

M. Ieda, W.-H. Zimmermann (eds.), *Cardiac Regeneration*, Cardiac and Vascular Biology 4, DOI 10.1007/978-3-319-56106-6_6

6.1 Introduction

Cardiomyocytes are terminally differentiated cells and are generally considered to have little regenerative potential. Therefore, after cardiomyocytes are damaged, the damaged areas are replaced with fibroblasts and fibrous tissue. Since fibroblasts have no contractile activity, cardiac function decreases, resulting in arrhythmia and heart failure. Cardiac regenerative medicine has arisen as a promising novel therapeutic field to overcome the naturally limited regenerative potential of the heart. The approaches for developing cardiac regeneration techniques are currently focused on cell transplantation using cardiomyocytes induced ex vivo from stem cells, such as embryonic and induced pluripotent stem cells (iPSCs) (Chong et al. 2014; Lalit et al. 2014; Shiba et al. 2012). These approaches, however, have multiple challenges, including tumorigenicity derived from residual pluripotent stem cells, long-term cell engraftment, and high costs, requiring further improvements (Okano et al. 2013; Yang et al. 2014). In contrast to cell transplantation, we have been developing a novel myocardial regeneration technique for direct reprogramming of cardiac non-myocytes (fibroblasts) into cardiomyocytes (Ieda et al. 2010; Inagawa and Ieda 2013; Sadahiro et al. 2015; Wada et al. 2013; Yamakawa et al. 2015). If pre-existing non-cardiomyocytes can be converted to cardiomyocytes within patient's heart, this new approach would be an attractive approach for pharmacologically induced remuscularization of the failing heart.

6.2 Discovery of Master Regulators for Direct Cardiac Reprogramming

In 1987, MyoD was identified as the single master gene for skeletal muscle differentiation (Davis et al. 1987). Similar investigational efforts failed to identify a single master gene for myocardial differentiation. The discovery of the four "Yamanaka factors" in 2006 suggested that the co-transfection of multiple transcription factors could potentially enable the reprogramming of terminally differentiated cells (Takahashi et al. 2007a, b; Takahashi and Yamanaka 2006). Thus, we performed a similar trial and identified three transcription factors (Gata4, Mef2c, and Tbx5, hereafter referred to as GMT) crucial for direct cardiac reprogramming in mice. In order to identify the transcription factors, we isolated cardiomyocytes and cardiac fibroblasts (CFs) using fluorescence-activated cell sorting (FACS: flow cytometry) and screened 14 genes that are specifically expressed in embryonic cardiomyocytes and play a critical role in cardiac formation as candidate factors (Ieda et al. 2009, 2010). Next, we established a genetically modified mouse that specifically expressed green fluorescent protein (GFP) only in differentiated cardiomyocytes (αMHC-GFP transgenic mouse) and screened gene candidates by quantitatively evaluating GFP expression. When all 14 genes were transduced to the CFs via retrovirus vectors, GFP-positive cells were detected at a level of approximately 1.7% a week after infection with the virus. Based on further GFP quantitative evaluations after removing each of the 14 genes, we identified three transcription factors (GMT) that are minimally required for cardiomyocyte generation. In addition, we identified a higher reprogramming efficiency with GMT (17% of GFP+ cells and 5% of cTnT+ cells). We named the cells induced from CFs that express these three transcription factors induced cardiomyocytes (iCMs).

We also determined whether iCMs have a similar morphology and function as native cardiomyocytes. Immunocytochemistry and microarray methods were applied to analyze expression at the protein and gene levels. The immunocytochemistry results demonstrated that iCMs express the cardiomyocyte-specific proteins α-actinin, cardiac troponin T (cTnT), and atrial natriuretic factor (ANF) and possess cardiomyocyte-specific sarcomeric structures. The microarray results demonstrated that gene expression changed from a CF pattern to a cardiomyocyte-like pattern, and the epigenetic status of histone methylation and DNA methylation was similar to that of cardiomyocytes. Functionally, Ca^{2+} imaging demonstrated similar intracellular Ca^{2+} kinetics as mouse neonatal cardiomyocytes. Furthermore, we confirmed spontaneous beating in some iCMs.

Subsequently, we demonstrated the direct reprogramming of CFs to a cardiomyocyte fate. Mesp1 and Isl1 are specifically expressed in cardiac progenitor cells (CPCs) (Laugwitz et al. 2005; Saga et al. 1999). We generated genetically modified mice that expressed yellow fluorescent protein (YFP) in all cells derived from CPCs by crossing Mesp1-Cre mice or Isl1-Cre mice with YFP-flox mice (Mesp1-YFP mice or Isl1-YFP mice). The iCMs derived from the fibroblasts in these mice did not express YFP, providing proof of the direct conversion to iCMs without a step involving CPCs.

Lastly, we also determined whether non-cardiac fibroblasts can be reprogrammed into iCMs by GMT. Mouse tail-tip fibroblasts (TTFs) were reprogrammed using GMT retrovirus vectors. The TTFs were successfully able to differentiate into αMHC-GFP+ cardiomyocyte-like cells. However, the reprogramming efficiency was low, as TTF-derived iCMs accounted for only ~2.5% according to a quantification cTnT+ cells, approximately half of the number of CF-derived iCMs. In addition, although TTF-derived iCMs possessed some cardiomyocyte functions such as Ca^{2+} regulation, they expressed higher levels of H3K27me3, a histone marker for gene suppression, in the promoters of cardiac genes than CF-derived iCMs and neonatal cardiomyocytes. These lines of evidence suggest that although various cells can be reprogrammed into iCMs, the induction rate varies depending on the original cell type. We confirmed cardiomyocyte conversion 1 week after the GMT infection by cTNT staining; few of the converted cells started to beat 4–5 weeks later. These results indicate that iCMs represent a heterogeneous cell population and that only 0.01–0.1% of CFs were reprogrammed completely into functional (i.e., beating) cardiomyocytes, warranting further improvement. Nevertheless, we identified the master regulators for cardiomyocytes that are minimally required for cardiac induction from other somatic cell types.

6.3 New Cardiac Reprogramming Factors and Stoichiometry of Transcription Factors for Efficient Reprogramming

Although GMT was identified to be a combination of critical transcription factors and enabled the reprogramming of mouse fibroblasts into functional cardiomyocytes, the poor induction rate warranted further improvements (Srivastava and Ieda 2012). Multiple studies on cardiac reprogramming have been performed since our report on the three "GMT transcription factors" and production of iCMs (Fig. 6.1).

Song et al. explored the optimal combination of core cardiac transcription factors necessary for efficient reprogramming of adult TTFs into functional cardiomyocytes (Song et al. 2012). Six cardiac transcription factors were screened using αMHC-GFP reporter mice to determine the core factors involved in cardiac reprogramming. When Hand2, a basic helix-loop-helix (bHLH) transcription factor, was added to GMT (GHMT—Gata4, Hand2, Mef2c, and Tbx5), adult CFs and TTFs were reprogrammed into functional cardiomyocyte-like cells more efficiently compared to other combinations. GMT alone resulted in only approximately 3% of the original fibroblasts becoming positive for both αMHC-GFP and cTnT. In contrast, GHMT induced approximately 9% of the adult fibroblasts to adopt a cTnT+ and αMHC-GFP+ phenotype. In addition, 5% of adult CF-derived iCMs and 1.8% of

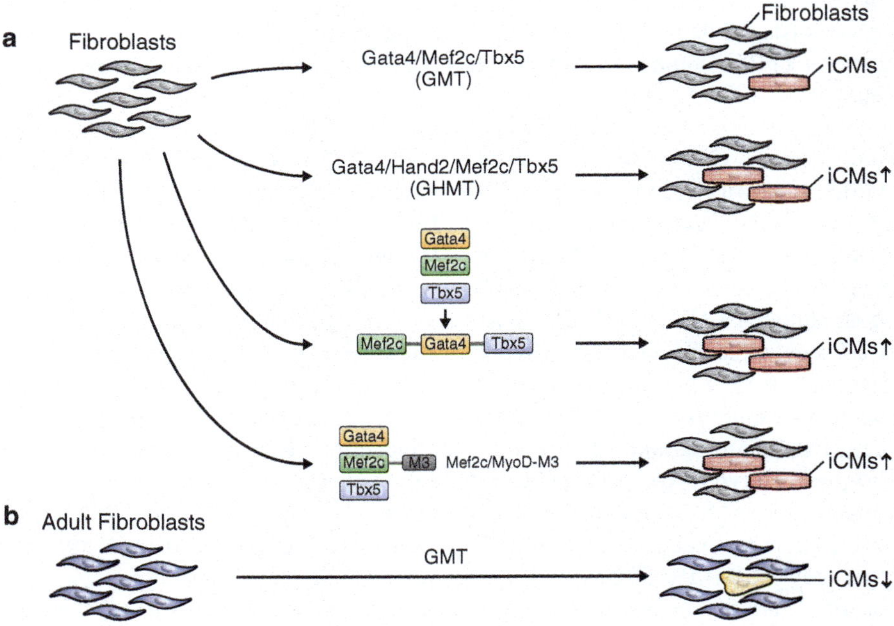

Fig. 6.1 Modification of reprogramming factors and fibroblast properties affect cardiac reprogramming efficiency. (**a**) Addition of Hand2 to Gata4, Mef2c, and Tbx5 (GMT), a polycistronic vector for GMT, and addition of the MyoD-M3 segment to Mef2c promoted cardiac reprogramming. (**b**) Cardiac reprogramming was inefficient in adult CFs and TTFs

adult TTF-derived iCMs possessed sarcomere-like structures. Microarray and quantitative RT-PCR (qPCR) analyses of gene expression patterns demonstrated the upregulation of a broad range of cardiac genes, indicative of a differentiated cardiac-like phenotype, and concomitant suppression of non-myocyte genes, including *Fsp1* (fibroblast-specific protein 1), in fibroblasts transduced with GHMT. Furthermore, when adult CFs or TTFs transduced with GHMT were continuously cultured for more than 5 weeks, Ca^{2+} transients, action potentials (APs), and spontaneous contractions were observed in the cell subsets, indicating that induced cardiomyocyte-like cells possessed similar functions as native cardiomyocytes (Nam et al. 2014). This study also demonstrated that a limited 10-day exogenous GHMT expression was sufficient to reprogram the fibroblasts toward a cardiomyocyte fate. This result suggests that GHMT may play a critical role in the onset of reprogramming, but thereafter, the reprogramming can continue to progress without exogenous GHMT expression. Based on these study results, the addition of Hand2 to GMT is considered to improve the efficiency of reprogramming of fibroblasts into cardiomyocytes compared to GMT alone (Nam et al. 2014; Song et al. 2012).

Protze et al. used a different approach to identify the optimal combination of transcription factors (Protze et al. 2012). They did not search a pool of transcription factor candidates for a single crucial gene but directly screened all triplet combinations of ten candidate factors combined with a qPCR assay to determine multiple cardiac-specific genes (MYH6, Myl2, Actc1, Nkx2.5, and SCN5A). Through this screening method, the combination of Mef2c, Myocd, and Tbx5 (MMT) was found to upregulate a broader spectrum of cardiac genes compared to other combinations. When neonatal CFs were transduced with MMT or GMT via lentivirus vectors, the expression of proteins involved in cardiac contractility and sodium and potassium ion channels were observed, and 0.08% of the CFs exhibited spontaneous contractions as well as action potential. These results indicate that the transduction of MMT as well as GMT can directly reprogram neonatal CFs into functional cardiomyocytes.

Addis et al. reported that Nkx2.5 is another important cardiac reprogramming factor that can improve the functionality of iCMs (Addis and Epstein 2013; Addis et al. 2013). They analyzed combinations of reprogramming factors using calcium activity in iCMs as a functional measure of cardiomyocytes. They constructed a GCaMP5 reporter lentivirus that allows for the real-time detection of calcium flux in live cells driven by the cardiomyocyte-specific troponin T (TNNT2) promoter. They transduced the GCaMP5 reporter virus containing the reprogramming factors into mouse fibroblasts. The results indicated that the addition of Nkx2.5 to GMT produced more functional cardiomyocytes, with $0.7 \pm 0.3\%$ of cells exhibiting GCaMP activity at 14 days post-induction, which was 22.5-fold higher than that with GMT alone ($0.03 \pm 0.02\%$). They also showed that the addition of Nkx2.5 to GHMT further increased the number of functional iCMs up to $1.6 \pm 0.3\%$ of the cells expressing GCaMP5, which is a 52-fold increase over GMT alone. Nkx2.5 did not increase the number of iCMs by promoting cell proliferation; instead, Nkx2.5 and Hand2 augmented the expression of cardiac genes related to excitation–contraction coupling, such as phospholamban (Pln) and calsequestrin (Casq2).

Hirai et al. reported that fusion of the MyoD transactivation domain to the pluripotency transcription factor Oct4 facilitated the transcriptional activity of Oct4, resulting in the highly efficient production of iPSCs (Hirai et al. 2010). They then applied this strategy to iCMs and showed that fusion of the MyoD transactivation domain to the cardiac reprograming factor Mef2c could greatly promote the direct reprogramming of fibroblasts into cardiomyocyte-like cells (Hirai et al. 2013). They fused the MyoD domain to Mef2c, Gata4, Hand2, and Tbx5 and transduced these four genes in various combinations into mouse fibroblasts. Transduction of the chimeric Mef2c with the wild-type forms of other three genes produced a much higher number of beating iCMs than with the other combinations of reprogramming factors. The induction efficiency of beating iCMs using chimeric Mef2c was 3.5%, which was 15-fold greater than when using the wild-type GHMT. Although the MyoD domain effectively increased the efficiency of iCM generation when fused at the carboxy terminus of Mef2c, it was not effective when fused to the other reprogramming factors Gata4, Hand2, and Tbx5. These results suggest that an optimal balance of the transcriptional activities of reprogramming factors is critical for successful cardiac induction; however, the exact molecular basis for this remains elusive.

Wang et al. also reported that a precise stoichiometry of GMT is critical for the efficiency and quality of iCM generation (Wang et al. 2014). They generated all possible combinations of G, M, and T with identical 2A sequences in a single transgene and transduced these viral vectors into mouse fibroblasts. They demonstrated that each combination of G, M, and T gave rise to distinct G, M, and T protein expression levels and that the MGT vector that expressed a higher protein level of Mef2c and lower levels of Gata4 and Tbx5 significantly enhanced reprogramming efficiency compared to the other GMT variants. Finally, the MGT vector resulted in more than a tenfold increase in the number of beating iCM loci than the separate GMT vectors, and the molecular characterization revealed that the optimal stoichiometry of G, M, and T correlated with the high expression of cardiac genes (Muraoka and Ieda 2015).

Given that a precise stoichiometry and the transcriptional activity of reprogramming factors are critical for reprogramming, it is conceivable that not all laboratories can reproduce cardiac reprogramming due to subtle differences in experimental conditions (Carey et al. 2011; Polo et al. 2012; Qian et al. 2013). Chen et al. reported that they could not produce functional iCMs using lentiviral GMT vectors from adult mouse fibroblasts (Chen et al. 2012). Although they were able to generate partially reprogrammed iCMs expressing some cardiac markers, the cells did not beat spontaneously or exhibit APs, suggesting that they were not fully reprogrammed iCMs.

In contrast to iCM generation, the induction of iPSCs has been widely reproduced by many laboratories. The iPSCs can expand indefinitely under standardized culture conditions, and the iPSC colonies are easily detected in the culture dish.

In contrast, iCMs do not proliferate because they are post-mitotic cells, which may hinder the detection of low numbers of beating iCMs (Yoshida and Yamanaka 2012b). Collectively, the available data highlight that cardiac reprogramming needs further optimization of the reprogramming factors and that defined culture conditions are necessary to establish a routine procedure that can be performed in a wide range of laboratories, as discussed below.

6.4 Identification of MicroRNAs for Cardiac Reprogramming

MicroRNAs (miRNAs) are ~22-nucleotide RNAs that modulate gene expression by inhibiting mRNA translation and promoting mRNA degradation (Liu and Olson 2010). Previous studies have revealed that MEF2 and SRF regulate the expression of two bicistronic muscle miRNA clusters encoding miR-133a-1/miR-1-2 and miR-133a-2/miR-1-1 in the embryonic and adult heart (Zhao et al. 2005). Loss-of-function mutation studies of these miRNAs in mice revealed that miR-1 and miR-133a regulate the gene expression programs required for normal cardiac growth and function (Zhao et al. 2007). For example, the absence of miR-133a expression results in the ectopic expression of smooth muscle genes and aberrant cardiomyocyte proliferation in the developing heart, leading to embryonic lethality in half of the mutant mice (Liu et al. 2008). These abnormal phenotypes can be attributed, at least in part, to the elevated expression of SRF and cyclin D2, which are direct targets of miR-133a. These results demonstrated that miRNAs and transcription factors orchestrate a complex network involved in normal heart development and cardiac cell fate determination.

There are four known muscle-specific miRNAs, miR-1, 133, 208, and 499, which regulate various stages of cardiac differentiation and development. Jayawardena et al. reported that a combination of these four miRNAs can reprogram mouse neonatal CFs into cardiomyocyte-like cells in vitro (Jayawardena et al. 2012). Their study demonstrated that a transduction of miR-1, 133, 208, and 499 resulted in 5% of adult fibroblasts from αMHC-CFP mice possessing an αMHC-CFP+ phenotype. In addition, the induced cells expressed cardiac-specific proteins and sarcomeric structures, indicative of iCMs. Furthermore, a JAK inhibitor improved the induction rate and quality of the miRNA-mediated reprogramming, leading to high induction rate of 1–2% of the initial fibroblast population into iCMs with spontaneous contractions. Although induction by miRNAs alone may not be sufficient to reprogram the fibroblasts from other origins such as tail-tip fibroblasts into cardiomyocytes, these results demonstrated miRNAs can reprogram at least CFs into functional cardiomyocytes. However, the molecular mechanisms of cardiac reprogramming by these miRNAs were not clarified.

We identified the molecular mechanisms and direct target of miR-133a in cardiac reprogramming (Muraoka and Ieda 2014; Muraoka et al. 2014). We also analyzed the cardiac reprogramming efficiency when using the miR-1, 133, 208, and 499 miRNAs in mouse embryonic fibroblasts (MEFs). We found that the miRNAs alone were not sufficient for cardiac reprogramming in MEFs, but the addition of miR-133a to GMT greatly increased the reprogramming efficiency compared with other miRNAs. Compared to GMT alone, miR-133a overexpression with GMT generated sevenfold more cTnT+ cells and beating iCMs from MEFs and shortened the duration to induce beating cells from 30 to 10 days. Microarray analyses revealed that more than 100 genes were downregulated by miR-133a, many of which were fibroblast-related genes, such as Col1a1, Col1a2, Fn1, and Postn. While the expression levels of Ccnd2, Cdc42, Hand2, RhoA, and Srf, which have been shown previously to be the direct targets of miR-133a, were not significantly altered, the expression of Snai1, a master regulator of epithelial-to-mesenchymal transformation, was significantly downregulated by miR-133a overexpression. Snai1 is a putative direct target of miR-133a and contains two conserved miR-133a-binding sites within its 3′-UTR as shown by bioinformatics analyses. Luciferase assays confirmed that miR-133a binds to both sites. The expression levels of Snai1 mRNA and protein were suppressed by miR-133a transduction, suggesting that Snai1 is a new direct target of miR-133a. To investigate the possible contribution of Snai1 to cardiac reprogramming, we suppressed Snai1 expression with siRNA in GMT-transduced MEFs. Snai1 knockdown suppressed fibroblast genes, upregulated cardiac gene expression, and promoted cardiac reprogramming, recapitulating the effects of miR-133a overexpression. In contrast, overexpression of Snai1 in GMT/miR-133a-transduced cells suppressed reprogramming by maintaining fibroblast signatures. MiR-133a-mediated Snai1 repression was also critical for cardiac reprogramming in adult mouse and human CFs. These results suggest that miR-133a-induced Snai1 and fibroblast gene suppressions are critical for cardiac reprogramming. Given that Snai1 suppression is also important for iPSC generation, which requires a mesenchymal-to-epithelial transition before reprogramming, Snai1 is a common target for cellular reprogramming from fibroblasts (Li et al. 2010; Unternaehrer et al. 2014). Moreover, the balance between master regulators of original cells (fibroblasts) and target cells (iCMs or iPSCs) may determine the success of the cell fate switch in general. Further studies might identify other new targets of the miRNAs involved in cardiac reprogramming, as miRNAs have numerous targets related to signal transduction, transcription factors, and epigenetic regulation. Nonetheless, this is the first study demonstrating a molecular circuit and mechanism of cardiac reprogramming (Fig. 6.2).

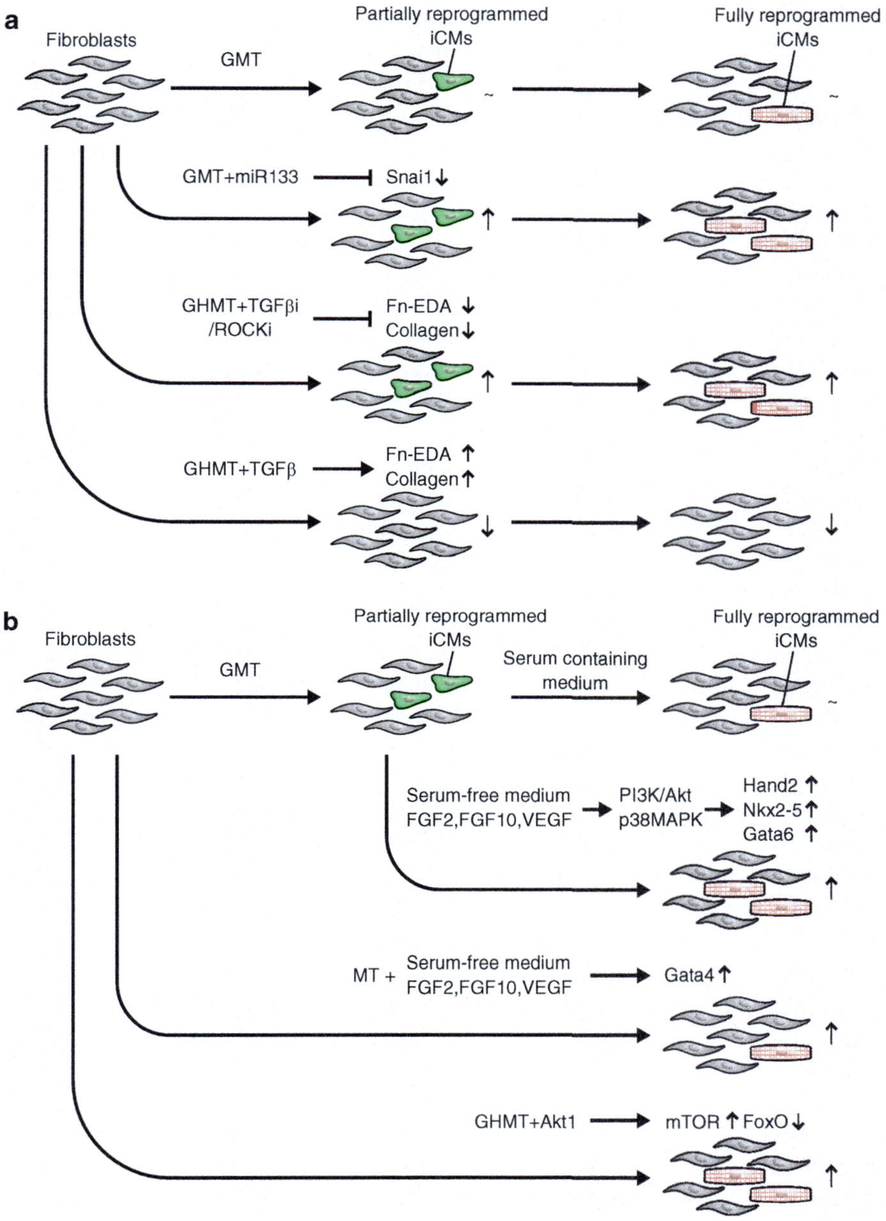

Fig. 6.2 Mechanisms of cardiac reprogramming. (**a**) Addition of miR-133, a TGFβ inhibitor, and a ROCK inhibitor increased cardiac reprogramming by repressing pro-fibrotic signaling, while TGFβ reduced it by maintaining fibroblast signatures. These interventions changed the generation of partially reprogrammed iCMs at the early stage of cardiac reprogramming. (**b**) Cultivation of the reprogrammed cells in serum-free medium with FGF2, FGF10, and VEGF promoted the conversion of partially reprogrammed iCMs into fully reprogrammed iCMs at the late stage of cardiac reprogramming. FGF2, FGF10, and VEGF increased the expression of cardiac reprogramming factors (Hand2, Nkx2.5, and Gata6) via the PI3K/Akt and p38MAPK pathways. The defined conditions increased the expression of Gata4 and enabled cardiac reprogramming with only MT. Activation of Akt1 promoted cardiac reprogramming and maturation of iCMs with alteration of mTOR and FoxO expression levels

6.5 Inhibition of Pro-fibrotic Signaling with Small Molecules Enhances Cardiac Reprogramming

Several small molecules have been utilized to enhance the differentiation of pluripotent stem and progenitor cells to cardiomyocytes and to promote the reprogramming of fibroblasts into iPSCs (Burridge et al. 2014; Kattman et al. 2011; Li et al. 2010). Ifkovits et al. reported that the inhibition of transforming growth factor β (TGFβ) signaling by small molecules increased the direct conversion of mouse fibroblasts to iCMs (Ifkovits et al. 2014). They overexpressed GHMT plus Nkx2.5 (GHMNT) with the calcium indicator GCaMP5, driven by the cTnT promoter, to quantify iCM yield in MEFs. They screened five small molecules, including a G9a histone methyltransferase inhibitor (BIX01294), canonical Wnt signaling activator (CHIR99021), Wnt inhibitor (XAV939), TGFβ/Activin/Nodal inhibitor (SB431542), and BMP inhibitor (DMH1), which were reported to promote directed differentiation from pluripotent stem cells to cardiomyocytes and iPSC reprogramming. Among them, only SB431542 promoted cardiac induction by GHMNT, and LY364947, a specific inhibitor of TGFβ, also showed similar effects. In contrast, addition of TGFβ to the culture media greatly reduced cardiac induction, suggesting that inhibition of TGFβ signaling increased the efficiency of iCM generation.

These results were confirmed and analyzed in more detail by Zhao et al. (2015). They performed RNA sequencing to identify the genes that were regulated by GHMT overexpression at day 7. Surprisingly, not only cardiac gene but also pro-fibrotic gene expression was activated during GHMT-mediated cardiac reprogramming at day 7. The expression of fibrotic genes was reduced 12 days post-GHMT infection, suggesting that transient activation of pro-fibrotic signaling at the early stage of reprogramming may inhibit the conversion of fibroblasts into cardiomyocytes. TGFβ signaling is an important pathway involved in controlling fibrotic events, and the expression levels of TGFβ signaling components, including phosphorylated Smad transcription factors, Smad2 and Smad3, and Tgfb2 and Tgfbr1, were all upregulated by GHMT transduction during the early stage of cardiac reprogramming. They showed that stimulation of pro-fibrotic signaling by TGFβ1 supplementation attenuated cardiac reprogramming by GHMT or GHMT plus miR-1 and 133 (GHMT2m). The addition of miR-1 and 133 to GHMT significantly decreased the expression of pro-fibrotic genes concomitant with the activation of cardiac genes, which was consistent with our previous data (Muraoka et al. 2014). In addition to TGFβ signaling, Rho triggers the formation of stress fibers and stimulates pro-fibrotic events via activation of its downstream effector, Rho-associated kinase (ROCK). Treatment with the ROCK inhibitor Y-27632 decreased the expression of the fibrotic genes Fn-EDA and aSMA and promoted cardiac reprogramming with GHMT and GHMT2m. Treatment of MEFs with A83-01, a selective inhibitor of TGFβ signaling, also decreased the phosphorylation of Smad2 and inhibited the expression of Fn-EDA, Col1a1, Col3a1, and SMA in GHMT- and GHMT2m-infected cells. Although A83-01 alone did not induce cardiac reprogramming in MEFs, the addition of A83-01 to GHMT or GHMT2m transduction greatly increased the cardiac reprogramming efficiency, with up to 60% of the MEFs reprogrammed into functional iCMs. The electrophysiological analysis revealed that APs were recorded from single spontaneously beating cells on day 9 of reprogramming. The APs of iCMs mimic those of fetal or nodal cardiomyocytes, as they show a high rate of spontaneous firing, short AP durations, and slow upstroke velocity, suggesting that they were an immature myocyte form. The frequency of cell contraction and spontaneous calcium transients in the iCMs were modulated by the administration of isoproterenol, a β-adrenergic agonist, and nifedipine, a blocker of L-type calcium channels, suggesting that functional excitation–contraction coupling machinery and β-adrenergic signaling components were developed in the iCMs. Inhibition of TGFβ signaling by A83-01 also enhanced the reprogramming of adult cardiac and dermal fibroblasts into functional cardiomyocytes, with frequencies of 2.5% and 4%, respectively. These results suggest that the inhibition of pro-fibrotic signaling by small molecules promotes cardiac reprogramming, but the manipulation of other molecules and signaling pathways will be necessary to further improve cardiac reprogramming in adult fibroblasts.

6.6 PI3K/Akt and p38MAPK Pathways Enhanced Cardiac Reprogramming Under Defined Culture Conditions

Our previous results demonstrated that transduction with GMT activated cardiac reporters and protein expression in ~20% of fibroblasts after 1 week; however, only 0.1% of the starting fibroblasts were fully reprogrammed into functional iCMs after 4 weeks under conventional serum-based culture conditions, suggesting that most cells remained partially reprogrammed or immature iCMs with the original method (Ieda et al. 2010). As discussed above, inhibition of fibroblast signatures by miR-133 and small molecules promoted cardiac reprogramming at the early stage of reprogramming; however, the molecular mechanisms underlying the conversion of partially reprogrammed cells into functional iCMs at the later stage remained unclear (Sadahiro et al. 2015). Moreover, the use of undefined serum-containing medium in the original method was associated with considerable batch-to-batch variation in cardiac reprogramming, leading to the variable and low reprogramming efficiencies in previous studies (Chen et al. 2012; Srivastava and Ieda 2012; Yoshida and Yamanaka 2012b). Recently, we were the first to describe the defined culture conditions that increased cardiac reprogramming by 100-fold compared with the conventional serum-based conditions (Yamakawa et al. 2015). We screened eight cardiogenic compounds and found that a combination of fibroblast growth factor (FGF) 2, FGF10, and vascular endothelial growth factor (VEGF), termed FFV, greatly improved the quality of cardiac reprogramming in mouse fibroblasts under serum-free culture conditions. FFV did not increase the generation of partially reprogrammed iCMs, while this treatment activated multiple cardiac transcriptional regulators, including Gata4/6, Hand2, and Nkx2.5, and converted partially reprogrammed iCMs into functional iCMs through the p38 mitogen-activated protein kinase (MAPK) and phosphoinositol 3-kinase (PI3K)/AKT pathways. Moreover, FFV enabled cardiac reprogramming with only Mef2c and Tbx5 without the need for Gata4, which may enable pharmacological reprogramming in the future. Thus, our results demonstrated that the FGF- and VEGF-mediated PI3K/Akt and p38MAPK signaling pathways are critical for the late stage of cardiac reprogramming, which has been a major hurdle for successful reprogramming. Identifying the efficient, reproducible, and defined culture conditions at least for mouse cells could allow for the standardization of the cardiac reprogramming procedure and lead to further improvements in the protocol in the future.

Consistent with our results, Zhou et al. reported that Akt dramatically enhanced cardiac reprogramming in three different types of fibroblasts (mouse embryo, adult cardiac, and tail tip) (Zhou et al. 2015). They used a retroviral expression library and screened 192 protein kinases that might augment the generation of functional iCMs by GHMT. They found that Akt1 overexpression induced Akt1 phosphorylation and increased the expression of cardiac genes, whereas a kinase-dead form of Akt1 abrogated the stimulatory activity on GHMT, suggesting that activation of Akt signaling is critical for cardiac reprogramming. Approximately 50% of the reprogrammed MEFs displayed spontaneous beating after 3 weeks of induction by Akt plus GHMT, while ~1% of the adult CFs and TTFs could be reprogrammed into beating iCMs with the same treatment, suggesting that some epigenetic barriers still exist in adult fibroblasts. Nevertheless, the iCMs generated by Akt plus GHMT displayed a more mature phenotype compared with those generated without Akt and that were polynucleated, hypertrophic, and responsive to β-adrenoceptor pharmacologic modulation. Mechanistically, insulin-like growth factor 1 (IGF1) and PI3K acted upstream of Akt, whereas the mitochondrial target of rapamycin complex 1 (mTORC1) and forkhead box o3 (Foxo3a) acted downstream of Akt to promote cardiac reprogramming. These findings provide new insights into the molecular mechanisms of cardiac reprogramming and might be valuable for future research on human cardiac reprogramming (Fig. 6.2).

6.7 Discovery of Human Cardiac Reprogramming Factors

We determined whether cardiomyocytes can be induced from human fibroblasts by direct reprogramming (Wada et al. 2013). First, we transduced GMT into human CFs in vitro; however, there was insufficient induction of cardiomyocytes. Thus, we explored new human cardiac reprogramming factors. When Mesp1 and Myocd, which are cardiomyocyte or CPC-specific transcription regulators, were added to GMT (GMT + Mesp1 + Myocd, hereafter referred to as GMTMM), the cardiac induction rate was improved. Microarray analysis demonstrated that the GMTMM-mediated human iCMs had increased expression levels of cardiomyocyte-specific genes and possessed sarcomeric structures. In addition, under co-culture conditions with primary cultured rat cardiomyocytes, the human iCMs demonstrated physical cardiac functions, indicating that the five factors of GMTMM can induce the reprogramming of human CFs to cardiomyocyte-like cells. We also demonstrated that the addition of miR-133a to GMTMM increased cardiac reprogramming in human fibroblasts via Snail repression. The induction of the cardiac markers α-actinin and cTnT increased from ~2% to ~27% by the addition of miR-133a (Muraoka et al. 2014).

Nam et al. also reported that the mouse reprogramming factors GHMT were ineffective in activating cardiac gene expression in human fibroblasts and that Myocd was required for human cardiac induction (Nam et al. 2013). Furthermore, the addition of miR-1 and miR-133a improved the myocardial conversion of human fibroblasts and eliminated the requirement for Mef2c in cardiac induction from human neonatal foreskin fibroblasts, adult CFs, and dermal fibroblasts. The induced human cardiomyocyte-like cells expressed multiple cardiac genes and developed sarcomere-like structures. Although the efficiency of inducing cTnT-expressing cells from human fibroblasts was 10–20%, only a small subset of the cells could exhibit spontaneous contractility after 11 weeks of culture.

Fu et al. reported that GMT plus Esrrg, Mesp1, Myocd, and Zfpm2 induced global cardiac gene expression and a phenotypic shift to a cardiac fate in human fibroblasts derived from human ESCs (Fu et al. 2013). While most cells were partially reprogrammed, a subset of human iCMs had sarcomere structures, calcium transients, and APs. They demonstrated that the epigenetic status of human iCMs resembled that of hESC-derived cardiomyocytes in terms of DNA and histone methylation status and that the iCMs were stably reprogrammed to a cardiac state without the need for the continuous expression of reprogramming factors.

Islas et al. reported that the transient overexpression of Ets2 and Mesp1, followed by activin A and BMP2 treatment, could reprogram human dermal fibroblasts into cardiac progenitor-like cells (Islas et al. 2012). The induced cardiac progenitor-like cells differentiated into immature cardiomyocytes that expressed several cardiac genes and exhibited sarcomeric structures and Ca^{2+} activities in a prolonged culture. Given that the induced progenitor-like cells formed colonies in a culture dish and expressed several cardiac progenitor genes, the route of cardiac induction by Ets2 and Mesp1 seemed to be different from that with the direct cardiac reprogramming performed by our group and other groups (Fu et al. 2013; Nam et al. 2013; Wada et al. 2013). These findings represent an important initial step toward potential therapeutic applications of the direct reprogramming approach in clinical situations. However, cardiac reprogramming of human fibroblasts was much slower and less efficient than that in mouse fibroblasts, and future studies are needed to optimize the necessary reprogramming factors and culture conditions for human cardiomyocyte induction and functional maturation.

6.8 Systemic Approach to Identify the Transcription Factors Required for Direct Reprogramming

Thus far, identification of the key transcription factors required for reprogramming has been performed by a process of exhaustive testing of large sets of plausible transcription factors (Han et al. 2014; Huang et al. 2011, 2014; Ieda et al.

2010; Riddell et al. 2014; Sekiya and Suzuki 2011; Vierbuchen et al. 2010; Zhou et al. 2008). As there are roughly 2000 different transcription factors, it is technically challenging to test all possible combinations of factors necessary for cardiac reprogramming. Recently, the Daley and Collins groups developed a network biology platform, CellNet, which can compare gene regulatory networks in engineered cell populations with those in in vivo counterparts and identify systematically the factors that can improve cellular reprogramming (Fig. 6.3) (Cahan et al. 2014; Morris et al. 2014). In the case of cardiac reprogramming, Cahan et al. used CellNet to analyze the gene regulatory network of the GMT-induced aMHC-GFP+ population derived from mouse fibroblasts, of which the vast majority of the cells were partially reprogrammed iCMs (Cahan et al. 2014). They found that the aMHC-GFP+ cells were exclusively classified as cardiomyocytes; however, multiple cardiac transcription factors, including Gata6, Tbx20, Hand2, and Nkx2.5, were incompletely activated in the GMT-mediated aMHC-GFP+ population. As discussed, at least some of the candidate factors, such as Hand2 and Nkx2.5, could improve the cardiac reprogramming by GMT, suggesting that this approach can be valuable for screening cardiac reprogramming factors (Addis et al. 2013; Song et al. 2012).

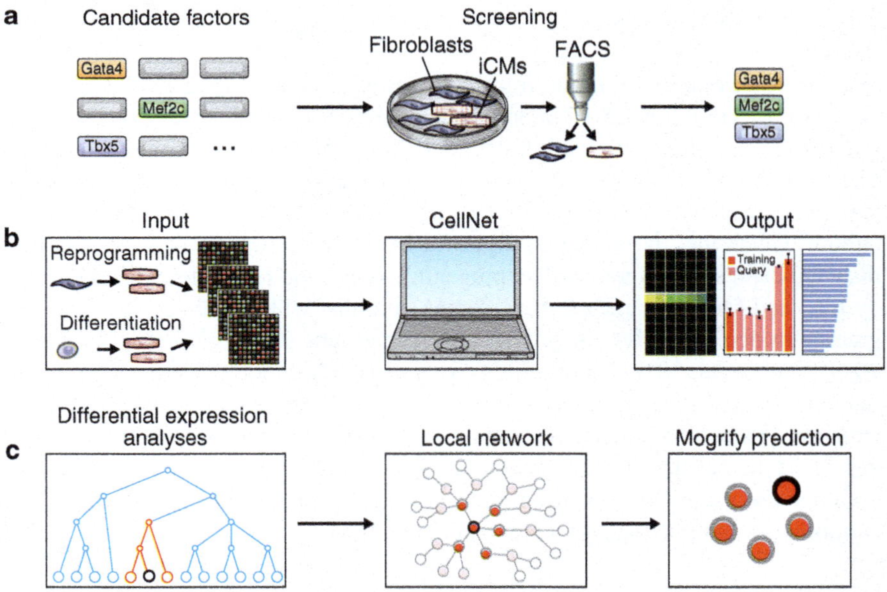

Fig. 6.3 Strategies to identify reprogramming factors. (**a**) Candidate approach for identification of reprogramming factors. Candidate factors defined empirically or experimentally are screened in in vitro experiments. (**b**) CellNet accurately assessed the fidelity of cellular engineering and identified the transcription factors that might be needed to enhance reprogramming. (**c**) The Mogrify algorithm for predicting transcription factors for cell conversion. The cell type ontology tree, gene expression data, and network analyses are used to predict the reprogramming factors necessary to induce cell conversion

More recently, Rackham et al. reported a predictive system (Mogrify) that combines human gene expression data with regulatory network information to predict the reprogramming factors necessary to induce cell conversion (Rackham et al. 2016). They applied Mogrify to 1173 human cell types and 1134 tissues and defined a cellular reprogramming atlas. To assess the predictive power of Mogrify, they first determined how Mogrify performed against previously published human cell conversions. Mogrify predicted NANOG, OCT4, and SOX2 as the top three transcription factors for iPSC conversion, a combination that was experimentally validated. For the conversion of human dermal fibroblasts into cardiomyocytes, Mogrify's predicted list included Gata4, Hand2, Mef2c, Nkx2.5, and Tbx5. Mogrify correctly predicted the transcription factors used in known transdifferentiations, and the average recovery rate of published transcription factors was 84%. Thus, these predictive computational programs can be useful to identify new transcription factors that could improve the efficiency of human cardiac reprogramming.

6.9 Cardiac Regeneration by In Vivo Direct Cardiac Reprogramming

For use in regenerative medicine, it would be ideal if cardiomyocytes could be reprogrammed from the endogenous CFs in situ. Based on our in vitro results, we determined whether the direct reprogramming approach could also be used to reprogram endogenous CFs into cardiomyocytes by applying cardiac reprogramming factors in vivo (Inagawa et al. 2012). We used a myocardial infarction (MI) mouse model, which was generated by coronary artery ligation. Retrovirus vectors were directly injected to transfer the transcription factors to the CFs on the day of coronary artery ligation. After 2 weeks, the hearts were removed, and the induction of cardiomyocytes was evaluated. In the negative control group that received a control retrovirus vector injection, there was no induction of cardiomyocytes from fibroblasts. In contrast, the group that received the GMT retrovirus vector injection exhibited induction of cardiomyocytes from fibroblasts. However, the majority of the inducted cells were immature cardiomyocyte-like cells, suggesting that all three transcription factors might not be transferred into the cells simultaneously. Thus, we constructed a single polycistronic GMT to ensure simultaneous delivery of the three factors. As a result, multiple cardiac genes were expressed, and the induction rate of sarcomere+ cardiomyocytes increased by twofold.

Similar outcomes were reported by multiple research groups. Qian et al. directly injected a GMT retrovirus into the mouse heart after coronary ligation and reported that 35% of the cardiomyocytes from the border/infarct zone were iCMs newly derived from endogenous CFs (Qian et al. 2012). Approximately 50% of the iCMs exhibited functional characteristics of adult ventricular cardiomyocytes, including an organized sarcomeric structure, cell contraction, electrophysiologic properties, and functional coupling to other cardiomyocytes. In order to determine whether iCMs were derived from CFs, genetic lineage tracing with a fibroblast-specific gene was applied using the transgenic mice (Postn-Cre and Fsp1-Cre with reporter mice) after direct injection with the GMT retrovirus into the heart. The post-MI myocardium contained iCMs, which expressed Postn-Cre- or Fsp1-Cre-induced reporter expression, confirming that these iCMs were derived from CFs and not from native cardiomyocytes. In addition, the functional evaluation after MI revealed that GMT retrovirus injection significantly improved cardiac function and suppressed fibrosis at least up to 3 months after the cardiac infarction.

Song et al. generated an ischemic cardiac disease mouse model and reported that GHMT retrovirus injection reprogrammed endogenous CFs into functional cardiomyocyte-like cells (Song et al. 2012). They also used transgenic mice with fibroblast lineage tracing and demonstrated that 2–6% of the cardiomyocytes from the border/infarct zone were newly induced cardiomyocyte-like cells. The newly differentiated cardiomyocyte-like cells possessed similar characteristics as the endogenous cardiomyocytes, such as a clear sarcomeric structure and functional properties. Their study also determined whether the newly induced cardiomyocyte-like cells might have arisen from the fusion of native cardiomyocytes with non-cardiomyocytes using mice with an inducible αMHC-MerCreMer transgene and Rosa26-*LacZ* reporter. The results demonstrated that cardiomyocyte-like cells were newly differentiated and independent of cell fusion. Lastly, GHMT-transduced mice had a twofold higher cardiac ejection fraction compared to the control mice. Furthermore, the damaged scar area was reduced by 50% at 12 weeks post-myocardial infarction.

Mathison et al. reported that in a rat cardiac infarction model, intramyocardial treatment with the proangiogenic VGEF together with GMT transduction enhanced the recovery of cardiac function and reduced the fibrosis area compared to GMT transduction alone (Mathison et al. 2012). This beneficial effect of VGEF treatment suggests that VEGF-mediated neovascularization may at least partially contribute to improved differentiation and survival of newly iCMs in the damaged myocardium, resulting in improved recovery after MI.

Jayawardena et al. used transgenic mice (Fsp1-Cre) with reporter mice and genetic tracing of fibroblast origin and demonstrated that direct injection of lentiviral miR-1, 133, 208, and 499 into the mouse heart after MI reprogrammed endogenous CFs into cardiomyocyte-like cells (Jayawardena et al. 2014). They used Fsp1-Cre mice for the linage tracing of non-myocytes and found that 12% of cardiomyocytes in the border/infarct area were newly generated iCMs. Serial cardiac echo mapping revealed that there was a progressive improvement in ventricular function following miRNA treatment, which began 1 month post-surgery and was enhanced at 3 months, similar to the period required for reprogramming using transcription factors. Thus, our group and other groups demonstrated that cardiac reprogramming can be achieved in vivo by efficiently transferring cardiac reprogramming factors into CFs (Fig. 6.4). Furthermore, in vivo cardiac reprogramming reduced scar size and improved cardiac function after MI, suggesting that in vivo cardiac reprogramming might be a promising approach for regenerative medicine. Given that endogenous CFs can be converted into more fully reprogrammed functional iCMs using in vivo reprogramming than using in vitro conditions, undefined factors in the microenvironment may improve the quality of cardiac reprogramming, which will be investigated in future studies (Yoshida and Yamanaka 2012a).

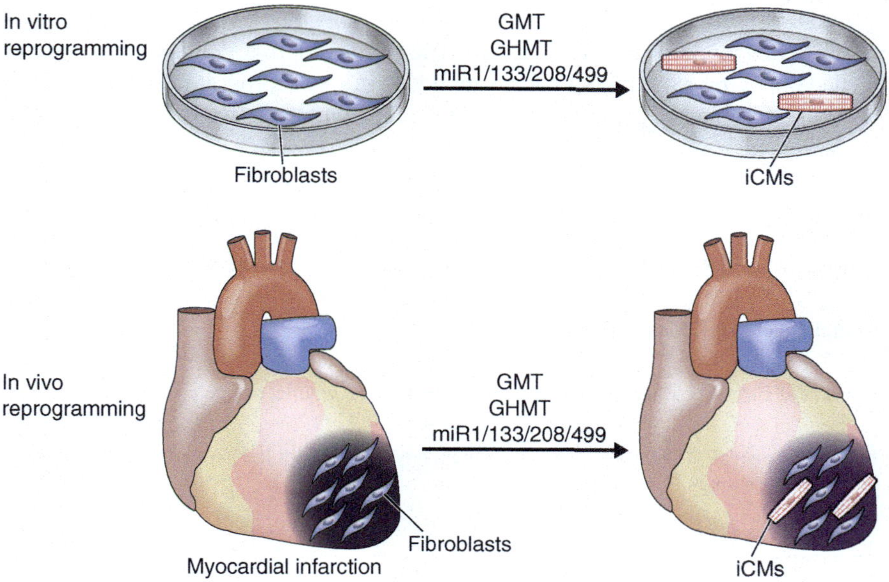

Fig. 6.4 In vitro and in vivo cardiac reprogramming

Conclusions

The heart consists of various types of cells, and cardiomyocytes account for only 30% of the total cells in the heart. Because of the low regenerative capability of adult cardiomyocytes, once they are damaged, they will be replaced by fibroblasts and fibrous tissue, resulting in impaired cardiac function and arrhythmia. Cardiac regenerative medicine has traditionally focused on procedures in which cardiomyocytes are prepared ex vivo and transplanted into the damaged heart. However, if a procedure for the direct reprogramming of endogenous CFs into cardiomyocytes using reprogramming factors is developed, it could become a promising therapeutic approach.

To date, our group and other groups have reported cardiac reprogramming in vitro and in vivo. However, the current reprogramming efficiency is not sufficient, especially in human cardiac reprogramming, and further improvement of the protocol and a better understanding of the molecular mechanisms are needed. In addition, the protocol using retrovirus vectors may potentially alter cellular function due to insertional mutagenesis by the integration of transgenes, requiring further investigation for safety concerns. Nevertheless, since our first discovery of cardiac reprogramming in 2010, there has been enormous progress in this direct reprogramming field as discussed in this chapter. We believe that future research can overcome these challenges, leading to the practical use of direct cardiac reprogramming in the regeneration of the failing heart.

Direct cardiac reprogramming converts endogenous CFs directly into cardiomyocytes by defined factors in vivo. The cardiac reprogramming factors identified in in vitro experiments can be applied to in vivo reprogramming. The fibrous tissue that mainly consists of CFs and extracellular matrix can be repaired by cardiac reprogramming.

Acknowledgments M.I. was supported by research grants from JST CREST, AMED PRIME, JSPS, Keio University Program for the Advancement of Next Generation Research Projects, Banyu Life Science, Senshin Medical Research Foundation, and Takeda Science Foundation.

Compliance with Ethical Standards

Conflict of Interest The authors declare that they have no conflict of interest.

Ethical Approval This article does not contain any studies with human participants performed by any of the authors.

References

Addis RC, Epstein JA (2013) Induced regeneration – the progress and promise of direct reprogramming for heart repair. Nat Med 19:829–836

Addis RC, Ifkovits JL, Pinto F, Kellam LD, Esteso P, Rentschler S, Christoforou N, Epstein JA, Gearhart JD (2013) Optimization of direct fibroblast reprogramming to cardiomyocytes using calcium activity as a functional measure of success. J Mol Cell Cardiol 60:97–106

Burridge PW, Matsa E, Shukla P, Lin ZC, Churko JM, Ebert AD, Lan F, Diecke S, Huber B, Mordwinkin NM et al (2014) Chemically defined generation of human cardiomyocytes. Nat Methods 11:855–860

Cahan P, Li H, Morris SA, Lummertz da Rocha E, Daley GQ, Collins JJ (2014) CellNet: network biology applied to stem cell engineering. Cell 158:903–915

Carey BW, Markoulaki S, Hanna JH, Faddah DA, Buganim Y, Kim J, Ganz K, Steine EJ, Cassady JP, Creyghton MP et al (2011) Reprogramming factor stoichiometry influences the epigenetic state and biological properties of induced pluripotent stem cells. Cell Stem Cell 9:588–598

Chen JX, Krane M, Deutsch MA, Wang L, Rav-Acha M, Gregoire S, Engels MC, Rajarajan K, Karra R, Abel ED et al (2012) Inefficient reprogramming of fibroblasts into cardiomyocytes using gata4, mef2c, and tbx5. Circ Res 111:50–55

Chong JJ, Yang X, Don CW, Minami E, Liu YW, Weyers JJ, Mahoney WM, Van Biber B, Cook SM, Palpant NJ et al (2014) Human embryonic-stem-cell-derived cardiomyocytes regenerate non-human primate hearts. Nature 510:273–277

Davis RL, Weintraub H, Lassar AB (1987) Expression of a single transfected cDNA converts fibroblasts to myoblasts. Cell 51:987–1000

Fu JD, Stone NR, Liu L, Spencer CI, Qian L, Hayashi Y, Delgado-Olguin P, Ding S, Bruneau BG, Srivastava D (2013) Direct reprogramming of human fibroblasts toward a cardiomyocyte-like state. Stem Cell Rep 1:235–247

Han JK, Chang SH, Cho HJ, Choi SB, Ahn HS, Lee J, Jeong H, Youn SW, Lee HJ, Kwon YW et al (2014) Direct conversion of adult skin fibroblasts to endothelial cells by defined factors. Circulation 130:1168–1178

Hirai H, Tani T, Kikyo N (2010) Structure and functions of powerful transactivators: VP16, MyoD and FoxA. Int J Dev Biol 54:1589–1596

Hirai H, Katoku-Kikyo N, Keirstead SA, Kikyo N (2013) Accelerated direct reprogramming of fibroblasts into cardiomyocyte-like cells with the MyoD transactivation domain. Cardiovasc Res 100:105–113

Huang P, He Z, Ji S, Sun H, Xiang D, Liu C, Hu Y, Wang X, Hui L (2011) Induction of functional hepatocyte-like cells from mouse fibroblasts by defined factors. Nature 475(7356):386–389

Huang P, Zhang L, Gao Y, He Z, Yao D, Wu Z, Cen J, Chen X, Liu C, Hu Y et al (2014) Direct reprogramming of human fibroblasts to functional and expandable hepatocytes. Cell Stem Cell 14:370–384

Ieda M, Tsuchihashi T, Ivey KN, Ross RS, Hong TT, Shaw RM, Srivastava D (2009) Cardiac fibroblasts regulate myocardial proliferation through beta1 integrin signaling. Dev Cell 16:233–244

Ieda M, Fu JD, Delgado-Olguin P, Vedantham V, Hayashi Y, Bruneau BG, Srivastava D (2010) Direct reprogramming of fibroblasts into functional cardiomyocytes by defined factors. Cell 142:375–386

Ifkovits JL, Addis RC, Epstein JA, Gearhart JD (2014) Inhibition of TGFbeta signaling increases direct conversion of fibroblasts to induced cardiomyocytes. PLoS One 9:e89678

Inagawa K, Ieda M (2013) Direct reprogramming of mouse fibroblasts into cardiac myocytes. J Cardiovasc Transl Res 6:37–45

Inagawa K, Miyamoto K, Yamakawa H, Muraoka N, Sadahiro T, Umei T, Wada R, Katsumata Y, Kaneda R, Nakade K et al (2012) Induction of cardiomyocyte-like cells in infarct hearts by gene transfer of Gata4, Mef2c, and Tbx5. Circ Res 111:1147–1156

Islas JF, Liu Y, Weng KC, Robertson MJ, Zhang S, Prejusa A, Harger J, Tikhomirova D, Chopra M, Iyer D et al (2012) Transcription factors ETS2 and MESP1 transdifferentiate human dermal fibroblasts into cardiac progenitors. Proc Natl Acad Sci U S A 109:13016–13021

Jayawardena TM, Egemnazarov B, Finch EA, Zhang L, Payne JA, Pandya K, Zhang Z, Rosenberg P, Mirotsou M, Dzau VJ (2012) MicroRNA-mediated in vitro and in vivo direct reprogramming of cardiac fibroblasts to cardiomyocytes. Circ Res 110:1465–1473

Jayawardena TM, Finch EA, Zhang L, Zhang H, Hodgkinson C, Pratt RE, Rosenberg PB, Mirotsou M, Dzau VJ (2014) MicroRNA induced cardiac reprogramming in vivo: evidence for mature cardiac myocytes and improved cardiac function. Circ Res 116(3):418–424

Kattman SJ, Witty AD, Gagliardi M, Dubois NC, Niapour M, Hotta A, Ellis J, Keller G (2011) Stage-specific optimization of activin/nodal and BMP signaling promotes cardiac differentiation of mouse and human pluripotent stem cell lines. Cell Stem Cell 8:228–240

Lalit PA, Hei DJ, Raval AN, Kamp TJ (2014) Induced pluripotent stem cells for post-myocardial infarction repair: remarkable opportunities and challenges. Circ Res 114:1328–1345

Laugwitz KL, Moretti A, Lam J, Gruber P, Chen Y, Woodard S, Lin LZ, Cai CL, Lu MM, Reth M et al (2005) Postnatal isl1+ cardioblasts enter fully differentiated cardiomyocyte lineages. Nature 433:647–653

Li R, Liang J, Ni S, Zhou T, Qing X, Li H, He W, Chen J, Li F, Zhuang Q et al (2010) A mesenchymal-to-epithelial transition initiates and is required for the nuclear reprogramming of mouse fibroblasts. Cell Stem Cell 7:51–63

Liu N, Olson EN (2010) MicroRNA regulatory networks in cardiovascular development. Dev Cell 18:510–525

Liu H, Zhu F, Yong J, Zhang P, Hou P, Li H, Jiang W, Cai J, Liu M, Cui K et al (2008) Generation of induced pluripotent stem cells from adult rhesus monkey fibroblasts. Cell Stem Cell 3:587–590

Mathison M, Gersch RP, Nasser A, Lilo S, Korman M, Fourman M, Hackett N, Shroyer K, Yang J, Ma Y et al (2012) In vivo cardiac cellular reprogramming efficacy is enhanced by angiogenic preconditioning of the infarcted myocardium with vascular endothelial growth factor. J Am Heart Assoc 1:e005652

Morris SA, Cahan P, Li H, Zhao AM, San Roman AK, Shivdasani RA, Collins JJ, Daley GQ (2014) Dissecting engineered cell types and enhancing cell fate conversion via CellNet. Cell 158:889–902

Muraoka N, Ieda M (2014) Direct reprogramming of fibroblasts into myocytes to reverse fibrosis. Annu Rev Physiol 76:21–37

Muraoka N, Ieda M (2015) Stoichiometry of transcription factors is critical for cardiac reprogramming. Circ Res 116:216–218

Muraoka N, Yamakawa H, Miyamoto K, Sadahiro T, Umei T, Isomi M, Nakashima H, Akiyama M, Wada R, Inagawa K et al (2014) MiR-133 promotes cardiac reprogramming by directly repressing Snai1 and silencing fibroblast signatures. EMBO J 33:1565–1581

Nam YJ, Song K, Luo X, Daniel E, Lambeth K, West K, Hill JA, Dimaio JM, Baker LA, Bassel-Duby R et al (2013) Reprogramming of human fibroblasts toward a cardiac fate. Proc Natl Acad Sci U S A 110:5588–5593

Nam YJ, Lubczyk C, Bhakta M, Zang T, Fernandez-Perez A, McAnally J, Bassel-Duby R, Olson EN, Munshi NV (2014) Induction of diverse cardiac cell types by reprogramming fibroblasts with cardiac transcription factors. Development 141:4267–4278

Okano H, Nakamura M, Yoshida K, Okada Y, Tsuji O, Nori S, Ikeda E, Yamanaka S, Miura K (2013) Steps toward safe cell therapy using induced pluripotent stem cells. Circ Res 112:523–533

Polo JM, Anderssen E, Walsh RM, Schwarz BA, Nefzger CM, Lim SM, Borkent M, Apostolou E, Alaei S, Cloutier J et al (2012) A molecular roadmap of reprogramming somatic cells into iPS cells. Cell 151:1617–1632

Protze S, Khattak S, Poulet C, Lindemann D, Tanaka EM, Ravens U (2012) A new approach to transcription factor screening for reprogramming of fibroblasts to cardiomyocyte-like cells. J Mol Cell Cardiol 53:323–332

Qian L, Huang Y, Spencer CI, Foley A, Vedantham V, Liu L, Conway SJ, Fu JD, Srivastava D (2012) In vivo reprogramming of murine cardiac fibroblasts into induced cardiomyocytes. Nature 485:593–598

Qian L, Berry EC, Fu JD, Ieda M, Srivastava D (2013) Reprogramming of mouse fibroblasts into cardiomyocyte-like cells in vitro. Nat Protoc 8:1204–1215

Rackham OJ, Firas J, Fang H, Oates ME, Holmes ML, Knaupp AS, Consortium F, Suzuki H, Nefzger CM, Daub CO et al (2016) A predictive computational framework for direct reprogramming between human cell types. Nat Genet 48(3):331–335

Riddell J, Gazit R, Garrison BS, Guo G, Saadatpour A, Mandal PK, Ebina W, Volchkov P, Yuan GC, Orkin SH et al (2014) Reprogramming committed murine blood cells to induced hematopoietic stem cells with defined factors. Cell 157:549–564

Sadahiro T, Yamanaka S, Ieda M (2015) Direct cardiac reprogramming: progress and challenges in basic biology and clinical applications. Circ Res 116:1378–1391

Saga Y, Miyagawa-Tomita S, Takagi A, Kitajima S, Miyazaki J, Inoue T (1999) MesP1 is expressed in the heart precursor cells and required for the formation of a single heart tube. Development 126:3437–3447

Sekiya S, Suzuki A (2011) Direct conversion of mouse fibroblasts to hepatocyte-like cells by defined factors. Nature 475:390–393

Shiba Y, Fernandes S, Zhu WZ, Filice D, Muskheli V, Kim J, Palpant NJ, Gantz J, Moyes KW, Reinecke H et al (2012) Human ES-cell-derived cardiomyocytes electrically couple and suppress arrhythmias in injured hearts. Nature 489:322–325

Song K, Nam YJ, Luo X, Qi X, Tan W, Huang GN, Acharya A, Smith CL, Tallquist MD, Neilson EG et al (2012) Heart repair by reprogramming non-myocytes with cardiac transcription factors. Nature 485:599–604

Srivastava D, Ieda M (2012) Critical factors for cardiac reprogramming. Circ Res 111:5–8

Takahashi K, Yamanaka S (2006) Induction of pluripotent stem cells from mouse embryonic and adult fibroblast cultures by defined factors. Cell 126:663–676

Takahashi K, Okita K, Nakagawa M, Yamanaka S (2007a) Induction of pluripotent stem cells from fibroblast cultures. Nat Protoc 2:3081–3089

Takahashi K, Tanabe K, Ohnuki M, Narita M, Ichisaka T, Tomoda K, Yamanaka S (2007b) Induction of pluripotent stem cells from adult human fibroblasts by defined factors. Cell 131:861–872

Unternaehrer JJ, Zhao R, Kim K, Cesana M, Powers JT, Ratanasirintrawoot S, Onder T, Shibue T, Weinberg RA, Daley GQ (2014) The epithelial-mesenchymal transition factor SNAIL paradoxically enhances reprogramming. Stem Cell Rep 3:691–698

Vierbuchen T, Ostermeier A, Pang ZP, Kokubu Y, Sudhof TC, Wernig M (2010) Direct conversion of fibroblasts to functional neurons by defined factors. Nature 463:1035–1041

Wada R, Muraoka N, Inagawa K, Yamakawa H, Miyamoto K, Sadahiro T, Umei T, Kaneda R, Suzuki T, Kamiya K et al (2013) Induction of human cardiomyocyte-like cells from fibroblasts by defined factors. Proc Natl Acad Sci U S A 110:12667–12672

Wang L, Liu Z, Yin C, Asfour H, Chen OM, Li Y, Bursac N, Liu J, Qian L (2014) Stoichiometry of Gata4, Mef2c, and Tbx5 influences the efficiency and quality of induced cardiac myocyte reprogramming. Circ Res 116(2):237–244

Yamakawa H, Muraoka N, Miyamoto K, Sadahiro T, Isomi M, Haginiwa S, Kojima H, Umei T, Akiyama M, Kuishi Y et al (2015) Fibroblast growth factors and vascular endothelial growth factor promote cardiac reprogramming under defined conditions. Stem Cell Rep 5:1128–1142

Yang X, Pabon L, Murry CE (2014) Engineering adolescence: maturation of human pluripotent stem cell-derived cardiomyocytes. Circ Res 114:511–523

Yoshida Y, Yamanaka S (2012a) An emerging strategy of gene therapy for cardiac disease. Circ Res 111:1108–1110

Yoshida Y, Yamanaka S (2012b) Labor pains of new technology: direct cardiac reprogramming. Circ Res 111:3–4

Zhao Y, Samal E, Srivastava D (2005) Serum response factor regulates a muscle-specific microRNA that targets Hand2 during cardiogenesis. Nature 436:214–220

Zhao Y, Ransom JF, Li A, Vedantham V, von Drehle M, Muth AN, Tsuchihashi T, McManus MT, Schwartz RJ, Srivastava D (2007) Dysregulation of cardiogenesis, cardiac conduction, and cell cycle in mice lacking miRNA-1-2. Cell 129:303–317

Zhao Y, Londono P, Cao Y, Sharpe EJ, Proenza C, O'Rourke R, Jones KL, Jeong MY, Walker LA, Buttrick PM et al (2015) High-efficiency reprogramming of fibroblasts into cardiomyocytes requires suppression of pro-fibrotic signalling. Nat Commun 6:8243

Zhou Q, Brown J, Kanarek A, Rajagopal J, Melton DA (2008) In vivo reprogramming of adult pancreatic exocrine cells to beta-cells. Nature 455:627–632

Zhou H, Dickson ME, Kim MS, Bassel-Duby R, Olson EN (2015) Akt1/protein kinase B enhances transcriptional reprogramming of fibroblasts to functional cardiomyocytes. Proc Natl Acad Sci U S A 112:11864–11869

7

Application of the Suspension Culture System for Scale-Up Manufacture of hPSCs and hPSC-Derived Cardiomyocytes

Vincent C. Chen, Larry A. Couture, and Joseph Gold

Abstract

Establishment of a scalable, robust, and GMP-compatible manufacturing process for human pluripotent stem cell-derived cardiomyocytes (hPSC-CMs) has been a bottleneck for the progress of cell therapy for heart diseases. The adherent cell culture platforms have been well developed for hPSC maintenance and cardiac differentiation. However, the two-dimensional culture system is limited by its scalability, hindering its application for scale-up cell production. Recent advances in development of suspension culture systems, which provide the advantage of scalability, have driven hPSC-CMs beyond bench research into preclinical development. With the suspension platform, the processes from hPSC expansion to cardiac differentiation have been streamlined for the hPSC-CM production. A fully suspension-based process avoids extensive labor associated with the conventional adherent culture and lowers production costs by reducing reagents, space, and operators. These advantages render this manufacturing process more manageable, cost-effective, labor-effective, and practical for large-scale cell production. In this chapter, we will review current status of the development of the suspension culture system for hPSC-CM production.

V.C. Chen (✉) • J. Gold
Center for Biomedicine and Genetics, Beckman Research Institute of City of Hope, 1500 E. Duarte Road, Duarte, CA 91010, USA
e-mail: cychen@coh.org

L.A. Couture
Arrogene, 8560 W. Sunset Blvd. Suite 424, Los Angeles, CA 90069, USA

Orbsen Therapeutics, Orbsen Building, NUIG, University Road Galway, Ireland, Ireland

M. Ieda, W.-H. Zimmermann (eds.), *Cardiac Regeneration*, Cardiac and Vascular Biology 4, DOI 10.1007/978-3-319-56106-6_7

7.1 Introduction

Human pluripotent stem cells (hPSCs), including human embryonic stem cells (hESCs) and human induced pluripotent stem cells (hiPSCs), are increasingly being used as sources to derive desired cell lineages for clinical trials of cell therapies. The availability of sufficient supplies of hPSCs and hPSC-derived cells to support preclinical and clinical studies is one of the key determinants for the advance of their therapeutic application. In the case of cell therapy for heart disease, it has been proposed that transplantation of a single dose on the order of 1 billion cells of hPSC-derived cardiomyocytes (hPSC-CMs) will be required to show clinical effects (Chong et al. 2014; Laflamme and Murry 2005). In order to meet the need for the enormous quantities of hPSCs and hPSC-CMs for IND-enabling large animal preclinical studies and clinical trials, it is essential to establish a robust and scalable process to produce the cells at a clinically meaningful scale.

The development of cell culture systems for manufacture of undifferentiated hPSCs and hPSC-CMs has made significant progress in the last decade. Two major types of cell culture platforms—2-dimensional (2D) adherent and 3-dimensional (3D) suspension cell culture systems—have been used to develop the processes for hPSC expansion and cardiac differentiation. The 3D suspension culture system is well-known for its scalability advantage and is commonly used in industry for commercial scales of production. In the following sections, we will provide an overview of current culture systems and focus on the process development of suspension culture systems for scale-up production of hPSCs and hPSC-CMs.

7.2 hPSC Cultures for Cell Expansion

7.2.1 Two-Dimensional Adherent Cell Culture System

The adherent cell culture platform is most commonly used for maintaining hPSCs. With advances in knowledge and technology, maintaining hPSC in adherent culture has become a relatively easy and routine laboratory process. For clinical purposes, recent developments in this culture system have been aimed at avoiding the use of non-defined materials such as feeder cells, Matrigel™, and serum-containing media. Tremendous progress in establishing culture systems suitable for cGMP compliance has been achieved by the development of chemically defined matrices, media, and reagents for adherent culture systems to meet the requirements for clinical quality (Li et al. 2005; Ludwig et al. 2006; Melkoumian et al. 2010; Rodin et al. 2010; Villa-Diaz et al. 2010; Wang et al. 2007). Regardless, scalability limitations of the adherent cell culture platform at some level still restrict its practical application for mass cell production. The requisite manipulations in handling large numbers of culture flasks for vessel coating, cell seeding, cell passaging, and harvesting are time-consuming and labor intensive. Such a scale-out approach for cell manufacturing not only can cause more variability among individual flasks but also compromise cell quality due to the lengthy process. While adherent cell culture has practical scalability limitations, multilayer culture vessels might potentially provide a solution for large-scale production. Current advances in operating multilayer vessels have achieved some degree of automation for assisting in vessel handling, media change, and cell detachment. However, there is very limited literature clearly demonstrating the feasibility of this system for large-scale production of hPSCs. Further evaluation is needed to determine if the multilayer culture system is practical for mass production of quality hPSCs. In addition to the restricted scalability inherent in the adherent culture system, another major obstacle for scale-up production is the current approach for hPSC passaging. In a majority of hPSC culture conditions, cells must be dissociated into small clusters at passaging in order to achieve better cell survival and plating efficiency. The dissociation method requires a good control of enzymatic treatment and mechanical dissociation of cells to generate clusters of the proper sizes. This process will be very challenging when operating with large amounts of culture vessels and surface area. Additionally, the cell-dissociation approach suffers from problems of uncontrollable cell cluster size and uneven distribution off cells during passaging and thus may engender undesired variability at subsequent stages.

7.2.2 Suspension Cell Culture System

In contrast to adherent culture, the suspension cell culture system is an ideal platform for cell production as it provides great scalability. Unlike the adherent 2D platform, the suspension platform in principle can be scaled up into one single culture vessel for large-scale production. A variety of agitation mechanisms, such as stirring, rocking, rotating, orbital shaking, and air sparging, have been employed to generate a dynamic suspension which can offer a homogeneous and scalable three-dimensional culture condition. Additionally, the nature of the suspension system makes culture sampling, media handling, and cell harvesting easier and more convenient. With the advantages mentioned above, the dynamic suspension culture system combined with bioreactors has been commonly used in industry to handle production scales of up to thousands of liters. Sophisticated bioreactors developed for suspension culture have the capacity to monitor parameters such as gas, pH, temperature, metabolites, and nutrients and control culture conditions. They also provide options for cell feeding, e.g., perfusion or fed batch, to replace manual media change and to achieve better production performance.

In the last decade, development of suspension culture for PSCs has made significant progress. One of the unique growth characteristics of hPSCs is that they need to grow in clusters, as they undergo apoptosis when dispersed to single cells. Development of hPSC suspension cultures has therefore been focused on growing the cells in the form of aggregates.

7.2.2.1 Microcarrier-Based Suspension Culture System

The early success of the hPSC suspension culture was achieved by growing cells as clusters on microcarriers (Phillips et al. 2008; Fernandes et al. 2009; Oh et al. 2009; Chen et al. 2011; Lock and Tzanakakis 2009; Kehoe et al. 2008). The microcarrier-based suspension culture system brings adherent cells cultured on surfaces of microcarriers into suspension for scalability and is therefore an intermediate between traditional adherent culture systems and true suspension cultures. The suspension system still requires significant manipulations that include the need for coating the microcarriers with matrix for cell attachment, the process of cell detachment, and the separation of the liberated cells from the microcarriers at cell passaging and harvesting. The complexity of the processes would be challenging for large-scale production. One concern inherent in this approach is that the effects of residual microcarriers and their allowable levels for clinical applications will need to be further evaluated.

7.2.2.2 Microcarrier-Free Suspension Culture Systems

In contrast, a matrix-independent cell suspension culture system which avoids the complexity of microcarrier-related processes is more favorable for cell production. One critical study showed that Y-27632, a Rho kinase (ROCK) inhibitor, could protect single cell-dissociated hESCs from apoptosis and the cells survived by forming aggregates (Watanabe et al. 2007). Based on this finding, several groups have successfully established microcarrier-free suspension cultures supporting hPSC growth in the form of cell aggregates by addition of Y-27632 (Chen and Couture 2015; Wang et al. 2013; Chen et al. 2012; Zweigerdt et al. 2011; Larijani et al. 2011; Amit et al. 2011; Steiner et al. 2010; Singh et al. 2010; Olmer et al. 2010; Krawetz et al. 2010; Amit et al. 2010). A broad spectrum of hPSC lines has been successfully adapted into suspension cultures with different media conditions and culture vessels by many research groups. Though the scalability of the hPSC suspension culture has been only demonstrated at a scale of 100 mL culture volume in published literature, we have been routinely applying the suspension cultures to produce GMP-grade hESC banks at the billion cell level with liter scale culture vessels.

Though a dynamic suspension system is favored for scale-up culture, some recent reports showed that media mixed with hydrogels can create a static suspension to grow and differentiate hPSCs as spheres in a 3D culture condition without sedimentation (Lei and Schaffer 2013; Zhang et al. 2013; Otsuji et al. 2014). hPSCs seeded in the hydrogel suspension culture either as single cells or in cell clusters also rely on the ROCK inhibitor Y-27632 to enhance cell survival, and the hPSCs form spheroids to expand. The hydrogel-based suspension culture system not only avoids potential shear damage to cells but also confers scalability on the static suspension. One study reported a remarkable 20-fold expansion rate and 2×10^7 cells/mL of cell yield after 5 days of the hPSC culture. While the hydrogel-based suspension culture system holds great potential for scale-up production, further testing in large scale for hydrogel preparation, handling viscous hydrogel-mixed media at media change, cell passaging, and harvesting need to be evaluated for practical application in cell production.

We will discuss the critical considerations for developing an hPSC suspension culture for production in the following sections.

Reagents

For GMP manufacture or clinical application of hPSC-related products, selection of appropriate media and reagents is critical to achieve predictable and consistent quality of cell products fitting therapeutic needs. Chemically defined reagents with documentation of proper quality testing are always preferred for GMP compliance.

Xeno-free reagents, though not required, are preferred for use in cell products aimed at therapeutic purposes. Although application of GMP or clinical grade reagents is not necessary at early stages of development, availability of equivalent reagents in GMP or clinical quality may need to be taken into consideration if the culturing method is intended to integrate into a GMP process. Acquiring reagents from qualified and reliable vendors is also important to ensure continuous supplies for reproducible cell manufacturing.

Media Conditions

As cells grow in aggregates in a dynamic suspension culture, the requirements of media components or concentrations of components could be different from cells growing in a static monolayer culture. Several common hPSC media, such as feeder conditioned-based media, Knockout Serum Replacement™ (KSR)-based media, mTeSR™, StemPro™ hESC SFM, and Essential 8 for adherent cell culture, have been used directly or modified to develop suspension cultures (Chen and Couture 2015; Wang et al. 2013; Chen et al. 2012; Zweigerdt et al. 2011; Larijani et al. 2011; Amit et al. 2011; Steiner et al. 2010; Singh et al. 2010; Olmer et al. 2010; Krawetz et al. 2010; Amit et al. 2010). As commercialized ready-to-use hPSC media provide advantages of convenience, quality control, and minimal lot-to-lot variation, they are favored for GMP production. However, those media are not specifically designed for hPSC suspension culture and are usually qualified based on adherent cell cultures. Qualification of new lots of media for an hPSC suspension culture would be necessary before large-scale manufacturing. Moreover, studies have shown that these media performed differently in support of hPSC suspension cultures. To improve maintenance of the undifferentiated state and expansion rates of hPSCs in suspension cultures, modification of media conditions in some cases is necessary. For instance, addition of rapamycin in mTeSR™ (Krawetz et al. 2010), IL-6 (interleukin-6), and IL-6 receptor chimera or LIF (leukemia inhibitory factor) in KSR-based medium (Amit et al. 2010; Amit et al. 2011) and increased concentrations of bFGF in StemPro™ hESC SFM (Chen et al. 2012) have been shown to provide better conditions for hPSC adaptation in suspension. Development of a common specialized medium for hPSC suspension culture would make the processes of adaptation and expansion more efficient.

Culture Vessels

A variety of vessel types, including plates, Petri dishes, Erlenmeyer flasks, T flasks, spinner flasks, and stirred bioreactors, have been used in static conditions with low attachment surfaces or dynamic platforms with different agitation mechanisms, such as orbital shaking, rocking, and stirring, to carry hPSC suspension cultures. Among these vessel types, spinner flasks and stirred bioreactors with the advantages mentioned above are more suitable for upscaling. We have been routinely using spinner flasks up to the 3 L scale for manufacturing hPSC-related cells, demonstrating scalability of this type of culture vessel. Another alternative is a bag-based bioreactor using wave-induced agitation on a rocking platform (Kalmbach et al. 2011; Oncul et al. 2010; Singh 1999). In contrast to stirred culture vessels, the bag-based vessel is designed to avoid the use of impellers and thus may significantly reduce mechanical force and uneven distribution of shear forces generated by impellers. The bag-based bioreactor has been used for culturing cell lines such as CHO and 293 in the form of single cells in large scale. While bag-based vessels may hold great potential for scale-up cell production, further evaluation to support hPSC culture is necessary.

Agitation Conditions

To generate 3D suspension environments, an agitation mechanism needs to be introduced into the cell culture. When cells are cultured in a hydrodynamic suspension, they will encounter shear force from fluid and mechanical forces caused by impact with solid surfaces in culture vessels. Shear and mechanical forces can have detrimental effects on cells and can also cause unwanted cell differentiation. Selection of proper agitation mechanisms, vessel designs, and optimization of agitation conditions to sufficiently suspend hPSC aggregates with minimal agglomeration and cell damage is essential for development of an efficient hPSC suspension culture system.

Cell Aggregate Size and Cell Passaging

Initial formation of small cell clusters is a critical step for hPSCs to grow in suspension culture. Cell aggregates gain in size due to cell expansion and agglomeration. Several studies suggested that the size of cell aggregates could affect cell quality and growth rates (Amit et al. 2010; Chen et al. 2012; Krawetz et al. 2010; Larijani et al. 2011; Olmer et al. 2010). As cell aggregates grow, at some point, they will reach sizes at which penetration of nutrients and metabolites inside the clusters will be less efficient and will cause cell differentiation and/or apoptosis. Therefore, monitoring cell aggregate size and passaging cell cultures at optimal intervals is critical for maintaining cell quality and expansion pace. The hPSC suspension cultures are usually passaged every 3–7 days when cell aggregate size reaches the range of 250–300 μM. The optimal subculture intervals for hPSCs may need to be determined individually for different cell lines and culture conditions.

Cell-Dissociation Methods for Passaging

Two approaches to dissociate cell aggregates have been used for cell passaging. One is to partially break down cell aggregates into small clusters by a strainer without enzyme treatment. The advantages of this approach are that it is free of enzyme and relatively easy. However, mechanical dissociation of cell aggregates could be detrimental to cells. Partially dissociated cell clusters are not only difficult to quantify but also result in poor control of cell aggregate sizes, which may lead to inconsistent cell cultures. For upscaling application, mechanical dissociation could also be challenging due to scale limit of strainers. The other common approach is enzymatic dissociation of cell aggregates into single cells. In contrast to mechanical dissociation, the single cell-dissociation approach is quantitative and can generate more controlled and homogeneous sizes of cell aggregates. Generating homogeneous sizes of cell aggregates is not only critical for reproducible quality of cell cultures but also critical for directed differentiation of cell clusters. With those advantages, the single cell-dissociation approach could be favorable for large-scale GMP production.

Stepwise Scale-Up of hPSC Suspension Cultures

In order to achieve large-scale cultures, the hPSC suspension culture usually requires stepwise expansion into larger-scale vessels. This cell expansion process may require using different types of vessels and agitation mechanisms. One study has shown inconsistency of cell expansion rates at scaling up hPSC suspension cultures from a static 6-well plate to an orbital-shaker Petri dish and finally to a stirred spinner, indicating the need for optimization for individual suspension platforms (Singh et al. 2010). Agitation condition and shear force are critical for the performance of an hPSC suspension culture and may need to be optimized for different scales and designs of culture vessels. To minimize the efforts involved in the optimization, use of similar or identical design of vessels that differ only in size and that use the same agitation mechanism during scale-up would simplify the conversion or estimation of agitation conditions among different scales.

7.3 Scale-Up Production of hPSC-CMs

7.3.1 Directed Differentiation of Cardiac Cells

Substantial advances in developing methods for cardiac differentiation from hPSCs have accelerated the application of hPSC-CMs to clinical translation. Several cardiac induction schemes have been designed based on knowledge of cardiomyogenesis during embryo development (Burridge et al. 2012). Temporal modulation of Activin/Nodal/TGF-β, BMP, and Wnt pathways to stepwise induce hPSCs through mesodermal and cardiac progenitor cells to cardiomyocytes (CMs) is a common strategy to direct cardiac differentiation. By this induction strategy, several procedures have been able to efficiently differentiate hPSCs into cardiomyocytes with more than 80% purity (Burridge et al. 2012; Fonoudi et al. 2015; Kattman et al. 2011; Kempf et al. 2014; Lian et al. 2012; Yang et al. 2008; Zhang et al. 2012; Zhu et al. 2011; Chen et al. 2015). Along with the advances in induction methods, materials and reagents used for cardiac differentiation have also been improved to comply with cGMP and clinical purposes (Burridge et al. 2014; Liang et al. 2013). While current differentiation procedures are able to generate high purity of hPSC-CMs with defined reagents, manufacture of the cells at a clinically meaningful scale is still challenging.

7.3.2 hPSC-CM Production with Two-Dimensional Differentiation Platforms

Several cardiac differentiation methods have been developed using monolayer cell culture on a 2D platform, in which cells could be induced efficiently and homogeneously (Burridge et al. 2014; Lian et al. 2013; Zhang et al. 2012; Zhu et al. 2011). Recent method developments have used defined matrices, such as vitronectin-based synthetic peptides and recombinant laminin, to replace Matrigel™ and feeders for cell attachment (Burridge et al. 2014). Growth factors used for cardiac induction have also been replaced by small-molecule compounds. One representative study has revealed a simple and efficient method for cardiac differentiation using a defined synthetic matrix (Synthemax™) to support cell attachment and sequential induction with two small molecules, CHIR99021 and IWP, modulating Wnt signaling to direct differentiation (Lian et al. 2012). Based on this method, a minimal media condition containing only RPMI 1640, ascorbic acid, and recombinant albumin was shown to be sufficient for cardiac differentiation (Burridge et al. 2014). Despite the improvement of reagents and simplification of the differentiation method, the 2D adherent culture system still has inherent limited scalability. As mentioned above, multilayer cell culture vessels hold great potential for large-scale production of hPSC-CMs. There is currently no report showing cardiac differentiation using this vessel format; it will require more investigation to evaluate whether the platform can serve for large-scale hPSC-CM production. There are some foreseeable concerns in using the multilayer vessel for cell manufacture, including homogeneity of media volume, cell seeding, and cell growth characteristics among different layers. These factors will affect cell density in different layers of the culture vessels. Cardiac induction on a 2D culture platform is highly dependent on cell density. Cell culture variability among different layers may therefore significantly affect differentiation efficiency and production consistency. In addition, cost and labor for vessel coating, plus difficulties in cell passaging and harvesting, also need to be considered in using the multilayer vessel for large-scale production.

7.3.3 hPSC-CM Production with Suspension Differentiation Platforms

7.3.3.1 Microcarrier-Based Suspension Culture for Cardiac Differentiation

Cardiac differentiation of hPSCs has been demonstrated with a dynamic microcarrier-based suspension culture system using spinner flasks (in 50 mL scale) or T25 flasks (in 10 mL scale) with rocking agitation (Lam et al. 2014; Ting et al. 2014). With an induction method based on Wnt regulation using the small molecules CHIR99021 and IWP2, the differentiation process was able to produce 1.2–1.9 x 10^6 cells/mL of hPSC-CMs with an average purity of 47–66% estimated by the cardiac troponin T-positive ($cTNT^+$) cell population. More importantly, this production process streamlined cell expansion and differentiation within a continuous suspension culture, making the cell production more efficient. However, as mentioned previously, processes associated with the microcarrier-based suspension culture, including matrix coating, cell attachment, and cell separation from the microcarriers, could be time-consuming and labor-intensive when operating a large-scale production.

7.3.3.2 Microcarrier-Free Suspension Culture for Cardiac Differentiation

The embryoid body (EB) formation assay is commonly used to demonstrate pluripotency of hPSCs. Cardiac induction procedures based on EB formation method with stepwise treatments of growth factors and small molecules have been shown to efficiently generate >80% purity of cardiomyocytes in small-scale static suspension and in dynamic suspension using a stirred bioreactor with volumes up to 100 mL (Kattman et al. 2011; Yang et al. 2008; Hemmi et al. 2014). By the conventional method, EB formation efficiency is low and the sizes of EBs are uncontrollable when set up from hPSC adherent cultures. These disadvantages potentially affect process reproducibility and product consistency. To improve the EB method, the ROCK inhibitor Y27632 has been used to enhance EB formation (Hemmi et al. 2014). In addition, the use of spinner flasks with proper agitation rates has also been applied to the process of EB formation to reduce agglomeration and to make the EB size more controllable (Rungarunlert et al. 2009). While the reports successfully demonstrated efficient cardiac differentiation in suspension cultures, consistency and reproducibility of the EB method for cardiac differentiation will need to be further tested. Moreover, those reported EB methods are based on cells derived from hPSCs in adherent culture, which may limit the scalability of the EB method in the hPSC-CM production.

Recently three reports, including our own, have successfully demonstrated integration of undifferentiated hPSC expansion and cardiac differentiation into a cell aggregate-based suspension culture system (Fonoudi et al. 2015; Chen et al. 2015; Kempf et al. 2014). In these studies, undifferentiated cell aggregates generated from the hPSC suspension culture were used as a cell format for cardiac induction in a continuous suspension culture. Taking advantage of the sizes of small molecules, which may penetrate the cell aggregate and induce cells inside more efficiently than growth factors, CHIR99021 and IWP were used to stepwise regulate Wnt pathways to generate hPSC-CMs. By this induction approach, one report showed a purity of up to ~60% hPSC-CMs generated in static suspension culture in 12-well plates and in dynamic suspension culture in rotated Erlenmeyer flasks (Kempf et al. 2014). The report further demonstrated a scale-up differentiation process to 100 mL suspension culture with spinner bioreactors. The results showed a purity of 68.6 ± 8.7% (ranging 53.9–84.1%) $cTNT^+$ cells from one hESC line and 58% ± 30% (ranging 27.2–88.3%) $cTNT^+$ cells from one hiPSC line, with hPSC-CM yield average of ~4 x 10^5 cells/mL. Another research group used the same Wnt modulation method but also inhibited TGF-β and activated sonic hedgehog pathways by adding small molecules at the stage of Wnt inhibition to generate hPSC-CMs (Fonoudi et al. 2015). With the modified induction method, the report demonstrated that the undifferentiated cell aggregates generated from the suspension culture can be induced to nearly 100% of beating aggregates in static suspension culture for 9 hPSC lines. The report analyzed four of these cell lines and showed up to 90% $cTnT^+$ cells. The authors further scaled up the differentiation process to 100 mL with a spinner bioreactor for one cell line and achieved purity of up to 85% alpha myosin heavy chain $(\alpha MHC)^+$ cells, though no cell yield was reported. With a small-molecule induction method that modulates only the Wnt pathway, we also reported that hPSCs expanded in suspension culture in the form of cell aggregates can efficiently differentiate into cardiac cells in suspension using spinner flasks for culture scales from 100 mL, to 500mL, to 1 L (Chen et al. 2015). Using the H7 hESC line, we were able to produce 90 ± 6% $cTnT^+$ cells with cell yields of 1.8 ± 0.2 × 10^6 cells/mL in 125 mL spinner flasks (n = 12), 92 ± 3% $cTnT^+$ cells with 2.3 ± 0.3 × 10^6 cells/mL in 500 mL spinner flasks (n = 3), and 96 ± 3% $cTnT^+$ cells with 1.4 ± 0.4 × 10^6 cells/mL in 1 L spinner flasks (n = 19). We also showed that the ESI-017 hESC line and a hiPSC line differentiated in 100 mL suspension cultures can generate 91 ± 4% $cTnT^+$ cells with a cell yield of 7.5 ± 2.5 × 10^5 cells/mL and 94 ± 4% cTnT with a cell yield of 2.0 ± 0.5 × 10^6 cells/mL, respectively. Recently, we successfully expanded H7 cells to 3 L culture scale in a 3 L spinner flask and produced approximately 3 billion hPSC-CMs with >90% purity (Table 7.1). The manufacturing process we established is able to expand hPSCs from 125 mL, to 500 mL, to 1 L, to 3 L spinner flasks and to produce hPSC-CMs at a desired scale (Fig. 7.1). The process has been

Table 7.1 Purity and cell yield for large-scale production of hPSC-CMs in 1 L and 3 L suspension culture

Culture scale	cTnT	Cell yield (cells/ml)
1 L	96 ± 4% (n = 26)	1.4 ± 0.4x10^6 (n = 26)
3 L	96 ± 4% (n = 3)	9.5 ± 0.8x10^5 (n = 3)

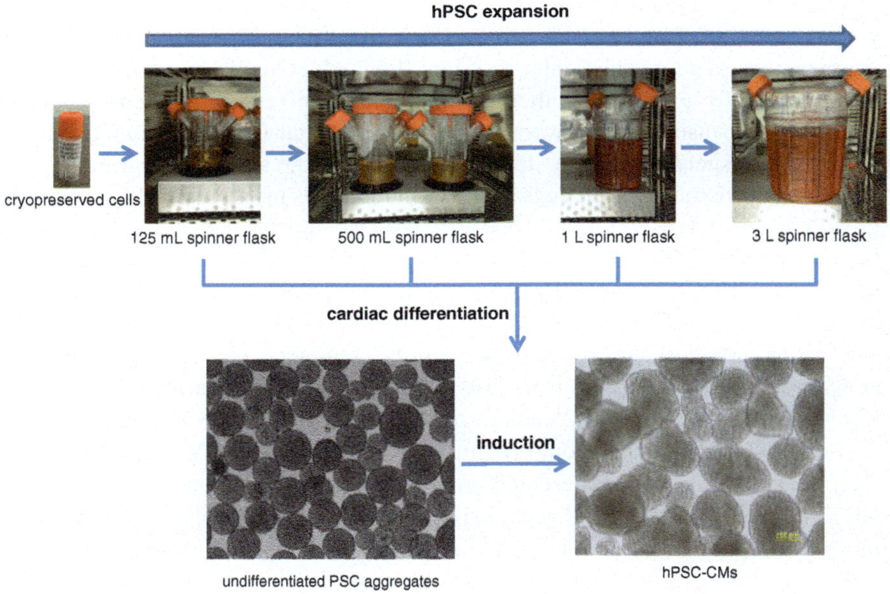

Fig. 7.1 *Scale-up manufacture of hPSC-CMs with the suspension culture system using spinner flasks*. hPSCs thawed from a cell bank were first grown and expanded in a suspension culture. The cell culture was stepwise scaled up from 125 mL spinner flasks to 500 mL, 1 L, and 3 L spinner flasks. The undifferentiated hPSC aggregates generated in the suspension culture at a target scale were then directly induced for cardiac differentiation to produce hPSC-CMs. In our manufacturing process, typically the undifferentiated hPSCs were passaged every 3 days with three- to fivefold of cell expansion and would require 2–3 weeks to expand into 3 L suspension culture for cardiac differentiation, which would take another 3 weeks to generate hPSC-CMs for a production

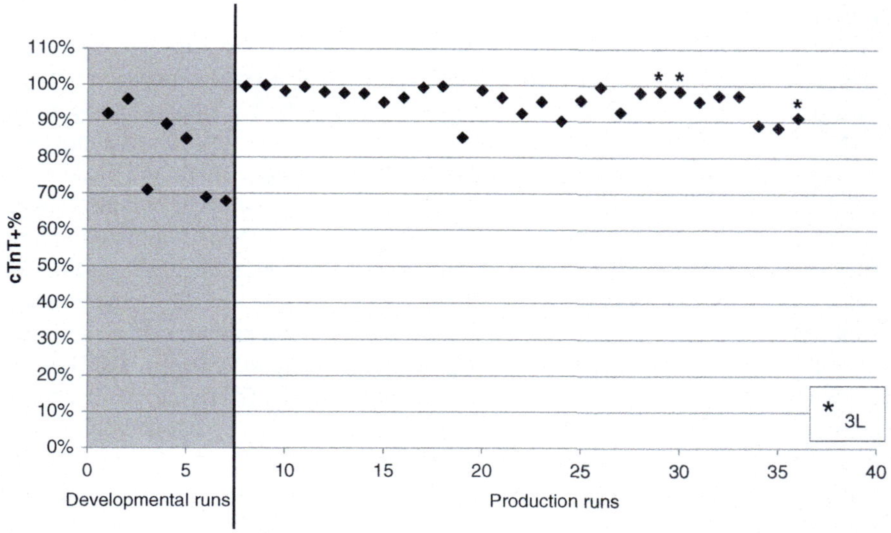

Fig. 7.2 *Reproducibility and consistency of hPSC-CM manufacture with liter scale of suspension cultures*. With culture conditions optimized in developmental runs, the process for large-scale hPSC-CM production has been validated by production runs using 1 L and 3 L spinner flasks. The purity of cell products was evaluated by percentage of cardiac troponin T (cTnT)-positive cells. H7 cells were used in the developmental and production runs (*x*-axis, percentage of $cTnT^{+}$ cells; *y*-axis, developmental and production runs; the asterisk symbol indicates the 3 L production run)

validated by a total of 26 runs of 1 L and 3 runs of 3 L hPSC-CM production generating cell products with superior cell purity, yield, and consistency (Fig. 7.2, Table 7.1). Taken together, these studies strongly indicate that the continuous processes of hPSC expansion and cardiac differentiation in the cell aggregate-based suspension culture system are scalable and can be applied to mass production of hPSC-CMs.

7.3.3.3 Considerations and Challenges for Development of the Cell Aggregate-Based Suspension Culture for Cardiac Differentiation

Conventionally, cardiac induction is performed in a static culture with adherent cells or EBs, and several induction methods for cardiac differentiation in the dynamic suspension culture have been developed based on those devised for static culture conditions. However, in hydrodynamic conditions, the biology of hPSCs; homogeneity of pH, gas, nutrients, and metabolites in the culture; and differentiation kinetics may not be the same as in the static culture condition. Moreover, induction of cells in the form of cell aggregates appears to be more complex than induction of the monolayer cell culture. These factors could potentially change the scheme of the optimal differentiation conditions. We will discuss the considerations for developing the method for cardiac differentiation in the cell aggregate-based suspension culture in the following sections.

Agitation Conditions

In the dynamic suspension culture, agitation generates a homogeneous physicochemical culture condition and avoids cell aggregates/EBs agglomeration but also causes shear and mechanical forces, which may affect hPSC differentiation. Shear and mechanical forces play a critical role in directing cell fate and morphogenesis from gastrulation to cardiogenesis during embryogenesis (Paluch and Heisenberg 2009; Patwari and Lee 2008; Ingber 2006). While effects of shear and mechanical forces on cardiac differentiation of hPSCs are not well studied, some reports have suggested agitation conditions can affect cardiac differentiation in the suspension culture (Ting et al. 2014; Geuss and Suggs 2013; Shafa et al. 2011; Niebruegge et al. 2009; Chen et al. 2015). By comparing different agitation mechanisms using suspension culture vessels such as roller bottles and spinner flasks with bulb or paddle types of impellers, one study showed that hESCs differentiated in these suspension platforms exhibited evident variation in cell yield and cardiac differentiation even at the same agitation speed (Niebruegge et al. 2009). In our previous study, we also found a significant difference in cardiac differentiation efficiency when the same induction scheme was applied to different scales and designs of the spinner flasks. These results imply that the agitation condition needs to be adjusted individually for different designs of the suspension culture vessels to support cardiac differentiation.

Optimization of agitation conditions is rarely mentioned in studies of cardiac differentiation in the dynamic suspension culture. One study using a rocking platform for cardiac differentiation in the microcarrier-based suspension culture showed that an intermittent rocking condition rather than a static or a continuous rocking supported better cardiac differentiation. Similarly, our previous study using spinner flasks demonstrated that optimal cardiac differentiation in dynamic suspension culture was achieved only within a limited range of stirring rates (Chen et al. 2015). These studies indicated that agitation condition is a critical parameter for optimization of cardiac differentiation in the suspension culture, though the mechanism for the effects of agitation or shear/mechanical forces on cardiac differentiation is not clear.

One challenge for cardiac differentiation in dynamic suspension is the conversion or estimation of a proper agitation condition among different scales or designs of culture vessels. In our previous study, we noticed that for individual scales and designs of suspension vessels, a fine adjustment of differentiation conditions, such as induction concentration of the small molecule, might be required to achieve consistent cardiac differentiation, likely due to variability of the agitation conditions (Chen et al. 2015). The observation suggested that agitation conditions or shear/mechanical forces unique to the culture vessels may affect the determination of the optimal differentiation conditions. Selection of an identical or similar design of vessel for scaling up would simplify the conversion of agitation conditions and may minimize the necessity of re-optimization of differentiation conditions.

Cell Aggregate Sizes

Control of cell aggregates in a homogeneous and proper size range is critical to ensure efficient and consistent cardiac induction. Unlike monolayer cell culture which allows even exposure of cells to the surrounding media condition, the hPSC aggregates are in a multilayer cell structure, which could potentially affect diffusion of molecules, such as cytokines, gases, nutrients, and metabolites, into the cell aggregate. Some reports used diffusion models paired with the experimental data to study the consumption, transport, and spatial distribution of oxygen and cytokines within the hESC aggregates and EBs in suspension (Wu et al. 2014; Van Winkle et al. 2012; Gassmann et al. 1996). Their models showed that the diffusing molecules distributed as a concentration gradient within cell aggregates/EBs and the molecule concentrations in the center regions decreased when the sizes of aggregates/EBs increased, suggesting the mass diffusion efficiency is dependent on the sizes of the cell aggregates and EBs. Consistent with the models, our previous study showed that the optimal concentrations of the small molecules for cardiac differentiation were dependent on the cell aggregate size (Chen et al. 2015). We found the efficiency of small molecule-induced cardiac differentiation declined when large cell aggregates were used for induction, and thus a higher concentration of the small molecule was required to drive cardiac differentiation more efficiently. Therefore, control of the cell aggregates within a size range that allows soluble molecules to diffuse throughout the cell aggregates is essential for an efficient cardiac differentiation.

Cell Line Variability

The hPSC lines are derived from individuals with distinct genetic variation, which may result in unique responses to culture conditions. The variability among the hPSC lines may complicate the development of cardiac induction methods. One study using the EB formation method for cardiac differentiation showed that different hPSC lines required individual optimization of induction schemes, including differential manipulation of Nodal/Activin and BMP4 pathways, titration of the inducers, and induction timing (Kattman et al. 2011). In our previous study using small-molecule modulation of the Wnt pathway to drive cardiac differentiation in suspension culture, we found that individual hPSC lines required re-titration of CHIR99021 to achieve an optimal cardiac induction (Chen et al. 2015). We also noted that an agitation condition optimized for one cell line is not necessarily applicable to another and therefore needed to re-optimize the agitation condition for each line. These observations suggest that individual optimization of the differentiation conditions may be necessary for some hPSC lines and that it will be essential to identify critical culture parameters in the cardiac differentiation method so that the differentiation condition can be optimized for a variety of cell lines with minimal efforts. In contrast, one recent study claimed the universality of a cardiac differentiation method for the suspension culture system by evaluating percentage of beating spheroids for nine cell lines differentiated in the static suspension culture (Fonoudi et al. 2015). The authors further performed flow cytometry analysis on four of the cell lines and showed approximately 90% differentiated cells positive for cTnT. Whether it is possible to establish a universal procedure for cardiac differentiation in suspension culture will need further investigation and testing.

7.4 Future Directions

Significant progress has been made in the development of hPSC expansion and differentiation that make it feasible to envision clinical use of stem cell-derived cardiomyocytes. There are still important issues that must be addressed before the routine therapeutic use of these cells becomes a reality:

- Advances in the process development for scale-up hPSC-CM production have integrated hPSC expansion and cardiac differentiation into a streamlined suspension culture system. Suspension cultures for hPSC production have been developed with defined reagents and media conditions; GMP-grade and xeno-free versions of the reagents for the hPSC expansion and the cardiac differentiation in the suspension culture need to be developed and tested to facilitate clinical applications.
- The suspension culture system has been practically applied for large-scale manufacture of hPSC-CMs at scale up to 1–3 L culture and generated 1–3 billion cells with >95% purity in one single vessel. While the current status of the suspension culture system is sufficient to produce meaningful scales of high-quality cells for large animal and phase I studies, a more sophisticated manufacturing process still needs to be developed when advancing toward clinical trials. Several hPSC culture and differentiation procedures have been developed using stirred-type suspension culture vessels. Alternative culture platforms such as rotary and bag-based rocking bioreactors, which can provide more homogeneous shear distribution and less mechanical stress, may prove beneficial when considering transferability of the agitation conditions during scale-up.
- It will be valuable to establish the kinetic profiles of gases, pH, nutrients, and metabolites for the suspension culture. With a better understanding of these dynamics, those parameters can be further studied, optimized, and controlled to achieve a consistent and robust culture condition. It is also critical to develop in-process analyses to evaluate and monitor cell quality, including induction efficiency and cell purity at different differentiation stages. Particularly for large-scale hPSC-CM production, evaluation of the intermediate cardiac cells during manufacturing is essential to assure and even predict the product quality and to assist in identification of problematic productions during early stages to minimize unnecessary costs and labor.

Given the unique ability of hPSCs to generate multiple cell types with therapeutic potential, advances in large-scale expansion of hPSCs and technologies to assess and control a differentiation process such as cardiomyocyte production should have impacts that extend well beyond a single-cell therapy target and further enable the deployment of these remarkable cells in the clinical arena.

Compliance with Ethical Standards

Conflict of Interest The authors declare that they have no conflict of interest.

Ethical approval This article does not contain any studies with human participants performed by any of the authors.

References

Amit M, Chebath J, Margulets V, Laevsky I, Miropolsky Y, Shariki K, Peri M, Blais I, Slutsky G, Revel M, Itskovitz-Eldor J (2010) Suspension culture of undifferentiated human embryonic and induced pluripotent stem cells. Stem Cell Rev 6(2):248–259

Amit M, Laevsky I, Miropolsky Y, Shariki K, Peri M, Itskovitz-Eldor J (2011) Dynamic suspension culture for scalable expansion of undifferentiated human pluripotent stem cells. Nat Protoc 6(5):572–579

Burridge PW, Keller G, Gold JD, JC W (2012) Production of de novo cardiomyocytes: human pluripotent stem cell differentiation and direct reprogramming. Cell Stem Cell 10(1):16–28

Burridge PW, Matsa E, Shukla P, Lin ZC, Churko JM, Ebert AD, Lan F, Diecke S, Huber B, Mordwinkin NM, Plews JR, Abilez OJ, Cui B, Gold JD, JC W (2014) Chemically defined generation of human cardiomyocytes. Nat Methods 11(8):855–860

Chen AK, Chen X, Choo AB, Reuveny S, SK O (2011) Critical microcarrier properties affecting the expansion of undifferentiated human embryonic stem cells. Stem Cell Res 7(2):97–111

Chen VC, Couture LA (2015) The suspension culture of undifferentiated human pluripotent stem cells using spinner flasks. Methods Mol Biol 1283:13–21

Chen VC, Couture SM, Ye J, Lin Z, Hua G, Huang HI, Wu J, Hsu D, Carpenter MK, Couture LA (2012) Scalable GMP compliant suspension culture system for human ES cells. Stem Cell Res 8(3):388–402

Chen VC, Ye J, Shukla P, Hua G, Chen D, Lin Z, Liu JC, Chai J, Gold J, Wu J, Hsu D, Couture LA (2015) Development of a scalable suspension culture for cardiac differentiation from human pluripotent stem cells. Stem Cell Res 15(2):365–375

Chong JJ, Yang X, Don CW, Minami E, Liu YW, Weyers JJ, Mahoney WM, Van Biber B, Cook SM, Palpant NJ, Gantz JA, Fugate JA, Muskheli V, Gough GM, Vogel KW, Astley CA, Hotchkiss CE, Baldessari A, Pabon L, Reinecke H, Gill EA, Nelson V, Kiem HP, Laflamme MA, Murry CE (2014) Human embryonic-stem-cell-derived cardiomyocytes regenerate non-human primate hearts. Nature 510(7504):273–277

Fernandes AM, Marinho PA, Sartore RC, Paulsen BS, Mariante RM, Castilho LR, Rehen SK (2009) Successful scale-up of human embryonic stem cell production in a stirred microcarrier culture system. Braz J Med Biol Res 42(6):515–522

Fonoudi H, Ansari H, Abbasalizadeh S, Larijani MR, Kiani S, Hashemizadeh S, Zarchi AS, Bosman A, Blue GM, Pahlavan S, Perry M, Orr Y, Mayorchak Y, Vandenberg J, Talkhabi M, Winlaw DS, Harvey RP, Aghdami N, Baharvand H (2015) A universal and robust integrated platform for the scalable production of human cardiomyocytes from pluripotent stem cells. Stem Cells Transl Med 4(12):1482–1494

Gassmann M, Fandrey J, Bichet S, Wartenberg M, Marti HH, Bauer C, Wenger RH, Acker H (1996) Oxygen supply and oxygen-dependent gene expression in differentiating embryonic stem cells. Proc Natl Acad Sci U S A 93(7):2867–2872

Geuss LR, Suggs LJ (2013) Making cardiomyocytes: how mechanical stimulation can influence differentiation of pluripotent stem cells. Biotechnol Prog 29(5):1089–1096

Hemmi N, Tohyama S, Nakajima K, Kanazawa H, Suzuki T, Hattori F, Seki T, Kishino Y, Hirano A, Okada M, Tabei R, Ohno R, Fujita C, Haruna T, Yuasa S, Sano M, Fujita J, Fukuda K (2014) A massive suspension culture system with metabolic purification for human pluripotent stem cell-derived cardiomyocytes. Stem Cells Transl Med 3(12):1473–1483

Ingber DE (2006) Mechanical control of tissue morphogenesis during embryological development. Int J Develop Biol 50(2–3):255–266

Kalmbach A, Bordas R, Oncul AA, Thevenin D, Genzel Y, Reichl U (2011) Experimental characterization of flow conditions in 2-and 20-L bioreactors with wave-induced motion. Biotechnol Prog 27(2):402–409

Kattman SJ, Witty AD, Gagliardi M, Dubois NC, Niapour M, Hotta A, Ellis J, Keller G (2011) Stage-specific optimization of activin/nodal and BMP signaling promotes cardiac differentiation of mouse and human pluripotent stem cell lines. Cell Stem Cell 8(2):228–240

Kehoe DE, Lock LT, Parikh A, Tzanakakis ES (2008) Propagation of embryonic stem cells in stirred suspension without serum. Biotechnol Prog 24(6):1342–1352

Kempf H, Olmer R, Kropp C, Ruckert M, Jara-Avaca M, Robles-Diaz D, Franke A, Elliott DA, Wojciechowski D, Fischer M, Roa Lara A, Kensah G, Gruh I, Haverich A, Martin U, Zweigerdt R (2014) Controlling expansion and cardiomyogenic differentiation of human pluripotent stem cells in scalable suspension culture. Stem cell reports 3(6):1132–1146

Krawetz R, Taiani JT, Liu S, Meng G, Li X, Kallos MS, Rancourt DE (2010) Large-scale expansion of pluripotent human embryonic stem cells in stirred-suspension bioreactors. Tissue Eng 16(4):573–582

Laflamme MA, Murry CE (2005) Regenerating the heart. Nat Biotechnol 23(7):845–856

Lam AT, Chen AK, Li J, Birch WR, Reuveny S, SK O (2014) Conjoint propagation and differentiation of human embryonic stem cells to cardiomyocytes in a defined microcarrier spinner culture. Stem Cell Res Therap 5(5):110

Larijani MR, Seifinejad A, Pournasr B, Hajihoseini V, Hassani SN, Totonchi M, Yousefi M, Shamsi F, Salekdeh GH, Baharvand H (2011) Long-term maintenance of undifferentiated human embryonic and induced pluripotent stem cells in suspension. Stem Cells Dev 20(11):1911–1923

Lei Y, Schaffer DV (2013) A fully defined and scalable 3D culture system for human pluripotent stem cell expansion and differentiation. Proc Natl Acad Sci U S A 110(52):E5039–E5048

Li Y, Powell S, Brunette E, Lebkowski J, Mandalam R (2005) Expansion of human embryonic stem cells in defined serum-free medium devoid of animal-derived products. Biotechnol Bioeng 91(6):688–698

Lian X, Hsiao C, Wilson G, Zhu K, Hazeltine LB, Azarin SM, Raval KK, Zhang J, Kamp TJ, Palecek SP (2012) Robust cardiomyocyte differentiation from human pluripotent stem cells via temporal modulation of canonical Wnt signaling. Proc Natl Acad Sci U S A 109(27):E1848–E1857

Lian X, Zhang J, Azarin SM, Zhu K, Hazeltine LB, Bao X, Hsiao C, Kamp TJ, Palecek SP (2013) Directed cardiomyocyte differentiation from human pluripotent stem cells by modulating Wnt/beta-catenin signaling under fully defined conditions. Nat Protoc 8(1):162–175

Liang P, Lan F, Lee AS, Gong T, Sanchez-Freire V, Wang Y, Diecke S, Sallam K, Knowles JW, Wang PJ, Nguyen PK, Bers DM, Robbins RC, JC W (2013) Drug screening using a library of human induced pluripotent stem cell-derived cardiomyocytes reveals disease-specific patterns of cardiotoxicity. Circulation 127(16):1677–1691

Lock LT, Tzanakakis ES (2009) Expansion and differentiation of human embryonic stem cells to endoderm progeny in a microcarrier stirred-suspension culture. Tissue Eng A 15(8):2051–2063

Ludwig TE, Levenstein ME, Jones JM, Berggren WT, Mitchen ER, Frane JL, Crandall LJ, Daigh CA, Conard KR, Piekarczyk MS, Llanas RA, Thomson JA (2006) Derivation of human embryonic stem cells in defined conditions. Nat Biotechnol 24(2):185–187

Melkoumian Z, Weber JL, Weber DM, Fadeev AG, Zhou Y, Dolley-Sonneville P, Yang J, Qiu L, Priest CA, Shogbon C, Martin AW, Nelson J, West P, Beltzer JP, Pal S, Brandenberger R (2010) Synthetic peptide-acrylate surfaces for long-term self-renewal and cardiomyocyte differentiation of human embryonic stem cells. Nat Biotechnol 28(6):606–610

Niebruegge S, Bauwens CL, Peerani R, Thavandiran N, Masse S, Sevaptisidis E, Nanthakumar K, Woodhouse K, Husain M, Kumacheva E, Zandstra PW (2009) Generation of human embryonic stem cell-derived mesoderm and cardiac cells using size-specified aggregates in an oxygen-controlled bioreactor. Biotechnol Bioeng 102(2):493–507

Oh SK, Chen AK, Mok Y, Chen X, Lim UM, Chin A, Choo AB, Reuveny S (2009) Long-term microcarrier suspension cultures of human embryonic stem cells. Stem Cell Res 2(3):219–230

Olmer R, Haase A, Merkert S, Cui W, Palecek J, Ran C, Kirschning A, Scheper T, Glage S, Miller K, Curnow EC, Hayes ES, Martin U (2010) Long term expansion of undifferentiated human iPS and ES cells in suspension culture using a defined medium. Stem Cell Res 5(1):51–64

Oncul AA, Kalmbach A, Genzel Y, Reichl U, Thevenin D (2010) Characterization of flow conditions in 2 L and 20 L wave bioreactors (R) using computational fluid dynamics. Biotechnol Prog 26(1):101–110

Otsuji TG, Bin J, Yoshimura A, Tomura M, Tateyama D, Minami I, Yoshikawa Y, Aiba K, Heuser JE, Nishino T, Hasegawa K, Nakatsuji N (2014) A 3D sphere culture system containing functional polymers for large-scale human pluripotent stem cell production. Stem Cell Rep 2(5):734–745

Paluch E, Heisenberg CP (2009) Biology and physics of cell shape changes in development. Curr Biol 19(17):R790–R799
Patwari P, Lee RT (2008) Mechanical control of tissue morphogenesis. Circ Res 103(3):234–243
Phillips BW, Horne R, Lay TS, Rust WL, Teck TT, Crook JM (2008) Attachment and growth of human embryonic stem cells on microcarriers. J Biotechnol 138(1–2):24–32
Rodin S, Domogatskaya A, Strom S, Hansson EM, Chien KR, Inzunza J, Hovatta O, Tryggvason K (2010) Long-term self-renewal of human pluripotent stem cells on human recombinant laminin-511. Nat Biotechnol 28(6):611–615
Rungarunlert S, Techakumphu M, Pirity MK, Dinnyes A (2009) Embryoid body formation from embryonic and induced pluripotent stem cells: benefits of bioreactors. World J Stem Cells 1(1):11–21
Shafa M, Krawetz R, Zhang Y, Rattner JB, Godollei A, Duff HJ, Rancourt DE (2011) Impact of stirred suspension bioreactor culture on the differentiation of murine embryonic stem cells into cardiomyocytes. BMC Cell Biol 12:53
Singh H, Mok P, Balakrishnan T, Rahmat SN, Zweigerdt R (2010) Up-scaling single cell-inoculated suspension culture of human embryonic stem cells. Stem Cell Res 4(3):165–179
Singh V (1999) Disposable bioreactor for cell culture using wave-induced agitation. Cytotechnology 30(1–3):149–158
Steiner D, Khaner H, Cohen M, Even-Ram S, Gil Y, Itsykson P, Turetsky T, Idelson M, Aizenman E, Ram R, Berman-Zaken Y, Reubinoff B (2010) Derivation, propagation and controlled differentiation of human embryonic stem cells in suspension. Nat Biotechnol 28(4):361–364
Ting S, Chen A, Reuveny S, Oh S (2014) An intermittent rocking platform for integrated expansion and differentiation of human pluripotent stem cells to cardiomyocytes in suspended microcarrier cultures. Stem Cell Res 13(2):202–213
Van Winkle AP, Gates ID, Kallos MS (2012) Mass transfer limitations in embryoid bodies during human embryonic stem cell differentiation. Cells Tissues Organs 196(1):34–47
Villa-Diaz LG, Nandivada H, Ding J, Nogueira-de-Souza NC, Krebsbach PH, O'Shea KS, Lahann J, Smith GD (2010) Synthetic polymer coatings for long-term growth of human embryonic stem cells. Nat Biotechnol 28(6):581–583
Wang L, Schulz TC, Sherrer ES, Dauphin DS, Shin S, Nelson AM, Ware CB, Zhan M, Song CZ, Chen X, Brimble SN, McLean A, Galeano MJ, Uhl EW, D'Amour KA, Chesnut JD, Rao MS, Blau CA, Robins AJ (2007) Self-renewal of human embryonic stem cells requires insulin-like growth factor-1 receptor and ERBB2 receptor signaling. Blood 110(12):4111–4119
Wang Y, Chou BK, Dowey S, He C, Gerecht S, Cheng L (2013) Scalable expansion of human induced pluripotent stem cells in the defined xeno-free E8 medium under adherent and suspension culture conditions. Stem Cell Res 11(3):1103–1116
Watanabe K, Ueno M, Kamiya D, Nishiyama A, Matsumura M, Wataya T, Takahashi JB, Nishikawa S, Nishikawa S, Muguruma K, Sasai Y (2007) A ROCK inhibitor permits survival of dissociated human embryonic stem cells. Nat Biotechnol 25(6):681–686
Wu J, Rostami MR, Cadavid Olaya DP, Tzanakakis ES (2014) Oxygen transport and stem cell aggregation in stirred-suspension bioreactor cultures. PLoS One 9(7):e102486
Yang L, Soonpaa MH, Adler ED, Roepke TK, Kattman SJ, Kennedy M, Henckaerts E, Bonham K, Abbott GW, Linden RM, Field LJ, Keller GM (2008) Human cardiovascular progenitor cells develop from a KDR+ embryonic-stem-cell-derived population. Nature 453(7194):524–528
Zhang J, Klos M, Wilson GF, Herman AM, Lian X, Raval KK, Barron MR, Hou L, Soerens AG, Yu J, Palecek SP, Lyons GE, Thomson JA, Herron TJ, Jalife J, Kamp TJ (2012) Extracellular matrix promotes highly efficient cardiac differentiation of human pluripotent stem cells: the matrix sandwich method. Circ Res 111(9):1125–1136
Zhang R, Mjoseng HK, Hoeve MA, Bauer NG, Pells S, Besseling R, Velugotla S, Tourniaire G, Kishen RE, Tsenkina Y, Armit C, Duffy CR, Helfen M, Edenhofer F, de Sousa PA, Bradley M (2013) A thermoresponsive and chemically defined hydrogel for long-term culture of human embryonic stem cells. Nat Commun 4:1335
Zhu WZ, Van Biber B, Laflamme MA (2011) Methods for the derivation and use of cardiomyocytes from human pluripotent stem cells. Methods Mol Biol 767:419–431
Zweigerdt R, Olmer R, Singh H, Haverich A, Martin U (2011) Scalable expansion of human pluripotent stem cells in suspension culture. Nat Protoc 6(5):689–700

Purification of Pluripotent Stem Cell-Derived Cardiomyocytes for Safe Cardiac Regeneration

8

Shugo Tohyama and Keiichi Fukuda

Abstract

Human pluripotent stem cells (PSCs), including embryonic stem cells (ESCs) and induced PSCs (iPSCs), have the potential to differentiate into various cells types and may be used as cell sources for regenerative medicine in the context of various diseases, including severe heart failure. However, one of the biggest hurdles in the use of human PSCs for clinical applications is tumor formation due to contamination with residual tumor-forming cells, primarily undifferentiated PSCs. In addition, hundreds of millions of cardiomyocytes are required for heart repair. Two approaches have been developed for achievement of safer cardiac regenerative therapy using human PSCs: (1) selective elimination of residual tumor-forming cells before cell transplantation and (2) purification of PSC-derived cardiomyocytes. Many methodologies, including genetic and nongenetic modification, have been developed using these strategies. In this chapter, we focus on the current status of selective elimination of residual PSCs and purification of cardiomyocytes for safe stem cell therapy.

S. Tohyama
Department of Cardiology, Keio University School of Medicine, Tokyo, Japan

Department of Organ Fabrication, Keio University School of Medicine, Tokyo, Japan
e-mail: shugotohyama@gmail.com

K. Fukuda (✉)
Department of Cardiology, Keio University School of Medicine, Tokyo, Japan
e-mail: kfukuda@a2.keio.jp

M. Ieda, W.-H. Zimmermann (eds.), *Cardiac Regeneration*, Cardiac and Vascular Biology 4, DOI 10.1007/978-3-319-56106-6_8

8.1 Introduction

The prognosis of patients with severe heart failure is extremely poor, and heart transplantation is the only effective treatment (Lund et al. 2015). However, lack of donors is a major problem worldwide. Cardiac regenerative therapy using human pluripotent stem cells (PSCs) may represent an effective alternative treatment option for heart transplantation. Human induced PSCs (iPSCs) have the potential to differentiate into various types of cells, similar to human embryonic stem cells (ESCs) (Takahashi 2007; Thomson et al. 1998), and may have applications as a new cell source for regenerative medicine in the context of various diseases, including severe heart failure (Burridge et al. 2012; Passier et al. 2008).

Although cardiac differentiation protocols have dramatically improved (Laflamme et al. 2007; Burridge et al. 2014; Lian 2012; Zhang et al. 2012; Willems et al. 2011; Minami et al. 2012), it may be impossible to stably differentiate into only target cells because many factors, including the specific cell lines used, affect differentiation efficiency (Kattman et al. 2011; Elliott et al. 2011; Osafune et al. 2008). Moreover, cardiac regenerative medicine using human PSCs will require hundreds of millions of cardiomyocytes. The use of this many cells increases the risk of contamination with residual PSCs or noncardiac proliferating cells, which is a major cause of tumor formation (Hentze et al. 2009; Miura et al. 2009; Kawamura et al. 2016; Zhang et al. 2014). Thus, many technologies have been developed to prevent tumor formation in cardiac regenerative medicine, including selective elimination of residual PSCs (Fig. 8.1a) and complete purification of cardiomyocytes (Fig. 8.1b).

In this chapter, we introduce these two strategies and discuss the use of these approaches for safe cardiac regeneration.

8.2 Elimination of Residual Pluripotent Stem Cells

Many studies have described methods for selective elimination of residual PSCs that have the capacity for teratoma formation (Fig. 8.1a). This strategy could theoretically have applications in all fields and is discussed in more detail in the following sections.

8.2.1 Cell Sorting by Stem Cell Markers

Separation strategies based on cell sorting using fluorescent-activated cell sorting (FACS) or magnetic-activated cell sorting (MACS) have been reported to eliminate residual undifferentiated PSCs. In such cell sorting methods, human PSC surface markers, such as TRA1–60, SSEA-4, and SSEA-5, are used (Fong et al. 2009; Tang 2011). In addition, claudin 6, a tight-junction protein specific for human PSCs, is also a useful surface marker for selective elimination of residual human PSCs through FACS (Ben-David et al. 2013). While these strategies are simple, they are

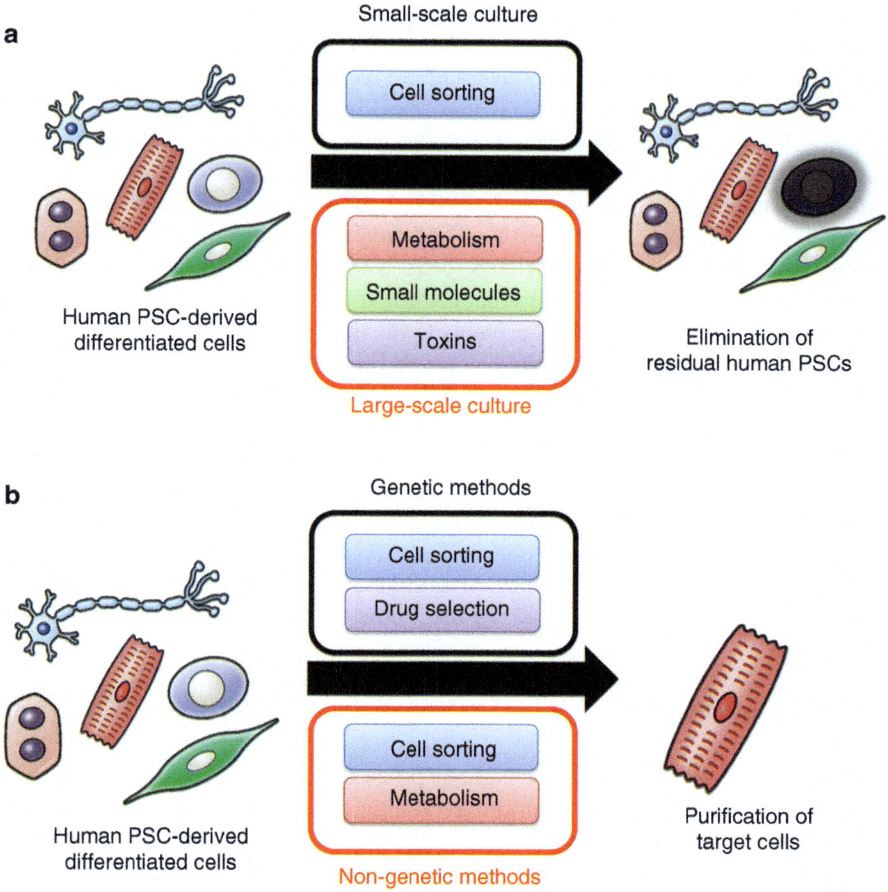

Fig. 8.1 *Strategies for prevention of tumor formation*. (**a**) Selective elimination of residual PSCs. (**b**) Purification of only cardiomyocytes from PSC-derived mixed cell populations

not suitable for large-scale culture because they require single-cell dissociation, which would be a time-consuming process when sorting a large number of cells.

8.2.2 Small Molecules or Toxins

Many studies have reported the elimination of undifferentiated human PSCs by utilization of toxins or small molecules. Of the toxins commonly used for this purpose, podocalyxin-like protein 1, a primary cytotoxic antibody for human PSCs, can eliminate residual PSCs (Choo 2008; Tan et al. 2009). In addition, *Clostridium perfringens* enterotoxin, which binds to claudin 6, has been reported to eliminate undifferentiated PSCs (Ben-David et al. 2013). Recently, Tateno et al. identified a human PSC-specific lectin (rBC2LCN) by glycome analysis and created a recombinant lectin-toxin fusion protein (rBC2LCN) using the catalytic domain of *Pseudomonas aeruginosa* exotoxin A (Tateno et al. 2015). This fusion protein could be specifically taken up into human PSCs and could kill residual PSCs.

Of the small molecules commonly used for elimination of undifferentiated PSCs, Bieberich et al. showed that ceramide analogs induce apoptosis and eliminate residual PSCs; indeed, treatment with ceramide analogs prevents teratoma formation after transplantation (Bieberich et al. 2004). Lee et al. targeted the hPSC-specific anti-apoptotic factor survivin and demonstrated that inhibition of survivin could selectively eliminate pluripotent stem cells with teratoma potential (Lee 2013). Using chemical screening, Dabir et al. identified a small molecule that inhibits the translocation of redox-regulated proteins to the mitochondria and showed that this small molecule induces apoptosis in human ESCs but not in differentiated cells (Dabir Deepa et al. 2013). Furthermore, Ben-David et al. identified a small molecule that inhibits the biosynthesis of oleic acid and specifically kills human PSCs using a screening library of more than 50,000 small molecules (Ben-David 2013).

These strategies using small molecules or toxins have many advantages because they are simple, efficient, and applicable for large numbers of cells and do not require single-cell dissociation. However, the cost of these strategies may be high owing to the need for large amounts of antibodies or small molecules. In addition, the components of these systems may affect other PSC-derived differentiated cells.

8.2.3 Metabolism

Improving our understanding of metabolic processes in human PSCs is necessary in order to remove undifferentiated tumor-forming cells by exploiting the metabolic environment (Fig. 8.2). Many studies have examined glucose metabolism in mouse and human PSCs (Kondoh et al. 2007; Panopoulos et al. 2011; Folmes Clifford et al. 2011). Folmes et al. reported that human iPSCs exhibit characteristics of elevated glucose utilization compared with mouse embryonic fibroblasts (MEFs) and that inhibition of glucose metabolism reduces the reprogramming efficiency of the cells (Folmes Clifford et al. 2011). Our group also showed that mouse and human PSCs mainly depend on activated glycolysis for ATP and biomass production and that glucose deprivation efficiently removes residual PSCs (Tohyama 2013).

In contrast, few studies have examined the effects of amino acid metabolism on mouse or human PSCs. Shyh-Chang et al. reported that mouse ESCs are critically dependent on threonine catabolism, which is important for synthesis of S-adenosylmethionine (SAM) and nucleotides (Shyh-Chang et al. 2013). Threonine starvation leads to decreased SAM levels, resulting in inhibition of histone H3K4 trimethylation and preventing mouse ESCs from maintaining pluripotency (Shyh-Chang et al. 2013; Wang et al. 2009). Additionally, Shiraki et al. evaluated the effects of essential amino acid deprivation on cell survival in human PSCs and found that methionine deprivation was the most effective inhibitor of human PSCs. They also reported that methionine is the main source of SAM production in human PSCs (Shiraki et al. 2014). Furthermore, Moussaieff et al. revealed that glucose-derived

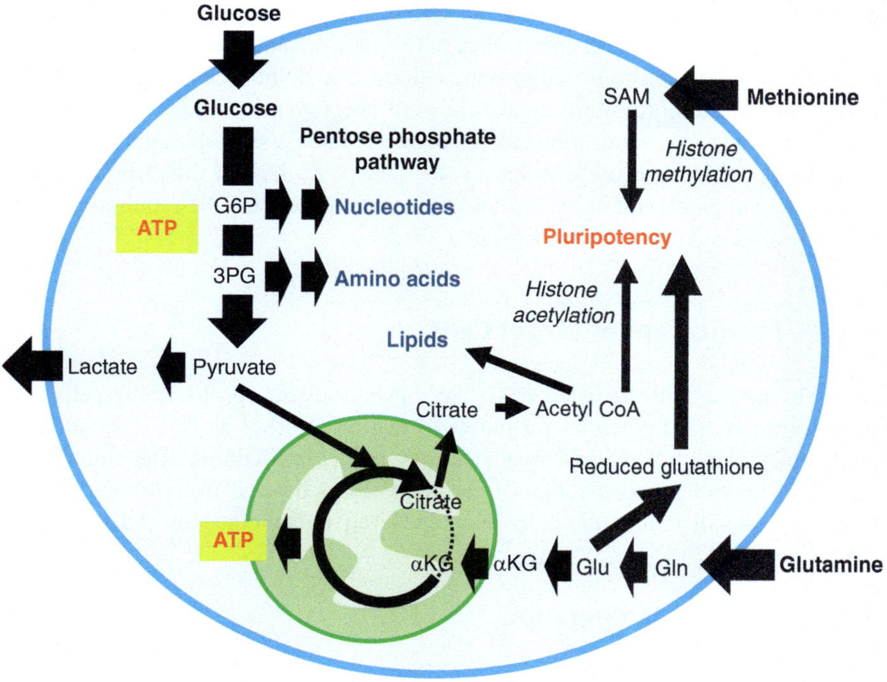

Fig. 8.2 *Metabolism in human PSCs*. Human PSCs depend on glucose and glutamine metabolism. Glycolysis contributes to ATP and biomass (amino acids and nucleotides) production. Glutamine metabolism contributes to not only ATP generation via OXPHOS but also to the maintenance of pluripotency via reduced glutathione synthesis. Methionine metabolism plays a role to produce *S*-adenosyl-methionine (SAM) that leads to maintain pluripotency via histone methylation. *G6P* glucose-6-phosphate, *3PG* glycerate 3-phosphate, *Gln* glutamine, *Glu* glutamate, *αKG* α-ketoglutarate

cytosolic acetyl-CoA contributes to the maintenance of pluripotency by induction of histone pan-acetylation (Moussaieff et al. 2015), and Carey et al. reported that naïve mouse ESCs utilize both glucose and glutamine catabolism to maintain a high level of intracellular α-ketoglutarate (αKG), which promotes histone and DNA demethylation and maintains pluripotency (Carey and Finley 2014). Recently, our group demonstrated that glutamine oxidation during the later steps of the tricarboxylic acid (TCA) cycle plays a key role in cell survival in human PSCs. In glucose-depleted conditions, glutaminolysis activation is increased, thereby promoting ATP production via oxidative phosphorylation (OXPHOS). Interestingly, human PSCs cannot utilize pyruvate efficiently because the expression levels of metabolic enzyme-related genes in the early steps of the TCA cycle are low, whereas those involved in the synthesis of cytosolic acetyl CoA are high. As a result, glucose deprivation and glutamine deprivation are most effective for elimination of residual human PSCs (Tohyama et al. 2016). Glutamine metabolism also contributes to synthesize reduced glutathione that plays a role in maintenance of pluripotency via prevention of OCT4 (Marsboom et al. 2016). Similar to approaches using small molecules and toxins, these approaches have many advantages, such as simplicity, efficiency, and suitability for large-scale culture, and do not require single-cell dissociation. Furthermore, these approaches are not expensive because they do not utilize antibodies or small molecules. However, these metabolic approaches also have the potential to cause damage to the other PSC-derived differentiated cells. Therefore, supplementation with alternative metabolites may be required to minimize the effects on other cells (Tohyama 2013).

8.3 Purification of Target Cells

Recently, both undifferentiated PSCs and other immature proliferating cells have been shown to have potential for tumor formation (Nori et al. 2015). In addition, contamination with noncardiomyocytes may induce arrhythmia after transplantation. Thus, complete purification of cardiomyocytes derived from human PSCs is necessary for safe realization of cardiac regenerative medicine (Fig. 8.1b).

8.3.1 Genetic Manipulation

Several studies have reported fluorescent protein expression-based purification of cardiomyocytes derived from mouse PSCs using various combinations of cardiomyocyte-specific promoters (e.g., αMHC, Mlc2v, Nkx2–5, and ANP) and reporters (e.g., green fluorescent protein [GFP]) (Gassanov et al. 2004; Anderson et al. 2007; Huber et al. 2007; van Laake et al. 2010). In humans, Elliott et al. introduced sequences encoding enhanced GFP (EGFP) into the NKX2–5 locus (NKX2–5-EGFP) by homologous recombination (Elliott et al. 2011). Furthermore, Ma et al. generated hiPSCs expressing a blasticidin-resistance gene under the control of the MYH6 promoter (MYH6-blasticidin) and obtain pure cardiomyocytes (Ma et al. 2011). While these methods are useful for basic research, they are not suitable for clinical application because they lack stability and safety. Therefore, it is necessary to establish nongenetic methods of purifying cardiomyocytes for clinical applications.

8.3.2 Nongenetic Cell Sorting

In nongenetic cell sorting strategies, some groups have attempted to obtain cardiac progenitor cells, whereas other groups, including ours, have attempted to isolate only cardiomyocytes. Yamashita et al. succeeded in obtaining mouse ESC-derived Flk-1-positive mesodermal cells, which could differentiate into several mesodermal lineages, including cardiomyocytes, smooth muscle cells, and endothelial cells (Yamashita et al. 2000). Hidaka et al. reported that prion protein and platelet-derived growth factor (PDGF) receptor α double-positive cells derived from mouse ESCs could differentiate into cardiomyocytes and smooth muscle cells (Hidaka et al. 2009).

On the other hand, it is difficult to isolate only cardiomyocytes because cardiomyocyte-specific surface markers have not yet been identified. Therefore, to isolate pure cardiomyocytes, we focused on the structural characteristics of cells rather than surface markers; using this approach, we succeeded in developing a nongenetic cardiomyocyte purification method (Hattori 2010). In short, because cardiomyocytes have many mature mitochondria and high mitochondrial membrane potential, we successfully purified cardiomyocytes (>99% purity) derived from mouse and human PSCs by a combination of FACS and the mitochondrial dye tetramethylrhodamine methyl ester perchlorate (TMRM). The fluorescence intensity of TMRM dye disappeared within 1 day, while that of other mitochondrial dyes was sustained over 5 days. Therefore, the effects of TMRM dye were suppressed. Other groups have also established nongenetic cardiomyocyte purification methods using FACS or MACS with antibodies against cell surface markers, including ALCAM (CD166) (Rust et al. 2009), signal-regulatory protein alpha (SIRPA) (Dubois et al. 2011), and vascular cell adhesion molecule 1 (VCAM1) (Uosaki 2011). Although cell sorting methods using antibodies or mitochondrial dyes are useful for the production of small numbers of cardiomyocytes, these methods are time consuming when using human PSC-derived mixed cell populations as the source cells (Fig. 8.3a). In addition, these methods require single-cell dissociation, which can damage the target cells, and transplantation of target cells with antibodies may result in immunogenicity. Therefore, further studies are needed to establish methods for scalable production of human PSC-derived pure cardiomyocytes for clinical applications.

8.3.3 Metabolic Selection

To establish an ideal method for scalable production of human cardiomyocytes for clinical applications, our group aimed to purify cardiomyocytes using specific metabolic culture conditions in which only cardiomyocytes and not residual PSCs can survive (Fig. 8.3b). To evaluate metabolic differences between PSCs and cardiomyocytes, we performed metabolome and transcriptome analyses. As mentioned above, we found that the PSCs mainly depended on activated glycolysis and that glucose deprivation could eliminate residual PSCs. However, because glucose-depleted conditions are also fatal for cardiomyocytes, supplementation with an alternative energy source is necessary for survival of cardiomyocytes. Interestingly, glucose and lactate are major energy substrates in fetal hearts, while fatty acids are major energy substrates in adult hearts based on the levels of energy substrates in the blood (Neely and Morgan 1974). Because PSC-derived cardiomyocytes show a fetal phenotype (Uosaki et al. 2015), we hypothesized that PSC-derived cardiomyocytes could efficiently utilize lactate for energy production and showed that mouse and human PSC-derived cardiomyocytes could survive under glucose-depleted and

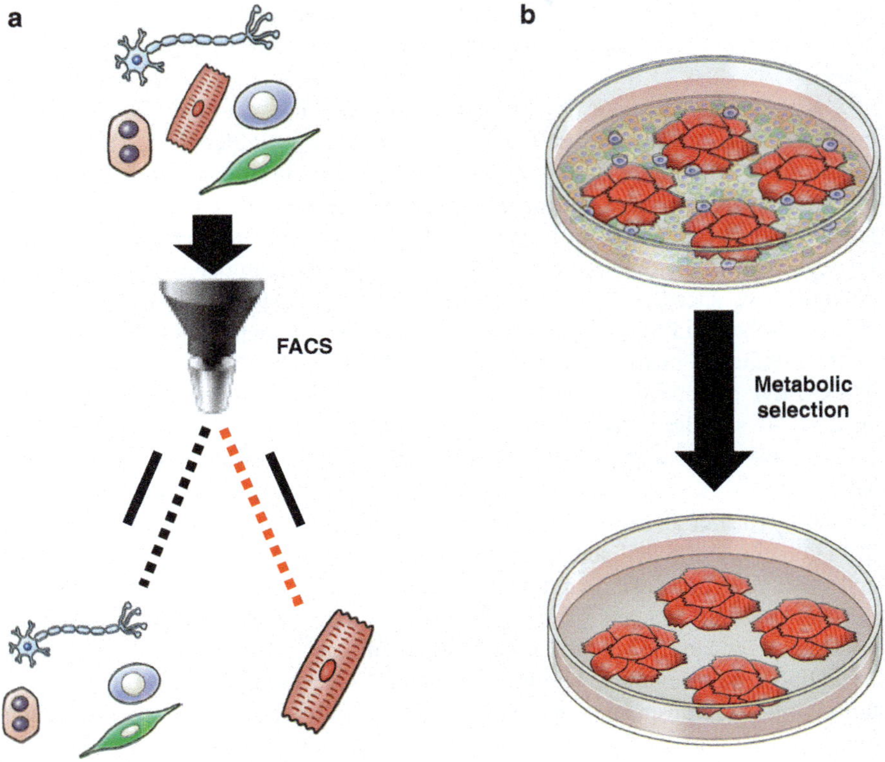

Fig. 8.3 *Purification of only cardiomyocytes from PSC-derived mixed cell populations.* (**a**), Purification of cardiomyocytes by a combination of FACS and antibodies or dyes. (**b**) Purification of cardiomyocytes by metabolic culture conditions

lactate-supplemented conditions. Moreover, because human PSC-derived noncardiac proliferating cells also depended on glycolysis like PSCs and cannot survive under these conditions, we were able to obtain pure cardiomyocytes (>95%) under these conditions (see Fig. 8.4).

As mentioned above, our group has recently reported that human PSCs depend on glycolysis and glutamine oxidation for ATP generation. Glucose and glutamine deprivation enabled complete removal of human PSCs in a much shorter period (Tohyama et al. 2016). Surprisingly, lactate supplementation could rescue only human PSC-derived cardiomyocytes because cardiomyocytes efficiently utilize lactate not only for ATP generation via OXPHOS but also for glutamate synthesis under glucose- and glutamine-depleted conditions. In short, lactate can compensate for the lack of intermediate metabolites and overcome the problem of cell damage in human PSC-derived cardiomyocytes, whereas residual human PSCs cannot utilize lactate-derived pyruvate, as mentioned above (Tohyama et al. 2016). In addition, most of the obtained pure cardiomyocytes were myosin light chain 2v (MLC2v)-positive ventricular cells. This metabolism-based method has the

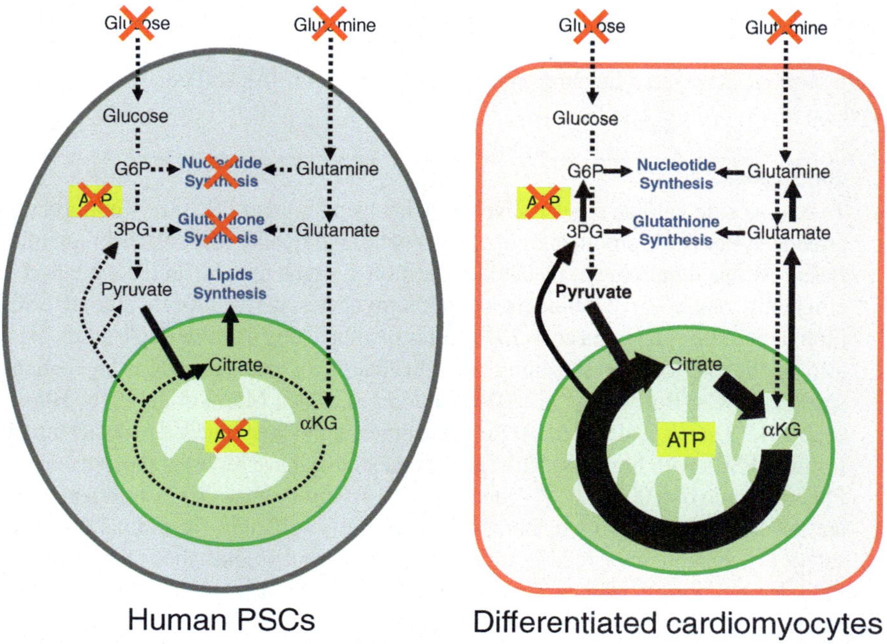

Fig. 8.4 *Metabolic differences in human PSCs and PSC-derived cardiomyocytes*. Under glucose- and glutamine-depleted conditions with pyruvate or lactate supplementation, human PSCs cannot utilize pyruvate efficiently because of low gene expression during the early steps in the TCA cycle. In contrast, human PSC-derived cardiomyocytes can efficiently use lactate-derived pyruvate because of high gene expression during the early steps in the TCA cycle

following advantages: (1) suitability for large-scale production of pure cardiomyocytes (Fonoudi et al. 2015; Hemmi et al. 2014), (2) simple procedure without specialized instrumentation, (3) low cost of culture medium, and (4) high yield of target cells (Aalto-Setala et al. 2015).

8.3.4 Other Nongenetic Methods

Xu succeeded in enriching human PSC-derived cardiomyocytes using a Percoll density gradient procedure (Xu et al. 2006), yielding cultures containing 35–66% cardiomyocytes. Nguyen et al. reported that the formation of cardiospheres derived from human PSCs enabled the enrichment of cardiomyocytes to over 80% (Nguyen Doan et al. 2014).

Ban et al. reported the purification of cardiomyocytes from mouse and human PSCs by a combination of FACS and molecular probes consisting of 15–30-bp dual-labeled oligonucleotides with a fluorophore and a quencher. In short, molecular probes could be used to identify and visualize cardiomyocyte-specific mRNA in live cells (Ban 2013). Recently, Miki et al. succeeded in establishing an efficient method for purifying cardiomyocytes based on endogenous microRNA (miRNA) activity (Miki et al. 2015). They utilized synthetic mRNAs encoding a fluorescent protein with sequences targeted by cardiomyocyte-specific miRNAs and purify cardiomyocytes with or without FACS.

Conclusions

To realize safe cardiac regenerative medicine using human PSCs, it is important to provide systems for producing target cells with high quality and sufficient quantity. Based on this requirement, metabolic selection systems may be an ideal method to efficiently obtain large numbers of cardiomyocytes derived from human PSCs (Tohyama 2013; Tohyama et al. 2016). This method using glucose-depleted media is also applicable to drug screening and elucidation of pathogenesis using patient-specific iPSCs (Burridge et al. 2016; Kodo et al. 2016; Matsa et al. 2016; Hinson et al. 2015; Dudek et al. 2015). At the same time, methods are needed to detect tumor-forming PSCs with higher sensitivity. Several studies have reported the detection of PSCs at a ratio of 0.001–0.01% (Tano et al. 2014; Kuroda et al. 2012). Further studies are needed to determine whether this sensitivity is sufficient for evaluation of the safety of techniques for regenerative medicine. Recent studies showed that human PSC-derived transplanted cardiomyocytes could electrically integrate with the host heart (Shiba et al. 2012; Gerbin et al. 2015) and mature over time (Hattori 2010; Chong et al. 2014; Funakoshi et al. 2016). Although the effectiveness of transplantation of human PSC-derived cardiomyocytes has been demonstrated in large animals (Chong et al. 2014; Ye et al. 2014; Kawamura et al. 2012), there is a risk of ventricular arrhythmia (Chong et al. 2014; Shiba et al. 2014). While the mechanism is unknown, there are two major possibilities: contamination with noncardiac cells derived from human PSCs and immaturity of PSC-derived cardiomyocytes. Large-scale purification methods for cardiomyocytes may yield solutions for overcoming both of these challenges in order to realize safe cardiac regenerative medicine.

Acknowledgments The present work was supported by the Highway Program for Realization of Regenerative Medicine from Japan Science and Technology Agency (to K.F.) and SENSHIN Medical Research Foundation (to S.T.).

Compliance with Ethical Standards

Conflict of Interest The Shugo Tohyama declare that they have no conflict of interest. Keiichi Fukuda is a cofounder of Heartseed Inc.

Ethical approval This article does not contain any studies with human participants performed by any of the authors.

References

Aalto-Setala K, Fuerstenau-Sharp M, Zimmermann ME, Stark K, Jentsch N, Klingenstein M et al (2015) Generation of highly purified human cardiomyocytes from peripheral blood mononuclear cell-derived induced pluripotent stem cells. PLoS One 10(5):e0126596

Anderson D, Self T, Mellor IR, Goh G, Hill SJ, Denning C (2007) Transgenic enrichment of Cardiomyocytes from human embryonic stem cells. Mol Ther 15(11):2027–2036

Ban K (2013) Purification of cardiomyocytes from differentiating pluripotent stem cells using molecular beacons that target cardiomyocyte-specific mRNA. Circulation 128:1897–1909. doi:10.1161/CIRCULATIONAHA.113.004228

Ben-David U (2013) Selective elimination of human pluripotent stem cells by an oleate synthesis inhibitor discovered in a high-throughput screen. Cell Stem Cell 12:162–179

Ben-David U, Nudel N, Benvenisty N (2013) Immunologic and chemical targeting of the tight-junction protein Claudin-6 eliminates tumorigenic human pluripotent stem cells. Nat Commun 4:1992

Bieberich E, Silva J, Wang G, Krishnamurthy K, Condie BG (2004) Selective apoptosis of pluripotent mouse and human stem cells by novel ceramide analogues prevents teratoma formation and enriches for neural precursors in ES cell-derived neural transplants. J Cell Biol 167:723–734. doi:10.1083/jcb.200405144

Burridge PW, Keller G, Gold JD, Wu JC (2012) Production of de novo cardiomyocytes: human pluripotent stem cell differentiation and direct reprogramming. Cell Stem Cell 10(1):16–28

Burridge PW, Matsa E, Shukla P, Lin ZC, Churko JM, Ebert AD et al (2014) Chemically defined generation of human cardiomyocytes. Nat Methods 11(8):855–860

Burridge PW, Li YF, Matsa E, Wu H, Ong SG, Sharma A et al (2016) Human induced pluripotent stem cell-derived cardiomyocytes recapitulate the predilection of breast cancer patients to doxorubicin-induced cardiotoxicity. Nat Med 22(5):547–556

Carey BW, Finley LW, Cross JR, Allis CD, Thompson CB (2014) Intracellular alpha-ketoglutarate maintains the pluripotency of embryonic stem cells. Nature 518(7539):413–416

Chong JJ, Yang X, Don CW, Minami E, Liu YW, Weyers JJ et al (2014) Human embryonic-stem-cell-derived cardiomyocytes regenerate non-human primate hearts. Nature 510(7504):273–277

Choo AB (2008) Selection against undifferentiated human embryonic stem cells by a cytotoxic antibody recognizing podocalyxin-like protein-1. Stem Cells 26:1454–1463. doi:10.1634/stemcells.2007-0576

Dabir Deepa V, Hasson Samuel A, Setoguchi K, Johnson Meghan E, Wongkongkathep P, Douglas Colin J et al (2013) A small molecule inhibitor of redox-regulated protein translocation into mitochondria. Dev Cell 25(1):81–92

Dubois NC, Craft AM, Sharma P, Elliott DA, Stanley EG, Elefanty AG et al (2011) SIRPA is a specific cell-surface marker for isolating cardiomyocytes derived from human pluripotent stem cells. Nat Biotechnol 29(11):1011–1018

Dudek J, Cheng IF, Chowdhury A, Wozny K, Balleininger M, Reinhold R et al (2015) Cardiac-specific succinate dehydrogenase deficiency in Barth syndrome. EMBO Mol Med 8(2):139–154

Elliott DA, Braam SR, Koutsis K, Ng ES, Jenny R, Lagerqvist EL et al (2011) NKX2-5eGFP/w hESCs for isolation of human cardiac progenitors and cardiomyocytes. Nat Methods 8(12):1037–1040

Folmes Clifford DL, Nelson Timothy J, Martinez-Fernandez A, Arrell DK, Lindor Jelena Z, Dzeja Petras P et al (2011) Somatic oxidative bioenergetics transitions into pluripotency-dependent glycolysis to facilitate nuclear reprogramming. Cell Metab 14(2):264–271

Fong CY, Peh GS, Gauthaman K, Bongso A (2009) Separation of SSEA-4 and TRA-1-60 labelled undifferentiated human embryonic stem cells from a heterogeneous cell population using magnetic-activated cell sorting (MACS) and fluorescence-activated cell sorting (FACS). Stem Cell Rev 5:72–80. doi:10.1007/s12015-009-9054-4

Fonoudi H, Ansari H, Abbasalizadeh S, Larijani MR, Kiani S, Hashemizadeh S et al (2015) A universal and robust integrated platform for the scalable production of human cardiomyocytes from pluripotent stem cells. Stem Cells Transl Med 4(12):1482–1494

Funakoshi S, Miki K, Takaki T, Okubo C, Hatani T, Chonabayashi K et al (2016) Enhanced engraftment, proliferation, and therapeutic potential in heart using optimized human iPSC-derived cardiomyocytes. Sci Rep 6:19111

Gassanov N, Er F, Zagidullin N, Hoppe UC (2004) Endothelin induces differentiation of ANP-EGFP expressing embryonic stem cells towards a pacemaker phenotype. FASEB J 18(14):1710–1712

Gerbin KA, Yang X, Murry CE, Coulombe KL (2015) Enhanced electrical integration of engineered human myocardium via intramyocardial versus epicardial delivery in infarcted rat hearts. PLoS One 10(7):e0131446

Hattori F (2010) Nongenetic method for purifying stem cell-derived cardiomyocytes. Nat Methods 7:61–66. doi:10.1038/nmeth.1403

Hemmi N, Tohyama S, Nakajima K, Kanazawa H, Suzuki T, Hattori F et al (2014) A massive suspension culture system with metabolic purification for human pluripotent stem cell-derived cardiomyocytes. Stem Cells Transl Med 3(12):1473–1483

Hentze H, Soong PL, Wang ST, Phillips BW, Putti TC, Dunn NR (2009) Teratoma formation by human embryonic stem cells: evaluation of essential parameters for future safety studies. Stem Cell Res 2(3):198–210

Hidaka K, Shirai M, Lee JK, Wakayama T, Kodama I, Schneider MD et al (2009) The cellular prion protein identifies bipotential cardiomyogenic progenitors. Circ Res 106(1):111–119

Hinson JT, Chopra A, Nafissi N, Polacheck WJ, Benson CC, Swist S et al (2015) Titin mutations in iPS cells define sarcomere insufficiency as a cause of dilated cardiomyopathy. Science 349(6251):982–986

Huber I, Itzhaki I, Caspi O, Arbel G, Tzukerman M, Gepstein A et al (2007) Identification and selection of cardiomyocytes during human embryonic stem cell differentiation. FASEB J 21(10):2551–2563

Kattman SJ, Witty AD, Gagliardi M, Dubois NC, Niapour M, Hotta A et al (2011) Stage-specific optimization of activin/nodal and BMP signaling promotes cardiac differentiation of mouse and human pluripotent stem cell lines. Cell Stem Cell 8(2):228–240

Kawamura M, Miyagawa S, Miki K, Saito A, Fukushima S, Higuchi T et al (2012) Feasibility, safety, and therapeutic efficacy of human induced pluripotent stem cell-derived cardiomyocyte sheets in a porcine ischemic cardiomyopathy model. Circulation 126(11 Suppl 1):S29–S37

Kawamura A, Miyagawa S, Fukushima S, Kawamura T, Kashiyama N, Ito E et al (2016) Teratocarcinomas arising from allogeneic induced pluripotent stem cell-derived cardiac tissue constructs provoked host immune rejection in mice. Sci Rep 6:19464

Kodo K, Ong SG, Jahanbani F, Termglinchan V, Hirono K, InanlooRahatloo K et al (2016) iPSC-derived cardiomyocytes reveal abnormal TGF-beta signalling in left ventricular non-compaction cardiomyopathy. Nat Cell Biol 18(10):1031–1042

Kondoh H, Lleonart ME, Nakashima Y, Yokode M, Tanaka M, Bernard D et al (2007) A high glycolytic flux supports the proliferative potential of murine embryonic stem cells. Antioxid Redox Signal 9(3):293–299

Kuroda T, Yasuda S, Kusakawa S, Hirata N, Kanda Y, Suzuki K et al (2012) Highly sensitive in vitro methods for detection of residual undifferentiated cells in retinal pigment epithelial cells derived from human iPS cells. PLoS One 7(5):e37342

van Laake LW, Qian L, Cheng P, Huang Y, Hsiao EC, Conklin BR et al (2010) Reporter-based isolation of induced pluripotent stem cell- and embryonic stem cell-derived cardiac progenitors reveals limited Gene expression variance. Circ Res 107(3):340–347

Laflamme MA, Chen KY, Naumova AV, Muskheli V, Fugate JA, Dupras SK et al (2007) Cardiomyocytes derived from human embryonic stem cells in pro-survival factors enhance function of infarcted rat hearts. Nat Biotechnol 25(9):1015–1024

Lee MO (2013) Inhibition of pluripotent stem cell-derived teratoma formation by small molecules. Proc Natl Acad Sci U S A 110:E3281–E3290. doi:10.1073/pnas.1303669110

Lian X (2012) Robust cardiomyocyte differentiation from human pluripotent stem cells via temporal modulation of canonical Wnt signaling. Proc Natl Acad Sci 109:E1848–E1857. doi:10.1073/pnas.1200250109

Lund LH, Edwards LB, Kucheryavaya AY, Benden C, Dipchand AI, Goldfarb S et al (2015) The registry of the international society for heart and lung transplantation: thirty-second official adult heart transplantation report–2015; focus theme: early graft failure. J Heart Lung Transplant 34(10):1244–1254

Ma J, Guo L, Fiene SJ, Anson BD, Thomson JA, Kamp TJ et al (2011) High purity human-induced pluripotent stem cell-derived cardiomyocytes: electrophysiological properties of action potentials and ionic currents. Am J Physiol Heart Circ Physiol 301(5):H2006–H2017

Marsboom G, Zhang G-F, Pohl-Avila N, Zhang Y, Yuan Y, Kang H et al (2016) Glutamine metabolism regulates the Pluripotency transcription factor OCT4. Cell Rep 16(2):323–332

Matsa E, Burridge PW, Yu KH, Ahrens JH, Termglinchan V, Wu H et al (2016) Transcriptome profiling of patient-specific human iPSC-cardiomyocytes predicts individual drug safety and efficacy responses in vitro. Cell Stem Cell 19(3):311–325

Miki K, Endo K, Takahashi S, Funakoshi S, Takei I, Katayama S et al (2015) Efficient detection and purification of cell populations using synthetic microRNA switches. Cell Stem Cell 16(6):699–711

Minami I, Yamada K, Otsuji TG, Yamamoto T, Shen Y, Otsuka S et al (2012) A small molecule that promotes cardiac differentiation of human pluripotent stem cells under defined, cytokine- and xeno-free conditions. Cell Rep 2(5):1448–1460

Miura K, Okada Y, Aoi T, Okada A, Takahashi K, Okita K et al (2009) Variation in the safety of induced pluripotent stem cell lines. Nat Biotechnol 27(8):743–745

Moussaieff A, Rouleau M, Kitsberg D, Cohen M, Levy G, Barasch D et al (2015) Glycolysis-mediated changes in acetyl-CoA and histone acetylation control the early differentiation of embryonic stem cells. Cell Metab 21(3):392–402

Neely JR, Morgan HE (1974) Relationship between carbohydrate and lipid metabolism and the energy balance of heart muscle. Annu Rev Physiol 36:413–459

Nguyen Doan C, Hookway Tracy A, Wu Q, Jha R, Preininger Marcela K, Chen X et al (2014) Microscale generation of cardiospheres promotes robust enrichment of cardiomyocytes derived from human pluripotent stem cells. Stem Cell Reports 3(2):260–268

Nori S, Okada Y, Nishimura S, Sasaki T, Itakura G, Kobayashi Y et al (2015) Long-term safety issues of iPSC-based cell therapy in a spinal cord injury model: oncogenic transformation with epithelial-mesenchymal transition. Stem Cell Reports. 4(3):360–373

Osafune K, Caron L, Borowiak M, Martinez RJ, Fitz-Gerald CS, Sato Y et al (2008) Marked differences in differentiation propensity among human embryonic stem cell lines. Nat Biotechnol 26(3):313–315

Panopoulos AD, Yanes O, Ruiz S, Kida YS, Diep D, Tautenhahn R et al (2011) The metabolome of induced pluripotent stem cells reveals metabolic changes occurring in somatic cell reprogramming. Cell Res 22(1):168–177

Passier R, van Laake LW, Mummery CL (2008) Stem-cell-based therapy and lessons from the heart. Nature 453(7193):322–329

Rust W, Balakrishnan T, Zweigerdt R (2009) Cardiomyocyte enrichment from human embryonic stem cell cultures by selection of ALCAM surface expression. Regen Med 4(2):225–237

Shiba Y, Fernandes S, Zhu WZ, Filice D, Muskheli V, Kim J et al (2012) Human ES-cell-derived cardiomyocytes electrically couple and suppress arrhythmias in injured hearts. Nature 489(7415):322–325

Shiba Y, Filice D, Fernandes S, Minami E, Dupras SK, Biber BV et al (2014) Electrical integration of human embryonic stem cell-derived cardiomyocytes in a guinea pig chronic infarct model. J Cardiovasc Pharmacol Ther 19(4):368–381

Shiraki N, Shiraki Y, Tsuyama T, Obata F, Miura M, Nagae G et al (2014) Methionine metabolism regulates maintenance and differentiation of human pluripotent stem cells. Cell Metab 19(5):p780–p794

Shyh-Chang N, Locasale JW, Lyssiotis CA, Zheng Y, Teo RY, Ratanasirintrawoot S et al (2013) Influence of threonine metabolism on S-adenosylmethionine and histone methylation. Science 339(6116):222–226

Takahashi K (2007) Induction of pluripotent stem cells from adult human fibroblasts by defined factors. Cell 131:861–872. doi:10.1016/j.cell.2007.11.019

Tan HL, Fong WJ, Lee EH, Yap M, Choo A (2009) mAb 84, a cytotoxic antibody that kills undifferentiated human embryonic stem cells via oncosis. Stem Cells 27:1792–1801. doi:10.1002/stem.109

Tang C (2011) An antibody against SSEA-5 glycan on human pluripotent stem cells enables removal of teratoma-forming cells. Nat Biotechnol 29:829–834. doi:10.1038/nbt.1947

Tano K, Yasuda S, Kuroda T, Saito H, Umezawa A, Sato Y (2014) A novel in vitro method for detecting undifferentiated human pluripotent stem cells as impurities in cell therapy products using a highly efficient culture system. PLoS One 9(10):e110496

Tateno H, Onuma Y, Ito Y, Minoshima F, Saito S, Shimizu M et al (2015) Elimination of tumorigenic human pluripotent stem cells by a recombinant lectin-toxin fusion protein. Stem Cell Rep 4(5):811–820

Thomson JA, Itskovitz-Eldor J, Shapiro SS, Waknitz MA, Swiergiel JJ, Marshall VS et al (1998) Embryonic stem cell lines derived from human blastocysts. Science 282(5391):1145–1147

Tohyama S (2013) Distinct metabolic flow enables large-scale purification of mouse and human pluripotent stem cell-derived cardiomyocytes. Cell Stem Cell 12:127–137. doi:10.1016/j.stem.2012.09.013

Tohyama S, Fujita J, Hishiki T, Matsuura T, Hattori F, Ohno R et al (2016) Glutamine oxidation is indispensable for survival of human pluripotent stem cells. Cell Metab 23(4):663–674

Uosaki H (2011) Efficient and scalable purification of cardiomyocytes from human embryonic and induced pluripotent stem cells by VCAM1 surface expression. PLoS One 6:e23657. doi:10.1371/journal.pone.0023657

Uosaki H, Cahan P, Lee Dong I, Wang S, Miyamoto M, Fernandez L et al (2015) Transcriptional landscape of cardiomyocyte maturation. Cell Rep 13(8):1705–1716

Wang J, Alexander P, Wu L, Hammer R, Cleaver O, McKnight SL (2009) Dependence of mouse embryonic stem cells on threonine catabolism. Science 325(5939):435–439

Willems E, Spiering S, Davidovics H, Lanier M, Xia Z, Dawson M et al (2011) Small-molecule inhibitors of the Wnt pathway potently promote cardiomyocytes from human embryonic stem cell-derived mesoderm. Circ Res 109(4):360–364

Xu C, Police S, Hassanipour M, Gold JD (2006) Cardiac bodies: a novel culture method for enrichment of cardiomyocytes derived from human embryonic stem cells. Stem Cells Dev 15(5):631–639

Yamashita J, Itoh H, Hirashima M, Ogawa M, Nishikawa S, Yurugi T et al (2000) Flk1-positive cells derived from embryonic stem cells serve as vascular progenitors. Nature 408(6808):92–96

Ye L, Chang Y-H, Xiong Q, Zhang P, Zhang L, Somasundaram P et al (2014) Cardiac repair in a porcine model of acute myocardial infarction with human induced pluripotent stem cell-derived cardiovascular cells. Cell Stem Cell 15(6):750–761

Zhang J, Klos M, Wilson GF, Herman AM, Lian X, Raval KK et al (2012) Extracellular matrix promotes highly efficient cardiac differentiation of human pluripotent stem cells: the matrix sandwich method. Circ Res 111(9):1125–1136

Zhang L, Pan Y, Qin G, Chen L, Chatterjee T, Weintraub N et al (2014) Inhibition of stearoyl-coA desaturase selectively eliminates tumorigenic Nanog-positive cells: improving the safety of iPS cell transplantation to myocardium. Cell Cycle 13(5):762–771

State of the Art in Cardiomyocyte Transplantation

9

Matthew E. Hartman, James J.H. Chong, and Michael A. Laflamme

Abstract

While other cell-based therapies for ischemic heart disease are generally understood to operate through indirect, so-called "paracrine" mechanisms of action (e.g., by promoting angiogenesis or attenuating adverse ventricular remodeling), cardiomyocyte transplantation aims to repopulate the infarct scar with functionally integrated new myocardium that will directly contribute to contractile function. Early proof of concept for this "direct remuscularization" strategy came from preclinical studies in rodent models in which neonatal and fetal cardiomyocytes were shown to stably engraft within the infarct scar and mediate beneficial

M.E. Hartman
Altius Institute for Biomedical Sciences, Seattle, WA 98121, USA

Division of Cardiology, Department of Medicine, University of Washington, Seattle, WA 98195, USA

Swedish Heart and Vascular, Seattle, WA 98122, USA
e-mail: mehartma@altius.org

J.J.H. Chong
Centre for Heart Research, The Westmead Institute for Medical Research, The University of Sydney, Hawkesbury Rd, Westmead, NSW 2145, Australia

Department of Cardiology, Westmead Hospital, Hawkesbury Rd, Westmead, NSW 2145, Australia
e-mail: james.chong@sydney.edu.au

M.A. Laflamme (✉)
Toronto General Hospital Research Institute, University Health Network, Toronto, ON M5G 1L7, Canada

McEwen Centre for Regenerative Medicine, University Health Network, Toronto, ON M5G 1L7, Canada

Department of Laboratory Medicine and Pathobiology, University of Toronto, Toronto, ON M5G 1L7, Canada
e-mail: michael.laflamme@uhnresearch.ca

M. Ieda, W.-H. Zimmermann (eds.), *Cardiac Regeneration*, Cardiac and Vascular Biology 4, DOI 10.1007/978-3-319-56106-6_9

effects. More recently, pluripotent stem cells (PSCs) have drawn considerable attention as a more practical, highly scalable source of cardiomyocytes, and these PSC-derived cardiomyocytes have shown considerable promise in both small- and large-animal models of post-infarct heart failure. In this chapter, we review the status of preclinical studies of cardiomyocyte transplantation, describe the significant remaining barriers to clinical translation, and consider potential target patient populations and study design for a first-in-human study of PSC-derived cardiomyocyte transplantation in ischemic heart disease.

9.1 Introduction

The transplantation of human cardiomyocytes has potential to fundamentally change the practice of cardiology by enabling a transition from treatments aimed at merely attenuating symptoms or disease progression to instead the large-scale replacement of damaged or poorly functioning heart muscle. Early work in the field demonstrated the feasibility of cardiomyocyte transplantation and showed that this intervention could result in the "remuscularization" of myocardial scar tissue in a variety of small-animal models of cardiac injury. While a variety of potentially cardiomyogenic cell types have been explored for this application, pluripotent stem cells (PSCs) have been the focus of particularly intense investigation in recent years, as these cells represent a reliable, essentially inexhaustible source of phenotypically unambiguous cardiomyocytes suitable for transplantation. There has been substantial progress toward understanding the factors that regulate the cardiac differentiation of PSCs, and these advances have translated into significant improvements in the reproducibility, yield, purity, and costs of generating PSC-derived cardiomyocytes (PSC-CMs). Much has been learned from PSC-CM transplantation studies in small- and large-animal models of cardiac injury, and investigators in France recently commenced the first clinical trial testing PSC-derived cardiac progenitors in patients with ischemic cardiomyopathy. That said, a number of important challenges remain that must be overcome if cardiomyocyte transplantation is to move from an experimental concept to mainstream practice. For example, currently available PSC-CMs have an immature phenotype, comparable to fetal cardiomyocytes, and this may limit their ability to safely integrate and improve contractile function. Other concerns that must be addressed include the risk of tumor formation and graft-related arrhythmias, the need to prevent graft cell death and immune rejection, and the requirement for large-scale, clinical-grade PSC-CM manufacturing. The most appropriate delivery mechanism has yet to be determined, with some investigators focusing on the direct intramyocardial injection of cardiomyocytes, while others focus on tissue-engineering approaches aimed at generating cellular patches or even whole organs for implantation. To optimize for successful translation and mainstream clinical adoption, it is essential that these and other uncertainties be resolved as much as possible before testing cardiomyocyte transplantation in human subjects.

9.2 How Might Cell Therapy Improve Cardiac Function?

There are multiple potential mechanisms by which intracardiac cell therapy might exert beneficial effects including (1) direct remuscularization (Fig. 9.1), (2) indirect remuscularization via the activation of host cardiomyocyte cell cycle activity or other endogenous reparative pathways, and (3) non-myogenic effects such as enhanced angiogenesis or attenuated adverse remodeling that do not involve remuscularization but might nonetheless enhance contractile function. Depending on the specific clinical scenario, some of these mechanisms may be more appropriate than others. For example, to replace the scar tissue left by myocardial infarction (MI) with functioning muscle, all three of these mechanisms might prove to be a useful component of therapy. Direct remuscularization might prove particularly useful in patients with end-stage post-MI heart failure and already established scar, because adverse remodeling has already plateaued and the window for indirect effects has

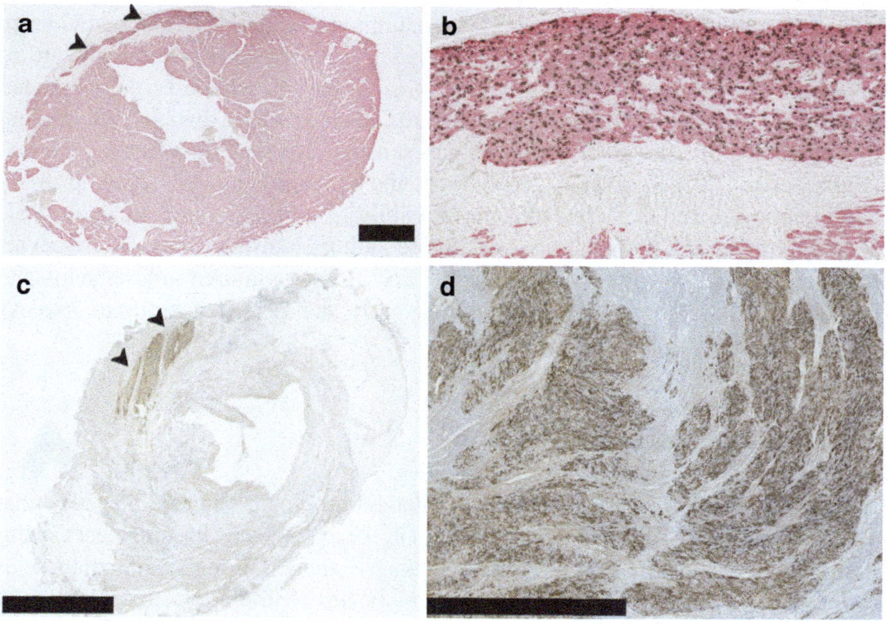

Fig. 9.1 The transplantation of human ESC-derived cardiomyocytes remuscularizes injured guinea pig and nonhuman primate hearts. (**a**) Human myocardium formed within the scar tissue of a cryoinjured guinea pig heart (*arrows*). Host and graft myocardium was identified by immunostaining with an antibody against the cardiac marker beta-myosin heavy chain (*red*), and human nuclei were identified by in situ hybridization with a human-specific pan-centromeric DNA probe (*brown*). Scale bar: 2 mm. (**b**) Inset of graft myocardium. (**c**) Human myocardium formed within the scar tissue of a nonhuman primate heart injured by left anterior descending coronary artery ligation (*arrows*). Graft human myocardium expressing green fluorescent protein (GFP) was identified by immunostaining with an antibody against GFP (*brown*), and host and graft nuclei were identified by hematoxylin stain (*purple*). Scale bar: 5 mm. (**d**) High-power image of graft human myocardium. Staining as in panel C. Scale bar: 4 mm

largely closed. Conversely, a pro-angiogenic cell therapy that increases myocardial perfusion might be expected to be more efficacious in patients suffering from chronic ischemia in the absence of significant scar tissue.

9.2.1 Direct Remuscularization

The delivery of exogenous cardiomyocytes is perhaps the most intuitive of the preceding mechanisms, and early work in the field was nearly exclusively focused on direct remuscularization. By this strategy, cardiomyocytes, cardiomyogenic progenitors, or engineered cardiac tissue are delivered to the injured heart, and they form nascent muscle that must become electromechanically integrated and contract synchronously with host myocardium to augment contractile function. As described in more detail below, early proof of principle for this approach was provided by animal studies in which fetal and neonatal cardiomyocytes were transplanted into a rat model of acute MI (Leor et al. 1996; Li et al. 1996; Scorsin et al. 1997; Scorsin et al. 2000). While these initial reports were limited to showing structural evidence for functional integration (e.g., shared gap junctions between host and graft cardiomyocytes in histological sections), host-graft coupling was later demonstrated directly by imaging genetically tagged grafts loaded with a dynamic fluorescent indicator of graft activation (e.g., calcium-sensitive dyes or fluorescent proteins) (Rubart et al. 2003; Roell et al. 2007). Field and colleagues moved beyond primary cardiomyocytes and demonstrated the possibility of using a more scalable PSC source (Klug et al. 1996). In this work, these authors transplanted cardiomyocytes derived from mouse embryonic stem cells (ESCs) into uninjured mouse hearts and showed that they formed stable intracardiac grafts that persisted for up to 7 weeks (Klug et al. 1996).

9.2.2 Indirect Remuscularization

An alternative approach to direct remuscularization would be to deliver cells that somehow stimulate endogenous cardiac repair, for example, by causing host cardiomyocytes to proliferate and repopulate damaged areas. During mammalian heart development, there is substantial cardiomyocyte proliferation that eventually declines in postnatal life (Pasumarthi and Field 2002; Bergmann et al. 2009). Work in lower vertebrates, which show far greater capacity for cardiomyocyte renewal in the adult heart, illustrates the potential utility of reactivating latent cardiomyocyte cell cycle activity. Indeed, the adult zebrafish heart is capable of large-scale regeneration following apical resection (Poss et al. 2002), and elegant fate-mapping studies have shown that this process involves the proliferation of already-committed cardiomyocytes rather than recruitment of cardiomyogenic progenitors (Jopling et al. 2010; Kikuchi et al. 2010). More recent studies suggest that mammalian hearts are capable of similarly dramatic regeneration in early postnatal life but only during a narrow time window that tracks closely with cardiomyocyte cell cycle activity.

Neonatal mice are able to fully regenerate the cardiac apex following resection at postnatal day 1; however, this ability is lost by postnatal day 7 (Porrello et al. 2011). Interestingly, adult hearts in transgenic mice with cardiomyocyte-restricted overexpression of the cell cycle regulator, cyclin D2, show sustained cardiomyocyte proliferation and robust regenerative capacity following experimental MI (Hassink et al. 2008). While the field continues to explore small-molecule and genetic approaches by which to reactivate cardiomyocyte proliferation in the adult heart, there is some preclinical evidence supporting the notion that the delivery of selected exogenous cell types may also upregulate host cardiomyocyte cycling (Tseliou et al. 2014).

9.2.3 Non-myogenic Mechanisms

Other cardiac cell therapies may be able to exert beneficial effects in the absence of direct remuscularization through indirect, so-called paracrine, mechanisms of action. A number of such effects have been attributed to cell transplantation including increased angiogenesis, decreased cell death, and reduced reverse remodeling (Mirotsou et al. 2011). The largest body of evidence for such indirect beneficial effects following intracardiac cell therapy comes from preclinical and clinical experience with various cell types that can be isolated from the bone marrow (BM) compartment. Indeed, while the field was initially captivated by reports of transdifferentiation and large-scale remuscularization following the transplantation of BM-derived cells (BMDCs), there is now general consensus that these observations can be accounted for by cell fusion and/or experimental artifact (Murry et al. 2004; Balsam et al. 2004; Norol et al. 2007; Iso et al. 2007). By contrast, there is strong evidence that BMDCs can stimulate angiogenesis (Tomita et al. 1999; Nagaya et al. 2004; Tang et al. 2004), reduce reverse remodeling post-MI (Fatkhudinov et al. 2015), reduce apoptosis (Yang et al. 2012), and help host cardiomyocytes re-enter the cell cycle (Matsuura et al. 2004).

9.2.4 Summary

Of course, it still remains to be determined whether any candidate cell type or any of these three broad potential mechanisms of action described above will actually prove efficacious in heart failure patients. Some support for a strategy based on direct remuscularization comes from a recent study that compared the intramyocardial transplantation of human BMDCs, human embryonic stem cell-derived cardiomyocytes (ESC-CMs) and human ESC-derived multipotent cardiovascular progenitors in a rat MI model (Fernandes et al. 2015). In that work, human ESC-CMs and ESC-derived progenitors formed comparable myocardial grafts with similar beneficial effects on left ventricular (LV) systolic function, while human BMDCs promoted angiogenesis and attenuated LV dilatation without engrafting long-term or improving contractility. Since multiple clinical trials suggest that BMDC-based therapies have only modest if any beneficial effects (Sadat et al. 2014), this preclinical study underscores the need for continued investigation into direct remuscularization as a therapeutic approach for post-MI heart failure.

9.3 What are the Potential Sources of Cardiomyocytes for Transplantation?

9.3.1 Primary Cardiomyocytes

In early proof-of-concept studies, investigators showed that fetal and neonatal cardiomyocytes transplanted into infarcted hearts could survive and mediate beneficial effects on left ventricular contractile function (Leor et al. 1996; Etzion et al. 2001; Scorsin et al. 1997; Li et al. 1996; Skobel et al. 2004; Rubart et al. 2003). In rodent models, fetal cardiomyocytes persist from at least 2 months (Leor et al. 1996; Etzion et al. 2001) to 6 months (Skobel et al. 2004) post-transplantation and couple with host myocardium (Rubart et al. 2003). Their transplantation improves contractile function in multiple models of cardiac injury including cryoinjury (Li et al. 1996), ischemia-reperfusion injury (Scorsin et al. 1997), and ligation of the left anterior descending artery (Etzion et al. 2001; Skobel et al. 2004). Fetal cardiomyocytes appear to improve left ventricular function more than neonatal cardiomyocytes, possibly a consequence of the former's higher proliferation rate post-transplantation (Fujimoto et al. 2011). Successful transplantation and engraftment have also been demonstrated with primary porcine (Watanabe et al. 1998) and human (Leor et al. 1996) fetal cardiomyocytes. While these preclinical studies provided excellent proof of principle for therapeutic remuscularization, it is generally agreed that neonatal and fetal cardiomyocytes are not a practical cell source for clinical application, given their obvious practical and ethical limitations.

9.3.2 Skeletal Muscle

Skeletal and cardiac muscle have many phenotypic properties in common, so skeletal muscle progenitors (myoblasts and satellite cells) drew considerable early attention in the field. This interest was only increased by early reports suggesting that they might be capable of transdifferentiation into functioning cardiomyocytes following intracardiac transplantation (Taylor et al. 1998; Atkins et al. 1999a; 1999b). The latter suggestion was exciting because skeletal and cardiac muscle have very different mechanisms of excitation-contraction coupling and electrical propagation, raising concerns about the capacity of skeletal muscle for electromechanical integration and the possibility of graft-related arrhythmias. Unfortunately, subsequent reports failed to confirm such transdifferentiation phenomena (Murry et al. 1996; Reinecke et al. 2002), and it appears that transplanted myoblasts remain committed to the skeletal muscle phenotype, including lack of expression of the gap junction protein connexin-43 (Cx43), which is required for function as an electrical syncytium (Scorsin et al. 2000). This outcome may explain in part the findings of the phase II MAGIC trial, in which the patients undergoing autologous skeletal myoblast transplantation showed a signal toward higher mortality and ventricular arrhythmias compared to those receiving placebo (Menasche et al. 2008).

9.3.3 Embryonic Stem Cells

Given their tremendous scalability and undisputed ability to differentiate into phenotypically unambiguous cardiomyocytes, a variety of PSC types have seen intense study in the field in recent years, including embryonic stem cells (ESCs), parthenogenetic stem cells (pSCs), and induced pluripotent stem cells (iPSCs). Human ESCs are isolated from the inner cell mass of blastocysts left over from in vitro fertilization (Thomson et al. 1998). As with all PSCs, ESCs are defined by their capacity for unlimited self-renewal and ability to differentiate into elements from all three embryonic germ layers (endoderm, mesoderm, and ectoderm). When placed into suspension culture in the presence of fetal bovine serum, human ESCs form three-dimensional aggregates called embryoid bodies that include multiple differentiated cell types, including cardiomyocytes (Itskovitz-Eldor et al. 2000; Kehat et al. 2001). Interestingly, different ESC lines show varying potential for "spontaneous" cardiac differentiation under these conditions (Leschik et al. 2015; Moore et al. 2008).

9.3.4 Parthenogenetic Stem Cells (pSCs)

Parthenogenesis is the process of oocyte activation without fertilization by a spermatocyte (Rogers et al. 2004; Cibelli et al. 2006). This process can be induced by chemical factors in the dish, and the resulting diploid cells can form blastocysts from which pSCs can be isolated. As with their counterparts isolated from normal blastocysts, human pSCs are capable of differentiating into multiple lineages from all the three embryonic germ layers (Lin et al. 2003; Revazova et al. 2007; Brevini et al. 2009; Li et al. 2014; Ahmad et al. 2012; Isaev et al. 2012). While the derivation of cardiomyocytes from human pSCs has not been reported to our knowledge, cardiomyocytes have been successfully obtained from mouse pSCs and employed in preclinical transplantation studies (Didie et al. 2013; Yang et al. 2015). The intriguing potential advantage of pSCs is that they have a largely haploidentical genotype, greatly simplifying their application in major histocompatibility complex (MHC)-matched allotransplantation. That said, a number of concerns have been raised about these cells including their propensity for dysregulated proliferation (Brevini et al. 2009), chromosomal instability (Liu et al. 2011a), and the effects of maternal gene imprinting on differentiation (Stelzer et al. 2011).

9.3.5 Induced Pluripotent Stem Cells

iPSCs are somatic cells that have been reprogrammed into an ESC-like state via the forced expression of transcription factors known to regulate pluripotency. As such, iPSCs offer a number of potential advantages over ESCs: they avoid the ethical concerns associated with the destruction of blastocysts and they could in principle be applied in autologous cell therapies. iPCS-CMs have already reached widespread application for in vitro disease modeling and drug screening (Itzhaki et al. 2011; Sun et al. 2012; Terrenoire et al. 2013; Bellin et al. 2013; Drawnel et al. 2014). That said, there are a number of concerns associated with iPSCs, particularly with regard to clinical applications. iPSCs were initially created by the overexpression of one of two different sets of transcription factors Oct3/Oct4, Sox2, c-Myc, and Klf4 (Takahashi et al. 2007) or Oct4, Sox2, Nanog, and LIN28 (Yu et al. 2007) that were delivered via integrating viral vectors, raising the possibility of random integration events and transformation. These concerns have been greatly allayed by safer reprogramming methods, including the use of non-integrating viral and nonviral vectors, cell-permeant proteins, modified mRNA, and/or small molecules (Zhou and Zeng 2013). Given the phenomenon of epigenetic memory, some investigators have wondered about the long-term phenotypic stability of the differentiation progeny of iPSCs. Finally, at present, individualized patient-specific iPSC-based therapeutics seem impractical from both a regulatory and an economic perspective, greatly undermining the rationale for the use of autologous iPSC derivatives.

9.3.6 Direct Reprogramming to Cardiomyocytes

Another intriguing possibility is the direct reprogramming of fibroblasts (or another readily available somatic cell type) into cardiomyocytes. Just as somatic cells can be reprogrammed into iPSCs via the forced expression of ESC-associated transcription factors, non-myocytes can be induced into a cardiomyocyte-like phenotype via the expression of cardiomyocyte-associated transcription factors. Mouse fibroblasts have been reprogrammed into cardiomyocyte-like cells by expression of Gata4, Mef2c, and Tbx5 (Ieda et al. 2010; Qian et al. 2012), a combination of miRNAs (Jayawardena et al. 2012), and through stimulation with small molecules (Fu et al. 2015). Human fibroblast reprogramming requires a different combination of transcription factors and microRNAs (Nam et al. 2013) but has also been reported. The major barriers to this approach currently are that the reprogramming is very inefficient and that the resultant cells have an incomplete cardiac phenotype (Chen et al. 2012; Qian et al. 2013).

9.4 How Are Human PSC-Derived Cardiomyocytes Obtained and What Is Their Phenotype?

Because PSC-CMs are generally regarded as the most practical, currently available source of cardiomyocytes for transplantation, we focus here on their derivation and phenotype:

9.4.1 Approaches to Cardiomyocyte Derivation from Human PSCs

Tremendous progress has been made toward improving the yield and purity of the cardiomyocytes that can be derived from human PSCs. Initial work in the field involved their differentiation in suspension cultures in the presence of high concentrations of fetal bovine serum. Under these conditions, PSCs can form three-dimensional aggregates, termed embryoid bodies (EBs), that include elements from all three embryonic germ layers (Itskovitz-Eldor et al. 2000). Some EBs contract rhythmically and contain cardiomyocytes, but cardiogenesis by this approach is very inefficient and the resultant cultures are typically comprised of <10% cardiomyocytes (Itskovitz-Eldor et al. 2000; Kehat et al. 2001; Xu et al. 2002). This situation drove the development of more efficient methods to direct the differentiation of PSCs into cardiomyocytes, and numerous protocols have been described that employ signaling molecules known from embryology to be involved in heart development, including Nodal, Wnt, and bone morphogenetic protein 4 (BMP4) (Kinder et al. 1999; Gadue et al. 2005; Naito et al. 2006; Ueno et al. 2007; Yang et al. 2008; Evans et al. 2010; Paige et al. 2010; Zhu et al. 2011; Zhang et al. 2012; Lian et al. 2012; Lian et al. 2013; Burridge et al. 2014). Additional refinements to such methods have been reported involving the manipulation of matricellular cues and/or differentiation under more biomimetic conditions, and there are now multiple protocols that can reach cardiomyocyte purities of >90% with yields of up to 10–40 cardiomyocytes per input undifferentiated cell (Zhang et al. 2012; Lian et al. 2012, 2013; Burridge et al. 2014). Interestingly, some of these methods enable the isolation of multipotent cardiovascular progenitor cells that are capable of giving rise to vascular elements in addition to cardiomyocytes making them potentially useful for cardiac tissue engineering (Yang et al. 2008). Recent efforts to optimize cardiomyocyte differentiation have focused on reducing costs and batch-to-batch variability through the use of small molecules (Lian et al. 2012, 2013; Burridge et al. 2014) while eliminating materials of nonhuman origin (Burridge et al. 2014). These advances have helped pave the way for the large-scale, clinical-grade cardiomyocyte production that would likely be required for therapeutic applications in humans.

9.4.2 Structure and Force Generation

In general, human PSC-CMs are immature and have a phenotype similar to that of cardiomyocytes in the early fetal human heart (Mummery et al. 2003). However, their structural and functional maturation does improve with duration in culture, so our laboratory has found it convenient to separately describe their phenotype at the two stages that have been most frequently examined in the literature (Lundy et al. 2013; Kim et al. 2010; Sartiani et al. 2007; Foldes et al. 2011; Robertson et al. 2013). By this admittedly imperfect schema, so-called "early-stage" cultures refer to human PSC-CMs after approximately 3–5 weeks, while "late-stage" cultures refer to those after >12 weeks of in vitro maturation (Lundy et al. 2013). Importantly, human PSC-CMs at both stages are quite distinct structurally from the large (~130 μm long), rod-shaped ventricular cardiomyocytes that populate the adult human heart (Gerdes et al. 1998). Early-stage human PSC-CMs are small (~10 x 15 μm), round or triangular cells (Mummery et al. 2003; Lundy et al. 2013; Otsuji et al. 2010; Snir et al. 2003), and they have an irregular myofibrillar structure with relatively short sarcomere lengths (Kehat et al. 2001; Mummery et al. 2003; Snir et al. 2003; Lundy et al. 2013). By contrast, late-stage human PSC-CMs are larger, more rectangular in shape (Snir et al. 2003; Lundy et al. 2013; Foldes et al. 2011) and have more and better organized sarcomeres (Lundy et al. 2013), enhanced mitochondrial-sarcomere alignment (Snir et al. 2003), and distinct Z-, A-, H-, and I-bands (Kamakura et al. 2013). Existing human PSC-CMs also have minimal force-generating capacity, with estimates in the range of 1–100 nN per cell (Liu et al. 2012; Taylor et al. 2013; Rodriguez et al. 2014; Yang et al. 2014) versus ~10–25 μN per cell for postnatal cardiomyocytes (van der Velden et al. 1999; Witjas-Paalberends et al. 2013). Somewhat improved measurements have been reported for human PSC-CMs organized into engineered heart tissues using decellularized human cardiac matrix as a scaffold (Guyette et al. 2016), so tissue engineering and/or in vitro electromechanical conditioning may provide a solution to the weak contractile forces of human PSC-CMs.

9.4.3 Electrophysiology

Human PSC-CMs also show gradual maturation in their electrophysiological properties with duration in culture (Mummery et al. 2003; Sartiani et al. 2007; Kim et al. 2010; Lundy et al. 2013), but even late-stage human PSC-CMs are still immature compared to adult cardiomyocytes. Further complicating their electrophysiological characterization, human PSC-CM cultures typically include an admixture of cardiac subtypes (i.e., nodal-, atrial-, and ventricular-like myocytes). As ventricular derivatives are the most appropriate for cardiac repair, we focus on their electrical phenotype here. Compared to adult human ventricular cardiomyocytes, late-stage human ventricular-type PSC-CMs have a less depolarized maximum diastolic potential (MDP), lower action potential (AP) amplitude, slower AP upstroke velocity, and shorter AP duration at 90% repolarization and retain some degree of automaticity (Nabauer et al. 1996; Drouin et al. 1998; Li et al. 1998; Piacentino et al. 2003; Zhang et al. 2009; He et al. 2003; Lundy et al. 2013). The amplitude and kinetics of some important ionic conductances are similar between human PSC-CMs and their adult counterparts, such as the fast sodium (I_{Na}) (Satin et al. 2004; Jonsson et al. 2012; Sheng et al. 2012), the L-type calcium(I_{CaL}) (Zhu et al. 2009; Jonsson et al. 2012), transient outward potassium (I_{to}) (Cordeiro et al. 2013), and delayed rectified potassium (I_{Kr} and I_{Ks}) (Jonsson et al. 2012) currents. However, other important currents show major differences that contribute to their immature AP phenotype. In particular, human ESC-CMs show relatively low levels of the inward rectifier potassium current (I_{K1}) (Sartiani et al. 2007; Cordeiro et al. 2013), which stabilizes the resting membrane potential and controls excitability, and relatively high levels of the hyperpolarization-activated pacemaker current (I_f) (Sartiani et al. 2007; Weisbrod et al. 2013). The net result is a propensity for automaticity or spontaneous electrical activity, and correcting these current densities may result in a more adult-like AP phenotype (Bett et al. 2013; Meijer van Putten et al. 2015). Human PSC-CMs also exhibit slow conduction velocities (only a fraction of that measured in monolayers of postnatal cardiomyocytes (Caspi et al. 2009; Mehta et al. 2011; Lee et al. 2012; Valderrábano 2007)), an outcome that may reflect impaired connexin expression and/or function, as well as isotropically distributed gap junctions.

9.4.4 Excitation-Contraction Coupling

Excitation-contraction coupling in adult ventricular cardiomyocytes involves the opening of voltage-dependent L-type calcium channels and trans-sarcolemmal calcium entry, which is then amplified by calcium-induced calcium release (CICR) from sarcoplasmic reticulum (SR) stores (Bers 2002). As in mature ventricular myocytes, human PSC-CMs exhibit CICR from SR calcium stores that operates by a stringent "local control" mechanism during excitation-contraction coupling (Zhu et al. 2009; Zhang et al. 2013a). However, human PSC-CMs lack transverse tubules (Lieu et al. 2009) and have slow SR calcium reuptake compared to adult cardiomyocytes (Lundy et al. 2013). One explanation for impaired calcium handling in early-stage PSC-CMs may be deficient calsequestrin expression, as overexpression of this calcium-binding protein of the SR has been reported to improve calcium handling in human PSC-CMs (Liu et al. 2009).

9.4.5 Metabolism and Mitochondria

During heart development, cardiomyocytes initially use glycolysis for primary energy production but then later transition to fatty acid metabolism (Fisher 1984; Lopaschuk et al. 1991), and this shift is accompanied by a substantial increase in mitochondrial content (Schaper et al. 1985). Mitochondrial number and volume increase as human PSCs differentiate into cardiomyocytes, but only about one-third of the levels seen in adult ventricular cardiomyocytes (Kerscher et al. 2015; Yang et al. 2014; Schaper et al. 1985). With prolonged duration in culture, human PSC-CMs do show increased fatty acid utilization (Kuppusamy et al. 2015) and show a more adult-like functional localization of mitochondria adjacent to myofibrils (Snir et al. 2003; Lundy et al. 2013).

9.5 How Can the Maturation of PSC-Derived Cardiomyocytes Be Improved?

The maturation of PSC-CMs improves with prolonged duration in culture, but this approach is tedious and expensive and would likely be impractical for large-scale cardiomyocyte manufacturing. Fortunately, there are a number of more scalable approaches that have shown to promote PSC-CM maturation, at least by some parameters. For example, PSC-CMs treated with the thyroid hormone, triiodothyronine (T3), hypertrophy, elongate, and exhibit longer sarcomere lengths, more mature calcium handling, greater force generation, and increased mitochondrial respiratory capacity (Yang et al. 2014). PSC-CMs also hypertrophy following stimulation with angiotensin II and phenylephrine (Foldes et al. 2011), while treatment with neuregulin promotes electrical remodeling toward a more mature ventricular phenotype (Zhu et al. 2010). Mitochondrial localization improves in PSC-CMs treated with isoproterenol and oleic acid (Chan et al. 2015). Overexpression of let-7-g or let-7i miRNAs accelerates maturation according to structural indices, gene expression profiles, and metabolic parameters (Kuppusamy et al. 2015).

Tissue-engineering approaches have also shown considerable promise for enhancing cardiomyocyte maturation. Human PSC-CMs cultured in three-dimensional environments exhibit improved structural and electrophysiological maturation relative to cells maintained in two-dimensional environments (Zhang et al. 2013b). Cyclical pacing or stretching of these three-dimensional constructs augments the maturation of human PSC-CMs in terms of gene expression, calcium kinetics, conduction velocity, sarcomere organization, and maximal force generation (Nunes et al. 2013; Mihic et al. 2014; Ruan et al. 2015; Hirt et al. 2014). In addition, culture of human PSC-CMs on certain scaffolds improves calcium handling and cell alignment (Khan et al. 2015). Tissue engineering in combination with the maturation methods described above may provide an economical means of maturing human PSC-CMs that could be incorporated into large-scale manufacturing.

9.6 What Have We Learned from Cardiomyocyte Transplantation in Small-Animal Models?

Inspired by previous successes in the field with the transplantation of fetal rodent cardiomyocytes (Leor et al. 1996; Etzion et al. 2001; Skobel et al. 2004; Rubart et al. 2003; Scorsin et al. 1997; Li et al. 1996) and mouse ESC-CMs (Naito et al. 2004; Kolossov et al. 2006; Cai et al. 2007), we and other investigators began human ESC-CM xenotransplantation studies into mice and rats (Kehat et al. 2004; Laflamme et al. 2005; Xue et al. 2005). In early work, we found that human ESC-CMs survived relatively poorly following transplantation into recently infarcted rat hearts but that their engraftment could be significantly enhanced by transiently heat-shocking the cells pre-transplantation and/or delivering them in a pro-survival "cocktail" of factors to improve retention and attenuate cell death (Laflamme et al. 2007a). With these interventions, we were able to remuscularize a significant fraction of the infarct scar in an athymic rat model of subacute MI and observed significant improvements in left ventricular dimensions and global contractile function (Laflamme et al. 2007a). Comparable salutary effects were subsequently reported by other investigators following the transplantation of human ESC-CMs in rodent models of acute or subacute MI (Caspi et al. 2007a, Cai et al. 2007; van Laake et al.

2007), although not all found these benefits to be durable to later time points (van Laake et al. 2007; 2008). Interestingly, in a rat model of chronic MI with already established scar tissue and ventricular dysfunction, we found that human ESC-CMs engraft but do not mediate significant beneficial effects on contractile function (Fernandes et al. 2010).

Less has been reported regarding the transplantation of human iPSC-CMs in animal models, but the best available data suggest that human iPSC-CMs and ESC-CMs have comparable phenotypes in vitro (Zhang et al. 2009; Lundy et al. 2013), so one might also expect comparable outcomes in vivo. A recent study in a rat MI model compared outcomes following the transplantation of human iPSC-CMs after 8, 20, and 30 days of in vitro differentiation (Funakoshi et al. 2016). These authors found that the optimal graft size was obtained with the transplantation of 20-day-old cultures, suggesting this may be an optimal time point for harvesting and implantation. Encouragingly, they found durable beneficial effects on left ventricular dilation and fractional shortening up to 12 weeks post-transplantation.

For true heart regeneration, the newly formed muscle must be capable of electromechanical integration and synchronous activation with host myocardium. Human ESC-CMs have the subcellular machinery necessary for such integration (e.g., gap junctions), and early mapping studies provided indirect evidence that they could couple with host tissue following transplantation in uninjured hearts (Kehat et al. 2004; Xue et al. 2005). To directly demonstrate whether human ESC-CMs were capable of 1:1 host-graft coupling following transplantation into injured hearts, our group turned to the guinea pig (Fig. 9.1 panels a, b), because this small-animal model has a sinus rate (Shiotani et al. 2007) that is much closer to that of humans than do mice and rats (Swoap et al. 2004). For these experiments, we created transgenic human ESC-CMs that stably expressed GCaMP3, a calcium-sensitive fluorescent protein. In vitro these myocytes show robust fluorescence transients with each cycle of contraction. Hence, by imaging the epicardial surface of hearts engrafted with GCaMP3^{+} human ESC-CMs, we can use the graft-autonomous GCaMP3 fluorescent signal to determine whether the graft tissue is electrically active and, if so, the temporal relationship of its activation to that in host myocardium. By this approach, we found that 100% of human GCaMP3+ ESC-CM grafts in uninjured guinea pig hearts were 1:1 coupled with host myocardium (as indicated by a 1:1 relationship between GCaMP3 fluorescent transients and QRS complexes of the host ECG). Applying the same approach to human GCaMP3^{+} ESC-CMs transplanted in a guinea model of subacute MI, we observed reliable 1:1 host-graft coupling in a majority (~60%) of recipient hearts, proving that the newly formed graft muscle is at least capable of electromechanical integration with host myocardium (Shiba et al. 2012). More recently, we found that foci of 1:1 host-graft coupling can also be found in a guinea pig model of chronic MI but that the extent of electromechanical integration is significantly reduced in this context (Shiba et al. 2014).

In summary, much important information was gleaned from the preceding work in small-animal models of MI. First, PSC-CMs, including human ESC-CMs, can engraft in injured hearts and form stable implants of electromechanically integrated new myocardium within the infarct scar, a sin qua non for true heart regeneration.

Second, teratoma formation has not been reported, at least when the implanted PSC-CMs were of sufficient cardiac purity. Finally, most studies have reported beneficial effects on left ventricular contractile function, at least attenuating the adverse remodeling and progressive loss of fractional shortening that otherwise occurs over time. Taken collectively, this body of work laid the groundwork for the large-animal studies discussed in the following section.

9.7 What Have We Learned from Large-Animal Studies?

Although not necessarily a prerequisite for regulatory approval prior to human clinical trials, large-animal models provide important information that cannot be obtained using small-animal or in vitro models. Large animals possess a heart size and rate similar to humans (Gandolfi et al. 2011). Differences also continue down to the cellular and molecular levels (for a detailed review, see Chong and Murry 2014). Large-animal models also enable the testing of clinical routes of cell delivery that are not feasible in small-animal models, such as intracoronary and trans-endocardial injection (discussed further below).

Kehat and colleagues were the first to use human ESC-CMs in a large-animal model study to test their capacity for electromechanical integration following transplantation into normal (i.e., uninjured) intact myocardium (Kehat et al. 2004). In a porcine model of heart block, human ESC-CMs engrafted, integrated with the host, and provided a new source of ventricular electrical activation (as evidenced by electrocardiographic recordings and electroanatomic mapping studies). This pioneering work heralded the possibility of "biological pacemakers" composed of human PSC-CMs that could 1 day replace the implantation of mechanical devices. Importantly, this large-animal model enabled the use of a highly specialized human clinical electroanatomical mapping system to localize the source of the new ventricular ectopic rhythm. Catheter-based mapping studies of this nature would be impractical in small-animal models.

Nonhuman primates have high genetic conservation with humans and therefore provide important information for what could be expected with human cardiomyocyte transplantation. Blin and colleagues investigated the intracardiac transplantation of allogeneic nonhuman primate ESC-derived cardiovascular progenitors defined by the expression of the cell surface marker stage-specific embryonic antigen-1 (SSEA-1 or CD15) and showed that these cells help repair and regenerate infarcted recipient hearts (Blin et al. 2010). After treatment with BMP2 and WNT3A to form cardiac mesoderm, 20 million SSEA-1-positive cardiovascular progenitors were purified by magnetic bead sorting and injected using direct visualization into rhesus monkey heart infarcts. Interestingly, only calcineurin inhibitors (FK506 or cyclosporine) were used for immunosuppression in this allogeneic approach. Small grafts that express cardiomyocyte, smooth muscle, and endothelial cell proteins were present at 2 months; however, no cardiac functional assessment was performed.

In a different approach, SSEA-1-positive nonhuman primate ESC-derived cells were combined with autologous adipose-derived mesenchymal stromal cells and arranged into sheets (Bel et al. 2010). These combined cell sheets were surgically implanted onto the epicardial surface of infarcted monkey hearts and showed donor-derived cardiomyocyte differentiation at 2 months. Although it was not technically possible to distinguish, the few cardiomyocytes most likely came from the ESC-derived cells rather than the mesenchymal cells within the delivered sheets. Following cell sheet delivery, there was a trend toward improved left ventricular function; however, the small number of animals examined precludes statistically robust conclusions.

Our group recently used a nonhuman primate model to demonstrate that human ESC-CMs can electromechanically couple and remuscularize the infarcted heart (Fig. 9.1 panels c, d) (Chong et al. 2014). In contrast to previous studies, we delivered a much higher cell number (1 billion cardiomyocytes per heart) via trans-epicardial injection under direct visualization following a mini-thoractomy. The immunosuppression regimen consisted of prednisolone, cyclosporine, and the T cell co-stimulatory blocking antibody, abatacept. Using this approach, we observed considerably larger cardiomyocyte grafts than other studies. To demonstrate electromechanical coupling, we used the same method as in the guinea pig model above to prove integration. Importantly, all infarcted monkeys receiving human ESC-CMs showed ventricular tachyarrhythmias by telemetric electrocardiographic monitoring, events that were not observed prior to cell transplantation or in infarcted animals receiving vehicle alone. The mechanistic basis for these arrhythmias remains undefined and will be crucial to determine prior to progression to clinical studies.

Recently, a porcine myocardial infarction model has been used to study reparative effects of human iPSC-derived cells (Ye et al. 2014). A unique tissue-engineering approach was employed involving fibrin patches containing insulin-like growth factor-1 (IGF-1) encapsulated microspheres. These patches were implanted alone or followed by trans-epicardial injection of a combination of 2 million human iPSC-CMs, endothelial cells, and smooth muscle cells (6 million cells in total). The fibrin patches enhanced cell engraftment and cardiac functional improvement. Notably, no ventricular arrhythmias were observed in this study. For the telemetry portion of the study, animals received the patch alone or followed by human iPSC-CM transplantation. It remains unclear if the lack of arrhythmias is due to a smaller graft size, treatment with IGF-1, or if arrhythmias would have been observed after cardiomyocyte transplantation without a fibrin patch.

The preceding large-animal studies provided excellent proof of concept for cardiomyocyte transplantation as well as information of high clinical relevance. Notably, the large-scale production of useful cardiovascular derivatives from PSCs is feasible, although additional refinements will be required to improve the efficiency and costs of cell manufacturing, as well as compliance with the good manufacturing practice guidelines required for clinical use. The lack of tumorgenicity seen in small-animal studies was replicated in larger animals despite the delivery of much larger cell doses in some studies. More ominously, our group observed graft-related arrhythmias following human ESC-CM transplantation in infarcted primate hearts. This phenomenon was not seen in early small-animal studies with the same cells, underscoring the need for additional work in large-animal models with slower heart rates and greater cardiac mass. These and other important issues that will need to be addressed before clinical trials can be efficiently run are considered in the following sections.

9.8 What Are the Remaining Challenges Associated with Cardiomyocyte Transplantation?

9.8.1 Cost, Scalability, and Production

The adult human heart contains ~4 billion cardiomyocytes (Murry et al. 2006), and the loss of one-quarter of these myocytes is known to result in heart failure (Caulfield et al. 1976). Hence, while detailed dose-response experiments will be required to empirically determine the precise number of cells required, we anticipate that a cell therapy predicated on the repopulation of infarcted myocardium would require the stable engraftment of ~1 billion cardiomyocytes. In the specific case of PSC-CM-based therapies, this cell number target would obviously need to be adjusted to account for losses due to graft cell death (Robey et al. 2008) and/or immune rejection (as discussed below). While these numbers are approximately two orders of magnitude larger than the number of cells that have been employed during earlier clinical experience with BMDCs, our group and others have shown the feasibility of human PSC-CM manufacturing at this scale, even with existing cardiac differentiation methods (Chong et al. 2014). One mitigating factor is that human PSC-CMs can be cryopreserved, banked, and then thawed at high viability just prior to transplantation (Xu et al. 2011).

It is generally accepted that suspension culture systems will likely be required for cardiomyocyte manufacturing on the scale needed for human patients. The large-scale growth of undifferentiated human PSCs has been demonstrated with up to 1 billion PSCs per flask (Chen and Couture 2015). Laminin microcarriers can potentially further improve the yield of undifferentiated human PSCs in culture and have the added benefit of significantly reducing media cost (Lam et al. 2015). While variability remains in cardiomyocyte yield, some microcarrier-based systems have reached 0.3–0.6 cardiomyocytes per input PSC (Lecina et al. 2010). Of course, cardiomyocyte purity is also a concern, and with cyclic perfusion feeding, human PSC-CMs can be produced at 85% purity (Kempf et al. 2014).

Any PSC line used in human patients will have to be cultured and differentiated using clinical-grade, good manufacturing practices (GMP). To date, there are only a handful of human ESC lines that have been produced under GMP conditions (Prowse et al. 2014). Only three human ESC lines (HAD-C 100, 102, and 106) have been reported that were derived using both GMP and xeno-free conditions (Tannenbaum et al. 2012). While it is not clear that animal-free conditions will be absolutely required for regulatory approval in humans, it is clearly desirable, given persistent concerns about the infectious risks of interspecies transplantation. Importantly, whichever human PSC line is selected for clinical-grade production of cardiomyocytes for human transplantation, robust and reproducible cardiomyocyte differentiation will have to be confirmed in that cell line along with repeating some basic preclinical studies to demonstrate the expected behavior following transplantation.

9.8.2 Graft Survival

It has been estimated that >90% of graft cardiomyocytes die shortly following intracardiac transplantation (Reinecke et al. 1999), an outcome that limits the extent of remuscularization obtained unless the initial cell dose can be increased proportionately. While some cell loss can be recovered through proliferation post-transplantation, human PSC-CMs show only modest cell cycle activity that tapers off over time (Laflamme et al. 2005, 2007a, Caspi et al. 2007a). This situation has led to intense interest in identifying the mechanisms of cell death and novel methods to overcome them. In addition to delivery into the poorly vascular and pro-inflammatory environment of the infarct scar, it is generally accepted that multiple pathways contribute to graft cell death and that a multifactorial approach will likely be required (Robey et al. 2008). For example, in our own early studies of human ESC-CM transplantation in infarcted rat hearts, we found that, while individual interventions were insufficient, engraftment could be improved by combining heat-shock, provision of matricellular attachment factors, and pharmacological inhibition of multiple mitochondrial and caspase-mediated cell death pathways (Laflamme et al. 2007a). A number of other effective pro-survival interventions have been reported with ESC-CMs, including delivery in aggregates rather than as dissociated single-cell suspensions (Moon et al. 2013), activation of cytoprotective heme oxygenase-1 activity by cobalt protoporphyrin pretreatment (Luo et al. 2014a; Luo et al. 2014b), and co-delivery with supportive stromal cells (Xi et al. 2011). Transplantation studies with primary cardiomyocytes and/or other myogenic cell types have suggested other possibilities including delivery in collagen hydrogels (Kutschka et al. 2006a), Bcl-2 overexpression (Kutschka et al. 2006b), overexpression of specific miRNAs (Hu et al. 2011; Liu et al. 2011b), and removal of the pro-fibrotic and antiangiogenic matrix protein, thrombospondin-2 (Reinecke et al. 2013).

9.8.3 Immune Rejection

It is well known that transplantation of cells or tissue, foreign to the host, results in an immune reaction and this has been observed with ESC-CM transplantation (Dai et al. 2007). Therefore, either preventing immune rejection via pharmacological immunosuppression and recipient-donor human leukocyte antigen (HLA) matching or evading the immune system altogether will be key components of a viable cardiomyocyte transplantation therapy. While the risks inherent in immunosuppression (including increased propensity for infection and malignancy) are deemed acceptable in end-stage heart failure patients being listed for whole-organ transplantation, this calculation would not necessarily apply in less severely ill individuals. Indeed, the need and nature of the immunosuppression associated with any given cell therapy will inevitably affect the type of patient for whom such therapy would be appropriate. Importantly, immunosuppression in heart transplant recipients is tailored over time based on pathological evaluation of surveillance biopsies obtained via catheter from the right ventricle. Because the graft myocardium formed following cell transplantation is unlikely to be similarly accessible to surveillance biopsy, it will be challenging to similarly "fine-tune" immunosuppressant therapy in cell therapy recipients.

At first glance, an attractive approach would be to use autologous cells for transplantation, for example, using patient-derived iPSCs in each and every patient. Unfortunately, in our opinion, this strategy is completely impractical at present. While the methods of reprogramming have become more standardized and efficient over time, the costs remain prohibitively high, and it takes weeks to months to generate iPSCs and their cardiac derivatives. Equally importantly, it is not clear that there is a regulatory structure in place to approve such individualized cell therapies, even if the economic and quality-control issues could be satisfactorily resolved (Condic and Rao 2008; Solomon et al. 2015).

An alternative approach is to create and maintain large banks of genetically diverse PSCs that would allow for HLA matching (Taylor et al. 2005; Nakajima et al. 2007), much as is done currently in solid-organ transplantation. Recognition of HLA class I molecules by T lymphocytes may be the primary mechanism for immune rejection of transplanted human ESCs (Deuse et al. 2011). HLA class I molecules are expressed on most nucleated cells (Karabekian et al. 2015), and HLA I matching could help reduce immunosuppression requirements following transplantation. To reduce the number of cell lines needed in such a repository for whole population matching, one could turn to HLA-homozygous human pESCs (Daughtry and Mitalipov 2014). That said, it is probably unnecessary to consider ABO blood group antigens for patient matching, since these antigens are lost after human ESCs differentiate into cardiomyocytes (Molne et al. 2008).

A more exotic but attractive alternative approach would be to create a universal donor line that has been engineered to evade the immune system. Targeting HLA class I molecule-mediated T lymphocyte activation holds some promise. Cytotoxic T lymphocyte antigen 4 (CTLA4) binds to CD80 and CD86 preventing T cell co-stimulation by CD28, whereas programmed death ligand-1 (PD-L1) directly inhibits T cell activity (Rong et al. 2014). Human ESC-CMs that express CTLA4-Ig and PD-L1 survive when transplanted into the hind limbs of non-immunosuppressed mice with humanized immune systems (Rong et al. 2014). Preventing T cell activation improves human ESC-CM graft survival in ischemic mouse myocardium (Huber et al. 2013). Beta-2 microglobulin is a component of the invariable, non-transmembrane light chain of HLA class I molecules (Karabekian et al. 2015). Human ESCs with disrupted beta-2 microglobulin have impaired surface expression of HLA class I molecules and reduced immunogenicity in vitro (Lu et al. 2013a), and beta-2 microglobulin-reduced human ESC-CMs do not stimulate allogeneic T cell proliferation in vitro

(Karabekian et al. 2015). If these strategies are successful in creating a universal donor human ESC, it would have tremendous impact on advancing the field of cardiomyocyte transplantation.

9.8.4 Integration and Arrhythmias

To contribute useful new force-generating units, graft myocardium must integrate and contract synchronously with host muscle. As described above, human ESC-CM grafts have been shown to be capable of such electromechanical integration in both intact and recently injured hearts (Kehat et al. 2004; Shiba et al. 2012; Shiba et al. 2014; Chong et al. 2014; Ye et al. 2014). Their capacity for integration in chronically injured hearts with established scar has been less well studied, but the best available evidence suggests that their integration is much more limited in this context, perhaps accounting for their failure to exert beneficial effects on contractile function in a guinea pig chronic MI model (Fernandes et al. 2010). If this observation is born out in subsequent large-animal studies, it would represent a significant barrier to successful translation in the large proportion of patients with remote MIs and established heart failure.

The capacity of PSC-CMs to integrate is a double-edged sword. While electrical coupling between graft and host tissue is required to mediate contractile benefits in a direct remuscularization strategy, it raises concerns about the risk of graft-related arrhythmias. Indeed, in the aforementioned work by our group in a nonhuman primate MI model, we observed nonlethal ventricular tachyarrhythmias in 100% of infarcted primates receiving hESC-CMs (Chong et al. 2014). There are several potential mechanisms by which PSC-CMs could contribute to electrical instability. One possibility is graft ectopy, i.e., a situation in which the graft tissue acts as an ectopic pacemaker. Consistent with this mechanism, most available PSC-CMs are comprised of a mixture of cardiac subtypes including a significant fraction of nodal/pacemaker myocytes (Zhang et al. 2012; Karakikes et al. 2014; He et al. 2003; Zhu et al. 2010; Moore et al. 2008; Kamakura et al. 2013; Weng et al. 2014). Moreover, ventricular myocytes in the developing heart show some degree of automaticity, so even purified ventricular PSC-CMs would be expected to retain some degree of automaticity unless matured (Zhang et al. 2002; Sartiani et al. 2007; Kim et al. 2010). Tachyarrhythmias may also arise from slowly firing ectopic foci within the infarcted area because of poor coupling, source-sink electrical mismatch, or network heterogeneity (Pumir et al. 2005; Biktashev et al. 2008). PSC-CMs have also been reported to have a propensity for afterdepolarizations and triggered arrhythmias (Zhang et al. 2002). Finally, PSC-CM graft tissue also shows low levels of isotropically distributed connexin-43, a situation that likely underlies the slow conduction velocities recently measured in graft tissue (Shiba et al. 2012, 2014; Chong et al. 2014). Slow conduction velocity, particularly through spatially heterogeneous islands of poorly connected graft tissue, seems likely to promote reentrant arrhythmias (Shiba et al. 2012, 2014). Overall, this is an understudied area, and the atypical ion channel expression and diverse mechanisms of graft-related arrhythmia may make them refractory to conventional antiarrhythmic therapies.

9.8.5 Tumorigenicity

PSCs are defined by their ability to form teratomas (a benign proliferation including differentiated elements from all three embryonic germ layers), so the risk of tumor formation is an inherent concern with any PSC-based cell therapy. The risk of tumor formation with any given therapy will obviously vary with cell dose, graft-cell autonomous factors (e.g., proliferative index), host factors (e.g., graft environment, host immune responses, inflammation), and the route of delivery. Cell dose is a particular issue for cardiovascular applications in which relatively large cell numbers (~10^9 per recipient) are likely to be required.

There are two sources from which tumors may arise following PSC-based therapies: (1) residual undifferentiated PSCs that might give rise to teratomas (Nussbaum et al. 2007; Fico et al. 2012; McLenachan et al. 2012; Menendez et al. 2012) and (2) malignant transformation of graft cells, which might lead to tumors even in the absence of undifferentiated cells (Ben-David et al. 2014). To minimize the risk of the first possibility, it is obviously desirable to maximize cardiomyocyte purity while minimizing or eliminating undifferentiated PSCs or incompletely committed PSC derivatives. Undifferentiated cells can be depleted by cytotoxic targeting of the cell surface marker claudin-6 (Ben-David et al. 2013), inhibition of stearoyl-coA desaturase (Zhang et al. 2014), or culture under methionine-free conditions (Matsuura et al. 2015). A number of strategies to increase PSC-CM purity have been described including Percoll gradient centrifugation (Lin et al. 2010), promoting differentiation of cardiac progenitor cells by TNFα overexpression (Behfar et al. 2007), sorting via expression of SIRPA (Dubois et al. 2011) or VCAM-1 (Ponten et al. 2013), mitochondrial labeling (Hattori et al. 2010), and metabolic selection with glucose depletion (Hemmi et al. 2014). Simply increasing the time in culture post-differentiation may also reduce or eliminate tumorigenicity as has been shown to be the case with neural precursors (Kozubenko et al. 2010; Doi et al. 2012a, b; Kobayashi et al. 2012). Collectively, the enrichment of human PSC-CM purity and time in culture post-derivation may account for their tumor-free engraftment observed in mouse (van Laake et al. 2008), rat (Caspi et al. 2007a; Citro et al. 2014), guinea pig (Shiba et al. 2012, 2014), pig (Ye et al. 2014), and nonhuman primate (Chong et al. 2014) models. With regard to the risk of malignant transformation of graft cells, the main factor is likely aneuploidy that can occur with prolonged cell culture (Ben-David et al. 2014). This risk can be mitigated through careful monitoring of human PSCs and their derivatives for karyotypic abnormalities during cell manufacturing. Cardiomyocytes themselves have a relatively low cell cycle activity (perhaps accounting for the relatively rarity of primary cardiac tumors), so the risk of transformation is probably relatively low once PSCs have become cardiac-committed.

9.9 How Should Cardiomyocytes Be Delivered to the Myocardium?

Depending on the clinical requirements, there are different therapeutic approaches to cardiomyocyte transplantation that could be considered. For example, if cell therapy must be administered emergently during an acute MI, then the transplant material has to be readily available, i.e., available "off the shelf". Many preclinical experiments performed thus far have used injection of whole-cell suspensions to repair damage from a MI, an approach that can be extrapolated to an allogeneic, "off-the-shelf" cell product. On the other hand, this approach, involving the local delivery of therapeutic cells, may be inappropriate in other clinical contexts, for example, a non-ischemic cardiomyopathy in which all or nearly all of the myocardium is dysfunctional. In this situation, tissue engineering or whole-organ fabrication may be needed to wholly replace the diseased tissue. These latter methods would also be amenable to incorporation of biomaterials aimed at structural support or providing cellular maturation cues.

The delivery mechanism that has been employed in the vast majority of preclinical studies of human PSC-CMs to date is direct, transepicardial cell injection via an open thoracotomy (Shiba et al. 2014; Chong et al. 2014). Because this approach requires exposure and direct visualization of the targeted region of the heart, it is necessarily invasive. Hence, unless less-invasive strategies (i.e., a mini-thoracotomy) can be demonstrated, this transepicardial route will require a sternotomy in humans, making individuals that already have an indication for cardiac surgery an attractive patient population for early clinical trials. A number of other less-invasive and even closed-chest approaches to cardiomyocyte delivery have been proposed, including steerable catheters with an injection needle for endocardial delivery or intracoronary catheters equipped with an injection needle (Wang et al. 2009). Steerable endocardial catheters could be guided using fluoroscopy (Amado et al. 2005), echocardiography, or voltage mapping as is done with cardiac electrophysiology procedures. Another benefit of direct myocardial injection over intracoronary delivery is that materials, such as hydrogels, can be delivered in addition to cardiomyocytes to provide a biodegradable scaffold during engraftment (Habib et al. 2011).

While a number of adult stem cell types have been delivered to the myocardium via an intracoronary route including autologous cardiosphere-derived cells (Makkar et al. 2012; Malliaras et al. 2014) and non-myogenic stem cells (Katritsis et al. 2005; Yang et al. 2010), this approach is not likely relevant to the delivery of PSC-CMs. First, the retention of cells following delivery via the intracoronary route is known to be relatively poor (with ~2% of the injected cells being retained) (Leiker et al. 2008), obviously problematic in a situation in which billions of stably engrafted cardiomyocytes are required. A more important consideration relates to cell size. Human PSC-CMs (≥25–40 μm long) (Lundy et al. 2013) are significantly larger than other cell types that have been delivery via the intracoronary route including BMDCs (~10 μm), mesenchymal stem cells (~20 μm) and cardiosphere-derived cells (~20 μm) (Johnston et al. 2009). It is very unlikely that differentiated cardiomyocytes would be capable of diapedesis and migration out of a vessel wall, so their larger size and predilection for clumping poses a major risk for vascular occlusion.

Another approach that has proven promising in animal studies is the delivery of human PSC-CMs after organization into engineered cardiac cell sheets. Early proof of principle for this approach came from studies using primary cardiomyocytes. Cardiac cell sheets are formed by culturing cardiomyocytes in monolayers on temperature-responsive polymers and then releasing the monolayer from the polymer by reducing the temperature (Shimizu et al. 2002). Monolayers can be stacked together to form multilayer cardiac cell sheets that can be transplanted onto the heart, integrate with host tissue, are angiogenic, and improve cardiac contractile function (Miyagawa et al. 2005; Sekiya et al. 2006; Furuta et al. 2006; Sekine et al. 2006). Importantly, engineered cardiac cells sheets can be made using human iPSC-CMs, and these have been engrafted into a porcine model of myocardial infarction with improved cardiac performance, even though the cells did not survive long term (Kawamura et al. 2012). The thickness of engineered cardiac cell sheets has typically been limited to three layers because of hypoxia; however, sequential grafting or alternating with layers of gelatin hydrogel microspheres can produce sheets between 9 and 15 layers thick (Komae et al. 2015; Matsuo et al. 2015). As an alternative to cardiac cell sheets, human ESC-CMs can be grown on polymer scaffolds or can be mixed with collagen and cast into three-dimensional constructs, termed engineered heart tissue (EHT) (Chen et al. 2010; Xu et al. 2013; Chen et al. 2015; Riegler et al. 2015). EHTs from human ESC-CMs and collagen have been successfully engrafted onto rat hearts in a chronic MI model, and while cardiac systolic function improved, the improvement did not correlate with cardiomyocyte viability (Riegler et al. 2015). Because cardiac cell sheets and EHTs are engineered, different cells and materials can be incorporated to change their function. Changing cellular composition by adding fibroblasts and/or vascular cells can improve vascularization, conduction velocity, or tension of the engineered tissue (Caspi et al. 2007b; Sekine et al. 2008; Lesman et al. 2010; Tulloch et al. 2011; Kreutziger et al. 2011; Liau et al. 2011; Hibino et al. 2012; Thavandiran et al. 2013). Some biodegradable scaffold materials have conductive properties (Baheiraei et al. 2015), are autologous like the omentum (Shevach et al. 2014) and urinary bladder (Turner et al. 2012), or structurally and mechanically tunable like silk combined with cardiac extracellular matrix (Stoppel et al. 2015). Engineered tissue may also present new challenges such as engineered cell sheets being capable of producing reentrant arrhythmias in vitro (Kadota et al. 2013); however, it is not known how host-graft integration will impact this.

Whole-organ tissue engineering is yet another approach to replacing damaged cardiac tissue, but this will be significantly more complicated than constructing small, three-dimensional EHTs. Building the scaffolds from the ground up for this type of fabrication has not yet been accomplished, but cell-free extracellular matrix scaffolds with preserved structure have been made from multiple organs, including the heart, using organ decellularization (Ott et al. 2008; Wainwright et al. 2010; Carvalho et al. 2012; Guyette et al. 2014; 2016; Weymann et al. 2015; Kawasaki et al. 2015). Human iPSC-CMs can grow on cell-free scaffolds made from both human (Oberwallner et al. 2014; Guyette et al. 2016) and mouse myocardium (Lu et al. 2013b). Decellularized porcine heart scaffolds have been reseeded with mixtures of cells including neonatal mouse cardiomyocytes and human umbilical vein endothelial cells (Weymann et al. 2014) as well as neonatal rat cardiomyocytes, fibroblasts, and endothelial cells in rat heart scaffolds (Yasui et al. 2014). For whole-organ tissue engineering to be a successful method of cardiomyocyte delivery, there are important hurdles that remain. One of the most significant problems is that acellular vascular grafts are thrombogenic; and while this phenomenon can be reduced somewhat by re-endothelialization (Robertson et al. 2014), graft intravascular thrombosis would clearly impair perfusion and graft viability. Seeding efficiency also remains very low, for example, despite the introduction of as many as 500 million human iPSC-CMs into whole human heart cardiac matrices, only 50% of a small (~5 cm^3) target region was successfully reseeded in one recent study (Guyette et al. 2016). In our opinion, these parameters would have to be greatly improved for this approach to compete successfully with the injection of cardiomyocyte cell suspensions, which have been shown to successfully form comparably sized volumes of host-perfused myocardium (Chong et al. 2014).

9.10 What Preclinical Information Is Still Needed to Facilitate Successful Translation?

The field has made tremendous progress in recent years, but, in the present authors' opinion, there are still many unanswered questions that should be addressed in preclinical studies to optimize cardiomyocyte transplantation for success in eventual clinical trials. Most importantly, we need to take every precaution to avoid an outcome akin to the one that derailed the gene therapy field for many years, and there are a number of major safety concerns (e.g., graft-related arrhythmias, tumorigenesis) that are inherent to cardiomyocyte transplantation. Moreover, we must be mindful that there are finite resources available to support the development and testing of novel cardiovascular therapeutics. Even putting aside the risk of patient harm, a premature move to the clinic and finding a null or insignificant beneficial effect could permanently derail progress in the field. Indeed, there may be recent precedent for this outcome from clinical trials testing BMDCs in ischemic heart disease. BMDCs moved relatively rapidly from small-animal studies to humans, and it has been suggested that their disappointing performance in randomized clinical trials may have resulted in part from an incomplete mechanistic understanding, suboptimal cell doses and timing of delivery, and an inappropriate target patient population (Laflamme et al. 2007b; Simari et al. 2014). Given that PSC-based therapies arguably represent a "higher-risk, higher-reward" cell source, a more systematic, methodical approach to preclinical optimization seems warranted and would help facilitate successful clinical translation.

First and foremost, the feasibility and efficacy of large-scale remuscularization must be firmly established in a large-animal model, and the specific dose and composition of the cardiomyocytes needs to be better defined. For example, doses of human ESC derivatives ranging from 2 million to 1 billion cells have been tested in large-animal models (Ye et al. 2014; Chong et al. 2014), suggesting that the field is far from a consensus in terms of what mechanism of action and cell numbers are most likely to mediate clinically relevant beneficial effects. The optimal cellular composition also needs to be firmly established. For example, stromal and endothelial cells make up the majority of the heart by nuclei (Bergmann et al. 2015), and so it makes intuitive sense that these cells should be incorporated in a heart regeneration strategy. However, human ESC-derived multipotent cardiovascular progenitors did not outperform committed cardiomyocytes in a recent head-to-head study (Fernandes et al. 2015).

Another issue that needs further optimization in large-animal studies is the approach to cardiomyocyte delivery. The various methods of cardiomyocyte delivery described above (e.g., surgical injection or catheter-based injection of cell suspensions versus implantation of an engineered muscle "patch") should be tested, ideally in a head-to-head comparison, and that approach resulting in the best retention of viable cells and/or greatest beneficial effects on LV contractile function should be prioritized in early clinical trials. To our knowledge, no one has reported the viability of PSC-CMs after passage through a cell-delivery catheter even in vitro, although this potential method of delivery has been frequently proposed. For any approach involving cell injection, large-animal studies are also required to inform the optimal number of injections, volume per injection site, and distribution of sites.

The risk of graft-related arrhythmias is a particularly important hurdle to successful translation for the reasons described above. Even if this risk can be greatly minimized by enhancing the input cell population (e.g., by employing more mature and/or ventricular-enriched PSC-CMs), additional preclinical studies seem warranted to determine the electrophysiological responses of graft tissue to antiarrhythmic drugs and other relevant conditions likely to be encountered in heart failure patients (e.g., ischemia). An informed decision will also have to be made as to whether all early PSC-CM recipients will be required to have an automatic implantable cardioverter-defibrillator (AICD), as was the case during clinical trials with skeletal myoblasts. Such a requirement would obviously provide a greater margin of safety and so may be entirely appropriate, but it would necessarily affect economic considerations and target patient population during trial design.

Finally, immunosuppression is another important issue and one that is particularly challenging to investigate based on xenotransplantation experiments involving human cardiomyocytes implanted into animal models. Allotransplantation experiments within a single species would likely better model responses to cardiomyocyte transplantation in humans. For example, monkey ESC-CMs could be transplanted into allogeneic monkeys and evaluated for the minimal level of immunosuppression needed to ensure graft survival.

9.11 What Is the Status of PSC-Derived Cardiomyocyte Transplantation in Human Trials?

To date there have been no clinical trials testing the transplantation of committed cardiomyocytes from PSCs or any potential source. However, investigators in France recently initiated a clinical trial to test human ESC-derived cardiovascular progenitor cells delivered via surgery in a fibrin scaffold, although their objective is not to achieve direct remuscularization. The Transplantation of Human Embryonic Stem Cell-Derived Progenitors in Severe Heart Failure (ESCORT, NCT02057900) trial is a phase 1 clinical trial that aims to recruit six patients undergoing cardiac surgery (e.g., coronary artery bypass grafting or mitral valve repair/replacement) to simultaneously receive an epicardial implant of multipotent cardiovascular progenitor cells. These cells are generated by treating undifferentiated human ESCs with the growth factor BMP2 and SU-5402, a small-molecule fibroblast growth factor receptor (FGF)-specific tyrosine kinase inhibitor. Somewhat paradoxically, these cells are defined by their expression of the surface marker SSEA-1, which is expressed by multipotent cardiovascular progenitors but is more commonly used as a pluripotency marker (Menasche et al. 2015a). A relatively small number of cells (up to 10 million) will be implanted in the ESCORT trial, with the expectation that cell implants will mediate beneficial effects via an indirect, paracrine mechanism of action rather than direct remuscularization (Menasche et al. 2015a). In addition to the aforementioned requirement that the enrolled patients must already be undergoing an open-chest procedure, other inclusion criteria include: (1) age 18–81 years, (2) LV ejection fraction ≤35%, (3) MI older than 6 months, (4) New York Heart Association class III or IV heart failure despite optimal medical care, (5) AICD placement with cardiac resynchronization therapy as indicated, and (6) not eligible for heart transplantation.

In 2015, the ESCORT team released a case report describing their first enrolled patient (Menasche et al. 2015b). The patient is a 68-year-old, insulin-dependent diabetic woman with New York Heart Association class III heart failure from an anterolateral MI with a LV ejection fraction of 26% on maximal medical therapy. She underwent coronary artery bypass grafting with left internal thoracic artery (left internal mammary artery) to the left anterior descending artery without suitable targets for revascularization of the left circumflex coronary arterial tree. The progenitor cell-containing patch made with 4 million cells was grafted onto the lateral LV in the region visually identified to have scar. She received immunosuppression with methylprednisolone, cyclosporine, and mycophenolate mofetil. At 3 months she has not had any adverse events and her LV ejection fraction improved to 36% by echocardiographic evaluation with the akinetic lateral wall improving to moderately hypokinetic. Her improvement in LV ejection fraction is in line with the effects of coronary artery bypass grafting for ischemic cardiomyopathy in the setting of anterior wall viability (Sharma et al. 2011); however, whether the functional improvement of the lateral wall was due to improved collateral flow or the overlying cellular patch is unknown.

9.12 How Should Future Clinical Trials with PSC-Derived Cardiomyocytes Be Designed?

Cardiomyocyte transplantation has the potential for both tremendous benefit and serious risks that will vary depending on the patient population selected. It will be important to recognize and account for these differences in planning the first clinical trials of human PSC-CM transplantation. In our opinion, the first patient population to be considered for study should have severe, symptomatic (NYHA class III or IV) ischemic cardiomyopathy, because patients with less severe cardiomyopathy are usually well managed with contemporary medical therapy alone. In our view, the most appropriate participants of such a trial can be further divided into four broad categories of patients: (1) individuals with another indication for cardiac surgery, (2) candidates for long-term mechanical circulatory support with a LV assist device (LVAD) with or without a subsequent heart transplant, (3) chronic heart failure patients who are not candidates for LVAD or heart transplantation, or (4) individuals with cardiogenic shock from acute MI. Each of these patient profiles has unique clinical characteristics that would affect trial design and outcomes.

Patients who are already undergoing cardiac surgery, as in the ESCORT trial, would allow for delivery of cardiomyocytes via transepicardial injection under direct visualization. As discussed above, this is the most studied delivery method to date and is arguably the approach by which the delivery of cardiomyocytes could be best controlled. Furthermore, since these patients will already have a cardiac surgery, the decision to transplant cardiomyocytes does not impose an additional sternotomy or surgery.

Another potential population that has been proposed for early clinical trials testing cardiomyocyte transplantation are end-stage heart failure patients receiving a LVAD. In this instance, the patients at question are obviously already undergoing an open-chest procedure (device implantation), and they will be receiving a device that is capable of taking over LV cardiac output, providing an additional margin of safety. That said, the support of an LVAD does not eliminate all risk; for example, a left-sided pump cannot compensate for right ventricular arrhythmias or right-sided failure. Importantly, the LVAD recipient population is further subdivided into two categories of patients: those being "bridged to transplantation" (i.e., supported with the device until a suitable donor heart becomes available) and those receiving the device as "destination therapy" (i.e., supported with the device permanently, as they are not candidates for heart transplantation). Each group has its pros and cons as potential subjects in a clinical trial of PSC-based cardiac therapy. Obviously, "bridge-to-transplant" patients have a "backup" plan in place, since they would be listed for heart transplantation. In this case, the engrafted heart would be removed at the time of transplantation, which would increase the margin of safety with regard to tumorigenesis while also providing invaluable information as to the structural fate of the graft cardiomyocytes. At the same time, a major disadvantage of "bridge-to-transplant" patients is that graft cells could allosensitize them, negatively affecting their transplant candidacy or outcomes post-transplant. A major drawback with both "bridge-to-transplant" and "destination therapy" patients is that human PSC-CM recipients would likely require chronic immunosuppression to avoid graft cell rejection, thereby greatly increasing their risk of serious LVAD or driveline infection. Finally, while the LVAD patient population has a number of safety advantages, it would be challenging to determine efficacy given the coexisting device, and there are obvious ethical issues inherent in cell transplantation in individuals that are unlikely to be benefit (and may be harmed).

On the other hand, end-stage heart failure patients who are not candidates for LVAD or heart transplantation would have to accept all of the risks of cardiomyocyte transplantation and immunosuppression without any "backup" strategy in the case of a bad outcome (e.g., graft-related tumor). Looking beyond early safety studies, an important consideration here would be whether such individuals might stand any chance of benefiting from cardiomyocyte transplantation. Given that human PSC-CMs have not mediated significant beneficial effects in chronic MI in the limited number of animal studies to date, it is possible that these patients would be accepting all of the risks with little or no likelihood of benefit.

By contrast, animal studies have so far suggested that human PSC-CM transplantation can have a significant beneficial effect in the acute phase post-MI. The risk-benefit ratio calculation here is very challenging, because most patients respond well to medical therapy following acute MI and go on to have only mild or moderate ischemic cardiomyopathy. Because it is not straightforward to identify and enroll those acute MI patients that will go on to have more serious dysfunction, one would likely need to have a large body of preclinical (and ideally clinical) safety data before contemplating cardiomyocyte transplantation in this context. One could conceivably target acute MI patients with cardiogenic shock that are anticipated to need long-term circulatory support (e.g., LVAD), but this is a relatively small proportion of patients and would make recruitment very challenging.

In summary, we think that currently there is a division between the clinical scenario where human PSC-CM transplantation will most likely be beneficial (e.g., acute MI) and the patient population with the greatest clinical need. Since patients with severe, chronic ischemic cardiomyopathy comprise the most likely target population for human PSC-CM transplantation, it seems important that the preclinical data demonstrate benefit in this disease context. As such, clinical benefit needs to be shown in small- and large-animal studies that best model patients with chronic ischemic cardiomyopathy. Once the efficacy of human PSC-CM transplantation is firmly established in chronic ischemic cardiomyopathy, then planning could begin for human clinical trials keeping in mind the different patient groups discussed above.

Conclusions

Cardiomyocyte transplantation is not yet ready to be tested in human clinical trials, but tremendous progress has been made in recent years through in vitro work and increasingly relevant animal models of MI. Cardiomyocyte manufacturing is now possible at a scale and cardiac purity that would have seemed impossible just a few years ago. Recent preclinical studies have demonstrated the feasibility of remuscularizing a significant fraction of the infarct scar with electrically integrated new myocardium, and impressive beneficial effects on LV contractile function have been reported, at least in small-animal MI models. Going forward, the field must address a number of practical uncertainties including the most appropriate dosage, cell composition, route of administration, and immunosuppressive therapy. These are issues that should be addressed first in large-animal studies, which are costly, tedious, and absolutely essential. The field also needs to demonstrate functional efficacy following cardiomyocyte transplantation in a more clinically relevant large-animal model and must satisfactorily overcome concerns about graft cell survival, arrhythmias, and tumor formation. Post-MI heart failure is a disease with high morbidity and mortality, so the need for new therapeutic options is acute. Cardiomyocyte transplantation is a highly promising approach, but its effective translation will require patience, hard work, cooperation, and a methodical approach in preclinical and first-in-human studies.

Compliance with Ethical Standards

Conflict of Interest The authors declare that they have no conflict of interest.

Ethical approval This article does not contain any studies with human participants performed by any of the authors. Animal research was performed in accordance with institutional, local, and federal regulations. This work was supported by T32 HL007312 (MEH), Future Leader Fellowship (ID 100463) from the National Heart Foundation of Australia (JJHC) and Sydney Medical School Foundation Fellowship (JJHC), RO1 NIH HL117991 (MAL), and funding from the McEwen Centre for Regenerative Medicine, the Toronto General Hospital Research Institute, the Ontario Institute for Regenerative Medicine, and the Peter Munk Cardiac Centre (MAL).

References

Ahmad R, Wolber W, Eckardt S et al (2012) Functional neuronal cells generated by human parthenogenetic stem cells. PLoS One 7(8):e42800. doi:10.1371/journal.pone.0042800

Amado LC, Saliaris AP, Schuleri KH et al (2005) Cardiac repair with intramyocardial injection of allogeneic mesenchymal stem cells after myocardial infarction. Proc Natl Acad Sci U S A 102(32):11474–11479

Atkins BZ, Lewis CW, Kraus WE et al (1999a) Intracardiac transplantation of skeletal myoblasts yields two populations of striated cells in situ. Ann Thorac Surg 67(1):124–129

Atkins BZ, Hueman MT, Meuchel JM et al (1999b) Myogenic cell transplantation improves in vivo regional performance in infarcted rabbit myocardium. J Heart Lung Transplant 18(12):1173–1180

Baheiraei N, Yeganeh H, Ai J et al (2015) Preparation of a porous conductive scaffold from aniline pentamer-modified polyurethane/PCL blend for cardiac tissue engineering. J Biomed Mater Res A 103(10):3179–3187. doi:10.1002/jbm.a.35447

Balsam LB, Wagers AJ, Christensen JL et al (2004) Haematopoietic stem cells adopt mature haematopoietic fates in ischaemic myocardium. Nature 428(6983):668–673. doi:10.1038/nature02460

Behfar A, Perez-Terzic C, Faustino RS et al (2007) Cardiopoietic programming of embryonic stem cells for tumor-free heart repair. J Exp Med 204(2):405–420

Bel A, Planat-Bernard V, Saito A et al (2010) Composite cell sheets: a further step toward safe and effective myocardial regeneration by cardiac progenitors derived from embryonic stem cells. Circulation 122(11 Suppl):S118–S123. doi:10.1161/CIRCULATIONAHA.109.927293

Bellin M, Casini S, Davis RP et al (2013) Isogenic human pluripotent stem cell pairs reveal the role of a KCNH2 mutation in long-QT syndrome. EMBO J 32(24):3161–3175. doi:10.1038/emboj.2013.240

Ben-David U, Nudel N, Benvenisty N (2013) Immunologic and chemical targeting of the tight-junction protein Claudin-6 eliminates tumorigenic human pluripotent stem cells. Nat Commun 4:1992. doi:10.1038/ncomms2992

Ben-David U, Arad G, Weissbein U et al (2014) Aneuploidy induces profound changes in gene expression, proliferation and tumorigenicity of human pluripotent stem cells. Nat Commun 5:4825. doi:10.1038/ncomms5825

Bergmann O, Bhardwaj RD, Bernard S et al (2009) Evidence for cardiomyocyte renewal in humans. Science 324(5923):98–102. doi:10.1126/science.1164680

Bergmann O, Zdunek S, Felker A et al (2015) Dynamics of cell generation and turnover in the human heart. Cell 161(7):1566–1575. doi:10.1016/j.cell.2015.05.026

Bers DM (2002) Cardiac excitation-contraction coupling. Nature 415(6868):198–205. doi:10.1038/415198a

Bett GC, Kaplan AD, Lis A et al (2013) Electronic "expression" of the inward rectifier in cardiocytes derived from human-induced pluripotent stem cells. Heart Rhythm 10(12):1903–1910. doi:10.1016/j.hrthm.2013.09.061

Biktashev VN, Arutunyan A, Sarvazyan NA (2008) Generation and escape of local waves from the boundary of uncoupled cardiac tissue. Biophys J 94(9):3726–3738. doi:10.1529/biophysj.107.117630

Blin G, Nury D, Stefanovic S et al (2010) A purified population of multipotent cardiovascular progenitors derived from primate pluripotent stem cells engrafts in postmyocardial infarcted nonhuman primates. J Clin Invest 120(4):1125–1139. doi:10.1172/JCI40120

Brevini TA, Pennarossa G, Antonini S et al (2009) Cell lines derived from human parthenogenetic embryos can display aberrant centriole distribution and altered expression levels of mitotic spindle check-point transcripts. Stem Cell Rev 5(4):340–352. doi:10.1007/s12015-009-9086-9

Burridge PW, Matsa E, Shukla P et al (2014) Chemically defined generation of human cardiomyocytes. Nat Methods 11(8):855–860. doi:10.1038/nmeth.2999

Cai J, Yi FF, Yang XC et al (2007) Transplantation of embryonic stem cell-derived cardiomyocytes improves cardiac function in infarcted rat hearts. Cytotherapy 9(3):283–291

Carvalho JL, de Carvalho PH, Gomes DA et al (2012) Characterization of decellularized heart matrices as biomaterials for regular and whole organ tissue engineering and initial recellularization with ips cells. J Tissue Sci Eng Suppl 11:002. doi:10.4172/2157-7552.S11-002

Caspi O, Huber I, Kehat I et al (2007a) Transplantation of human embryonic stem cell-derived cardiomyocytes improves myocardial performance in infarcted rat hearts. J Am Coll Cardiol 50(19):1884–1893

Caspi O, Lesman A, Basevitch Y et al (2007b) Tissue engineering of vascularized cardiac muscle from human embryonic stem cells. Circ Res 100(2):263–272

Caspi O, Itzhaki I, Kehat I et al (2009) In vitro electrophysiological drug testing using human embryonic stem cell derived cardiomyocytes. Stem Cells Dev 18(1):161–172. doi:10.1089/scd.2007.0280

Caulfield JB, Leinbach R, Gold H (1976) The relationship of myocardial infarct size and prognosis. Circulation 53(3 Suppl):I141–I144

Chan HY, Keung W, Li RA et al (2015) Morphometric analysis of human embryonic stem cell-derived ventricular cardiomyocytes: determining the maturation state of a population by quantifying parameters in individual cells. Stem Cells Int 2015:586908. doi:10.1155/2015/586908

Chen VC, Couture LA (2015) The suspension culture of undifferentiated human pluripotent stem cells using spinner flasks. Methods Mol Biol 1283:13–21. doi:10.1007/7651_2014_118

Chen QZ, Ishii H, Thouas GA et al (2010) An elastomeric patch derived from poly(glycerol sebacate) for delivery of embryonic stem cells to the heart. Biomaterials 31(14):3885–3893. doi:10.1016/j.biomaterials.2010.01.108

Chen JX, Krane M, Deutsch MA et al (2012) Inefficient reprogramming of fibroblasts into cardiomyocytes using Gata4, Mef2c, and Tbx5. Circ Res 111(1):50-55. doi:10.1161/CIRCRESAHA.112.270264

Chen Y, Wang J, Shen B et al (2015) Engineering a freestanding biomimetic cardiac patch using biodegradable poly(lactic-co-glycolic acid) (PLGA) and human embryonic stem cell-derived ventricular cardiomyocytes (hESC-VCMs). Macromol Biosci 15(3):426–436. doi:10.1002/mabi.201400448

Chong JJ, Murry CE (2014) Cardiac regeneration using pluripotent stem cells--progression to large animal models. Stem Cell Res 13(3 Pt B):654–665. doi:10.1016/j.scr.2014.06.005

Chong JJ, Yang X, Don CW et al (2014) Human embryonic-stem-cell-derived cardiomyocytes regenerate non-human primate hearts. Nature 510(7504):273–277. doi:10.1038/nature13233

Cibelli JB, Cunniff K, Vrana KE (2006) Embryonic stem cells from parthenotes. Methods Enzymol 418:117–135

Citro L, Naidu S, Hassan F et al (2014) Comparison of human induced pluripotent stem-cell derived cardiomyocytes with human mesenchymal stem cells following acute myocardial infarction. PLoS One 9(12):e116281. doi:10.1371/journal.pone.0116281

Condic ML, Rao M (2008) Regulatory issues for personalized pluripotent cells. Stem Cells 26(11):2753–2758. doi:10.1634/stemcells.2008-0421

Cordeiro JM, Nesterenko VV, Sicouri S et al (2013) Identification and characterization of a transient outward K+ current in human induced pluripotent stem cell-derived cardiomyocytes. J Mol Cell Cardiol 60:36–46. doi:10.1016/j.yjmcc.2013.03.014

Dai W, Field LJ, Rubart M et al (2007) Survival and maturation of human embryonic stem cell-derived cardiomyocytes in rat hearts. J Mol Cell Cardiol 43(4):504–516

Daughtry B, Mitalipov S (2014) Concise review: parthenote stem cells for regenerative medicine: genetic, epigenetic, and developmental features. Stem Cells Transl Med 3(3):290–298. doi:10.5966/sctm.2013-0127

Deuse T, Seifert M, Tyan D et al (2011) Immunobiology of naive and genetically modified HLA-class-I-knockdown human embryonic stem cells. J Cell Sci 124(Pt 17):3029–3037. doi:10.1242/jcs.087718

Didie M, Christalla P, Rubart M et al (2013) Parthenogenetic stem cells for tissue-engineered heart repair. J Clin Invest 123(3):1285–1298. doi:10.1172/JCI66854

Doi D, Morizane A, Kikuchi T et al (2012a) Prolonged maturation culture favors a reduction in the tumorigenicity and the dopaminergic function of human ESC-derived neural cells in a primate model of Parkinson's disease. Stem Cells 30(5):935–945. doi:10.1002/stem.1060

Doi D, Morizane A, Kikuchi T et al (2012b) Prolonged maturation culture favors a reduction in the tumorigenicity and the dopaminergic function of human ESC-derived neural cells in a primate model of Parkinson's disease. Stem Cells 30(5):935–945. doi:10.1002/stem.1060

Drawnel FM, Boccardo S, Prummer M et al (2014) Disease modeling and phenotypic drug screening for diabetic cardiomyopathy using human induced pluripotent stem cells. Cell Rep 9(3):810–821. doi:10.1016/j.celrep.2014.09.055

Drouin E, Lande G, Charpentier F (1998) Amiodarone reduces transmural heterogeneity of repolarization in the human heart. J Am Coll Cardiol 32(4):1063–1067

Dubois NC, Craft AM, Sharma P et al (2011) SIRPA is a specific cell-surface marker for isolating cardiomyocytes derived from human pluripotent stem cells. Nat Biotechnol 29(11):1011–1018. doi:10.1038/nbt.2005

Etzion S, Battler A, Barbash IM et al (2001) Influence of embryonic cardiomyocyte transplantation on the progression of heart failure in a rat model of extensive myocardial infarction. J Mol Cell Cardiol 33(7):1321–1330. doi:10.1006/jmcc. 2000.1391

Evans SM, Yelon D, Conlon FL et al (2010) Myocardial lineage development. Circ Res 107(12):1428–1444. doi:10.1161/CIRCRESAHA.110.227405

Fatkhudinov T, Bolshakova G, Arutyunyan I et al (2015) Bone marrow-derived multipotent stromal cells promote myocardial fibrosis and reverse remodeling of the left ventricle. Stem Cells Int 2015:746873. doi:10.1155/2015/746873

Fernandes S, Naumova AV, Zhu WZ et al (2010) Human embryonic stem cell-derived cardiomyocytes engraft but do not alter cardiac remodeling after chronic infarction in rats. J Mol Cell Cardiol 49(6):941–949. doi:10.1016/j.yjmcc.2010.09.008

Fernandes S, Chong JJ, Paige SL et al (2015) Comparison of human embryonic stem cell-derived cardiomyocytes, cardiovascular progenitors, and bone marrow mononuclear cells for cardiac repair. Stem Cell Reports 5(5):753–762. doi:10.1016/j.stemcr. 2015.09.011

Fico A, De Chevigny A, Egea J et al (2012) Modulating Glypican4 suppresses tumorigenicity of embryonic stem cells while preserving self-renewal and pluripotency. Stem Cells 30(9):1863–1874. doi:10.1002/stem.1165

Fisher DJ (1984) Oxygenation and metabolism in the developing heart. Semin Perinatol 8(3):217–225

Foldes G, Mioulane M, Wright JS et al (2011) Modulation of human embryonic stem cell-derived cardiomyocyte growth: a testbed for studying human cardiac hypertrophy? J Mol Cell Cardiol 50(2):367–376. doi:10.1016/j.yjmcc.2010.10.029

Fu Y, Huang C, Xu X et al (2015) Direct reprogramming of mouse fibroblasts into cardiomyocytes with chemical cocktails. Cell Res 25(9):1013–1024. doi:10.1038/cr.2015.99

Fujimoto KL, Clause KC, Liu LJ et al (2011) Engineered fetal cardiac graft preserves its cardiomyocyte proliferation within postinfarcted myocardium and sustains cardiac function. Tissue Eng Part A 17(5–6):585–596. doi:10.1089/ten.TEA.2010.0259

Funakoshi S, Miki K, Takaki T et al (2016) Enhanced engraftment, proliferation, and therapeutic potential in heart using optimized human iPSC-derived cardiomyocytes. Sci Rep 6:19111. doi:10.1038/srep19111

Furuta A, Miyoshi S, Itabashi Y et al (2006) Pulsatile cardiac tissue grafts using a novel three-dimensional cell sheet manipulation technique functionally integrates with the host heart, in vivo. Circ Res 98(5):705–712

Gadue P, Huber TL, Nostro MC et al (2005) Germ layer induction from embryonic stem cells. Exp Hematol 33(9):955–964. doi:10.1016/j.exphem.2005.06.009

Gandolfi F, Vanelli A, Pennarossa G et al (2011) Large animal models for cardiac stem cell therapies. Theriogenology 75(8):1416–1425. doi:10.1016/j.theriogenology.2011.01.026

Gerdes AM, Onodera T, Tamura T et al (1998) New method to evaluate myocyte remodeling from formalin-fixed biopsy and autopsy material. J Card Fail 4(4):343–348

Guyette JP, Gilpin SE, Charest JM et al (2014) Perfusion decellularization of whole organs. Nat Protoc 9(6):1451–1468. doi:10.1038/nprot.2014.097

Guyette JP, Charest J, Mills RW et al (2016) Bioengineering human myocardium on native extracellular matrix. Circ Res 118(1):56–72. doi:10.1161/CIRCRESAHA.115.306874

Habib M, Shapira-Schweitzer K, Caspi O et al (2011) A combined cell therapy and in-situ tissue-engineering approach for myocardial repair. Biomaterials 32(30):7514–7523. doi:10.1016/j.biomaterials.2011.06.049

Hassink RJ, Pasumarthi KB, Nakajima H et al (2008) Cardiomyocyte cell cycle activation improves cardiac function after myocardial infarction. Cardiovasc Res 78(1):18–25

Hattori F, Chen H, Yamashita H et al (2010) Nongenetic method for purifying stem cell-derived cardiomyocytes. Nat Methods 7(1):61–66. doi:10.1038/nmeth.1403

He JQ, Ma Y, Lee Y et al (2003) Human embryonic stem cells develop into multiple types of cardiac myocytes: action potential characterization. Circ Res 93(1):32–39. doi:10.1161/01.RES.0000080317.92718.99

Hemmi N, Tohyama S, Nakajima K et al (2014) A massive suspension culture system with metabolic purification for human pluripotent stem cell-derived cardiomyocytes. Stem Cells Transl Med 3(12):1473–1483. doi:10.5966/sctm.2014-0072

Hibino N, Duncan DR, Nalbandian A et al (2012) Evaluation of the use of an induced pluripotent stem cell sheet for the construction of tissue-engineered vascular grafts. J Thorac Cardiovasc Surg 143(3):696–703. doi:10.1016/j.jtcvs.2011.06.046

Hirt MN, Boeddinghaus J, Mitchell A et al (2014) Functional improvement and maturation of rat and human engineered heart tissue by chronic electrical stimulation. J Mol Cell Cardiol 74:151–161. doi:10.1016/j.yjmcc.2014.05.009

Hu S, Huang M, Nguyen PK et al (2011) Novel microRNA prosurvival cocktail for improving engraftment and function of cardiac progenitor cell transplantation. Circulation 124(11 Suppl):S27–S34. doi:10.1161/CIRCULATIONAHA.111.017954

Huber BC, Ransohoff JD, Ransohoff KJ et al (2013) Costimulation-adhesion blockade is superior to cyclosporine a and prednisone immunosuppressive therapy for preventing rejection of differentiated human embryonic stem cells following transplantation. Stem Cells 31(11):2354–2363. doi:10.1002/stem.1501

Ieda M, Fu JD, Delgado-Olguin P et al (2010) Direct reprogramming of fibroblasts into functional cardiomyocytes by defined factors. Cell 142(3):375–386. doi:10.1016/j.cell.2010.07.002

Isaev DA, Garitaonandia I, Abramihina TV et al (2012) In vitro differentiation of human parthenogenetic stem cells into neural lineages. Regen Med 7(1):37–45. doi:10.2217/rme.11.110

Iso Y, Spees JL, Serrano C et al (2007) Multipotent human stromal cells improve cardiac function after myocardial infarction in mice without long-term engraftment. Biochem Biophys Res Commun 354(3):700–706

Itskovitz-Eldor J, Schuldiner M, Karsenti D et al (2000) Differentiation of human embryonic stem cells into embryoid bodies compromising the three embryonic germ layers. Mol Med 6(2):88–95

Itzhaki I, Maizels L, Huber I et al (2011) Modelling the long QT syndrome with induced pluripotent stem cells. Nature 471(7337):225–229. doi:10.1038/nature09747

Jayawardena TM, Egemnazarov B, Finch EA et al (2012) MicroRNA-mediated in vitro and in vivo direct reprogramming of cardiac fibroblasts to cardiomyocytes. Circ Res 110(11):1465–1473. doi:10.1161/CIRCRESAHA.112.269035

Johnston PV, Sasano T, Mills K et al (2009) Engraftment, differentiation, and functional benefits of autologous cardiosphere-derived cells in porcine ischemic cardiomyopathy. Circulation 120(12):1075–1083., 7 p following 1083. doi:10.1161/CIRCULATIONAHA.108.816058

Jonsson MK, Vos MA, Mirams GR et al (2012) Application of human stem cell-derived cardiomyocytes in safety pharmacology requires caution beyond hERG. J Mol Cell Cardiol 52(5):998–1008. doi:10.1016/j.yjmcc.2012.02.002

Jopling C, Sleep E, Raya M et al (2010) Zebrafish heart regeneration occurs by cardiomyocyte dedifferentiation and proliferation. Nature 464(7288):606–609. doi:10.1038/nature08899

Kadota S, Minami I, Morone N et al (2013) Development of a reentrant arrhythmia model in human pluripotent stem cell-derived cardiac cell sheets. Eur Heart J 34(15):1147–1156. doi:10.1093/eurheartj/ehs418

Kamakura T, Makiyama T, Sasaki K et al (2013) Ultrastructural maturation of human-induced pluripotent stem cell-derived cardiomyocytes in a long-term culture. Circ J 77(5):1307–1314

Karabekian Z, Ding H, Gtybayeva G et al (2015) HLA class I depleted hESC as a source of hypoimmunogenic cells for tissue engineering applications. Tissue Eng Part A. doi:10.1089/ten.TEA.2015.0105

Karakikes I, Senyei GD, Hansen J et al (2014) Small molecule-mediated directed differentiation of human embryonic stem cells toward ventricular cardiomyocytes. Stem Cells Transl Med 3(1):18–31. doi:10.5966/sctm.2013-0110

Katritsis DG, Sotiropoulou PA, Karvouni E et al (2005) Transcoronary transplantation of autologous mesenchymal stem cells and endothelial progenitors into infarcted human myocardium. Catheter Cardiovasc Interv 65(3):321–329. doi:10.1002/ccd.20406

Kawamura M, Miyagawa S, Miki K et al (2012) Feasibility, safety, and therapeutic efficacy of human induced pluripotent stem cell-derived cardiomyocyte sheets in a porcine ischemic cardiomyopathy model. Circulation 126(11 Suppl 1):S29–S37. doi:10.1161/CIRCULATIONAHA.111.084343

Kawasaki T, Kirita Y, Kami D et al (2015) Novel detergent for whole organ tissue engineering. J Biomed Mater Res A 103(10):3364–3373. doi:10.1002/jbm.a.35474

Kehat I, Kenyagin-Karsenti D, Snir M et al (2001) Human embryonic stem cells can differentiate into myocytes with structural and functional properties of cardiomyocytes. J Clin Invest 108(3):407–414. doi:10.1172/JCI12131

Kehat I, Khimovich L, Caspi O et al (2004) Electromechanical integration of cardiomyocytes derived from human embryonic stem cells. Nat Biotechnol 22(10):1282–1289. doi:10.1038/nbt1014

Kempf H, Olmer R, Kropp C et al (2014) Controlling expansion and cardiomyogenic differentiation of human pluripotent stem cells in scalable suspension culture. Stem Cell Rep 3(6):1132–1146. doi:10.1016/j.stemcr.2014.09.017

Kerscher P, Bussie BS, DeSimone KM et al (2015) Characterization of mitochondrial populations during stem cell differentiation. Methods Mol Biol 1264:453–463. doi:10.1007/978-1-4939-2257-4_37

Khan M, Xu Y, Hua S et al (2015) Evaluation of changes in morphology and function of human induced pluripotent stem cell derived cardiomyocytes (HiPSC-CMs) cultured on an aligned-nanofiber cardiac patch. PLoS One 10(5):e0126338. doi:10.1371/journal.pone.0126338

Kikuchi K, Holdway JE, Werdich AA et al (2010) Primary contribution to zebrafish heart regeneration by gata 4(+) cardiomyocytes. Nature 464(7288):601–605. doi:10.1038/nature08804

Kim C, Majdi M, Xia P et al (2010) Non-cardiomyocytes influence the electrophysiological maturation of human embryonic stem cell-derived cardiomyocytes during differentiation. Stem Cells Dev 19(6):783–795. doi:10.1089/scd.2009.0349

Kinder SJ, Tsang TE, Quinlan GA et al (1999) The orderly allocation of mesodermal cells to the extraembryonic structures and the anteroposterior axis during gastrulation of the mouse embryo. Development 126(21):4691–4701

Klug MG, Soonpaa MH, Koh GY et al (1996) Genetically selected cardiomyocytes from differentiating embronic stem cells form stable intracardiac grafts. J Clin Invest 98(1):216–224. doi:10.1172/JCI118769

Kobayashi Y, Okada Y, Itakura G et al (2012) Pre-evaluated safe human iPSC-derived neural stem cells promote functional recovery after spinal cord injury in common marmoset without tumorigenicity. PLoS One 7(12):e52787. doi:10.1371/journal.pone.0052787

Kolossov E, Bostani T, Roell W et al (2006) Engraftment of engineered ES cell-derived cardiomyocytes but not BM cells restores contractile function to the infarcted myocardium. J Exp Med 203(10):2315–2327. doi:10.1084/jem.20061469

Komae H, Sekine H, Dobashi I et al (2015) Three-dimensional functional human myocardial tissues fabricated from induced pluripotent stem cells. J Tissue Eng Regen Med 11(3):926–935. doi:10.1002/term.1995

Kozubenko N, Turnovcova K, Kapcalova M et al (2010) Analysis of in vitro and in vivo characteristics of human embryonic stem cell-derived neural precursors. Cell Transplant 19(4):471–486. doi:10.3727/096368909X484707

Kreutziger KL, Muskheli V, Johnson P et al (2011) Developing vasculature and stroma in engineered human myocardium. Tissue Eng Part A 17(9–10):1219–1228. doi:10.1089/ten.TEA.2010.0557

Kuppusamy KT, Jones DC, Sperber H et al (2015) Let-7 family of microRNA is required for maturation and adult-like metabolism in stem cell-derived cardiomyocytes. Proc Natl Acad Sci U S A 112(21):E2785–E2794. doi:10.1073/pnas.1424042112

Kutschka I, Chen IY, Kofidis T et al (2006a) Collagen matrices enhance survival of transplanted cardiomyoblasts and contribute to functional improvement of ischemic rat hearts. Circulation 114(1 Suppl):I167–I173. doi:10.1161/CIRCULATIONAHA.105.001297

Kutschka I, Kofidis T, Chen IY et al (2006b) Adenoviral human BCL-2 transgene expression attenuates early donor cell death after cardiomyoblast transplantation into ischemic rat hearts. Circulation 114(1 Suppl):I174–I180. doi:10.1161/CIRCULATIONAHA.105.001370

van Laake LW, Passier R, Monshouwer-Kloots J et al (2007) Human embryonic stem cell-derived cardiomyocytes survive and mature in the mouse heart and transiently improve function after myocardial infarction. Stem Cell Res 1(1):9–24. doi:10.1016/j.scr.2007.06.001

van Laake LW, Passier R, Doevendans PA et al (2008) Human embryonic stem cell-derived cardiomyocytes and cardiac repair in rodents. Circ Res 102(9):1008–1010. doi:10.1161/CIRCRESAHA.108.175505

Laflamme MA, Gold J, Xu C et al (2005) Formation of human myocardium in the rat heart from human embryonic stem cells. Am J Pathol 167(3):663–671. doi:10.1016/S0002-9440(10)62041-X

Laflamme MA, Chen KY, Naumova AV et al (2007a) Cardiomyocytes derived from human embryonic stem cells in pro-survival factors enhance function of infarcted rat hearts. Nat Biotechnol 25(9):1015–1024. doi:10.1038/nbt1327

Laflamme MA, Zbinden S, Epstein SE et al (2007b) Cell-based therapy for myocardial ischemia and infarction: pathophysiological mechanisms. Annu Rev Pathol 2:307–339. doi:10.1146/annurev.pathol.2.010506.092038

Lam AT, Li J, Chen AK et al (2015) Improved human pluripotent stem cell attachment and spreading on xeno-free laminin-521-coated microcarriers results in efficient growth in agitated cultures. Biores Open Access 4(1):242–257. doi:10.1089/biores.2015.0010

Lecina M, Ting S, Choo A et al (2010) Scalable platform for human embryonic stem cell differentiation to cardiomyocytes in suspended microcarrier cultures. Tissue Eng Part C Methods 16(6):1609–1619. doi:10.1089/ten.TEC.2010.0104

Lee P, Klos M, Bollensdorff C et al (2012) Simultaneous voltage and calcium mapping of genetically purified human induced pluripotent stem cell-derived cardiac myocyte monolayers. Circ Res 110(12):1556–1563. doi:10.1161/CIRCRESAHA.111.262535

Leiker M, Suzuki G, Iyer VS et al (2008) Assessment of a nuclear affinity labeling method for tracking implanted mesenchymal stem cells. Cell Transplant 17(8):911–922

Leor J, Patterson M, Quinones MJ et al (1996) Transplantation of fetal myocardial tissue into the infarcted myocardium of rat. A potential method for repair of infarcted myocardium? Circulation 94(9 Suppl):II332–II336

Leschik J, Caron L, Yang H et al (2015) A view of bivalent epigenetic marks in two human embryonic stem cell lines reveals a different cardiogenic potential. Stem Cells Dev 24(3):384–392. doi:10.1089/scd. 2014.0345

Lesman A, Habib M, Caspi O et al (2010) Transplantation of a tissue-engineered human vascularized cardiac muscle. Tissue Eng Part A 16(1):115–125. doi:10.1089/ten.TEA.2009.0130

Li RK, Jia ZQ, Weisel RD et al (1996) Cardiomyocyte transplantation improves heart function. Ann Thorac Surg 62(3):654–660. discussion 660-1

Li GR, Feng J, Yue L et al (1998) Transmural heterogeneity of action potentials and Ito1 in myocytes isolated from the human right ventricle. Am J Phys 275(2 Pt 2):H369–H377

Li J, He J, Lin G et al (2014) Inducing human parthenogenetic embryonic stem cells into isletlike clusters. Mol Med Rep 10(6):2882–2890. doi:10.3892/mmr.2014.2588

Lian X, Hsiao C, Wilson G et al (2012) Robust cardiomyocyte differentiation from human pluripotent stem cells via temporal modulation of canonical Wnt signaling. Proc Natl Acad Sci U S A 109(27):E1848–E1857. doi:10.1073/pnas.1200250109

Lian X, Zhang J, Azarin SM et al (2013) Directed cardiomyocyte differentiation from human pluripotent stem cells by modulating Wnt/beta-catenin signaling under fully defined conditions. Nat Protoc 8(1):162–175. doi:10.1038/nprot.2012.150

Liau B, Christoforou N, Leong KW et al (2011) Pluripotent stem cell-derived cardiac tissue patch with advanced structure and function. Biomaterials 32(35):9180–9187. doi:10.1016/j.biomaterials.2011.08.050

Lieu DK, Liu J, Siu CW et al (2009) Absence of transverse tubules contributes to non-uniform ca(2+) wavefronts in mouse and human embryonic stem cell-derived cardiomyocytes. Stem Cells Dev 18(10):1493–1500. doi:10.1089/scd.2009.0052

Lin H, Lei J, Wininger D et al (2003) Multilineage potential of homozygous stem cells derived from metaphase II oocytes. Stem Cells 21(2):152–161. doi:10.1634/stemcells.21-2-152

Lin Q, Fu Q, Zhang Y et al (2010) Tumourigenesis in the infarcted rat heart is eliminated through differentiation and enrichment of the transplanted embryonic stem cells. Eur J Heart Fail 12(11):1179–1185. doi:10.1093/eurjhf/hfq144

Liu J, Lieu DK, Siu CW et al (2009) Facilitated maturation of Ca2+ handling properties of human embryonic stem cell-derived cardiomyocytes by calsequestrin expression. Am J Phys Cell Phys 297(1):C152–C159. doi:10.1152/ajpcell.00060.2009

Liu W, Yin Y, Jiang Y et al (2011a) Genetic and epigenetic X-chromosome variations in a parthenogenetic human embryonic stem cell line. J Assist Reprod Genet 28(4):303–313. doi:10.1007/s10815-010-9517-1

Liu J, van Mil A, Vrijsen K et al (2011b) MicroRNA-155 prevents necrotic cell death in human cardiomyocyte progenitor cells via targeting RIP1. J Cell Mol Med 15(7):1474–1482. doi:10.1111/j.1582-4934.2010.01104.x

Liu J, Sun N, Bruce MA et al (2012) Atomic force mechanobiology of pluripotent stem cell-derived cardiomyocytes. PLoS One 7(5):e37559. doi:10.1371/journal.pone.0037559

Lopaschuk GD, Spafford MA, Marsh DR (1991) Glycolysis is predominant source of myocardial ATP production immediately after birth. Am J Phys 261(6 Pt 2):H1698–H1705

Lu P, Chen J, He L et al (2013a) Generating hypoimmunogenic human embryonic stem cells by the disruption of beta 2-microglobulin. Stem Cell Rev 9(6):806–813. doi:10.1007/s12015-013-9457-0

Lu TY, Lin B, Kim J et al (2013b) Repopulation of decellularized mouse heart with human induced pluripotent stem cell-derived cardiovascular progenitor cells. Nat Commun 4:2307. doi:10.1038/ncomms3307

Lundy SD, Zhu WZ, Regnier M et al (2013) Structural and functional maturation of cardiomyocytes derived from human pluripotent stem cells. Stem Cells Dev 22(14):1991–2002. doi:10.1089/scd.2012.0490

Luo J, Weaver MS, Cao B et al (2014a) Cobalt protoporphyrin pretreatment protects human embryonic stem cell-derived cardiomyocytes from hypoxia/reoxygenation injury in vitro and increases graft size and vascularization in vivo. Stem Cells Transl Med 3(6):734–744. doi:10.5966/sctm.2013-0189

Luo J, Weaver MS, Dennis JE et al (2014b) Targeting survival pathways to create infarct-spanning bridges of human embryonic stem cell-derived cardiomyocytes. J Thorac Cardiovasc Surg 148(6):3180-8.e1. doi:10.1016/j.jtcvs.2014.06.087

Makkar RR, Smith RR, Cheng K et al (2012) Intracoronary cardiosphere-derived cells for heart regeneration after myocardial infarction (CADUCEUS): a prospective, randomised phase 1 trial. Lancet 379(9819):895–904. doi:10.1016/S0140-6736(12)60195-0

Malliaras K, Makkar RR, Smith RR et al (2014) Intracoronary cardiosphere-derived cells after myocardial infarction: evidence of therapeutic regeneration in the final 1-year results of the CADUCEUS trial (CArdiosphere-derived aUtologous stem CElls to reverse ventricUlar dySfunction). J Am Coll Cardiol 63(2):110–122. doi:10.1016/j.jacc.2013.08.724

Matsuo T, Masumoto H, Tajima S et al (2015) Efficient long-term survival of cell grafts after myocardial infarction with thick viable cardiac tissue entirely from pluripotent stem cells. Sci Rep 5:16842. doi:10.1038/srep16842

Matsuura K, Wada H, Nagai T et al (2004) Cardiomyocytes fuse with surrounding noncardiomyocytes and reenter the cell cycle. J Cell Biol 167(2):351–363. doi:10.1083/jcb.200312111

Matsuura K, Kodama F, Sugiyama K et al (2015) Elimination of remaining undifferentiated induced pluripotent stem cells in the process of human cardiac cell sheet fabrication using a methionine-free culture condition. Tissue Eng Part C Methods 21(3):330–338. doi:10.1089/ten.TEC.2014.0198

McLenachan S, Menchon C, Raya A et al (2012) Cyclin A1 is essential for setting the pluripotent state and reducing tumorigenicity of induced pluripotent stem cells. Stem Cells Dev 21(15):2891–2899. doi:10.1089/scd.2012.0190

Mehta A, Chung YY, Ng A et al (2011) Pharmacological response of human cardiomyocytes derived from virus-free induced pluripotent stem cells. Cardiovasc Res 91(4):577–586. doi:10.1093/cvr/cvr132

Meijer van Putten RM, Mengarelli I, Guan K et al (2015) Ion channelopathies in human induced pluripotent stem cell derived cardiomyocytes: a dynamic clamp study with virtual IK1. Front Physiol 6:7. doi:10.3389/fphys.2015.00007

Menasche P, Alfieri O, Janssens S et al (2008) The myoblast autologous grafting in ischemic cardiomyopathy (MAGIC) trial: first randomized placebo-controlled study of myoblast transplantation. Circulation 117(9):1189–1200. doi:10.1161/CIRCULATIONAHA.107.734103

Menasche P, Vanneaux V, Fabreguettes JR et al (2015a) Towards a clinical use of human embryonic stem cell-derived cardiac progenitors: a translational experience. Eur Heart J 36(12):743–750. doi:10.1093/eurheartj/ehu192

Menasche P, Vanneaux V, Hagege A et al (2015b) Human embryonic stem cell-derived cardiac progenitors for severe heart failure treatment: first clinical case report. Eur Heart J 36(30):2011–2017. doi:10.1093/eurheartj/ehv189

Menendez S, Camus S, Herreria A et al (2012) Increased dosage of tumor suppressors limits the tumorigenicity of iPS cells without affecting their pluripotency. Aging Cell 11(1):41–50. doi:10.1111/j.1474-9726.2011.00754.x

Mihic A, Li J, Miyagi Y et al (2014) The effect of cyclic stretch on maturation and 3D tissue formation of human embryonic stem cell-derived cardiomyocytes. Biomaterials 35(9):2798–2808. doi:10.1016/j.biomaterials.2013.12.052

Mirotsou M, Jayawardena TM, Schmeckpeper J et al (2011) Paracrine mechanisms of stem cell reparative and regenerative actions in the heart. J Mol Cell Cardiol 50(2):280–289. doi:10.1016/j.yjmcc. 2010.08.005

Miyagawa S, Sawa Y, Sakakida S et al (2005) Tissue cardiomyoplasty using bioengineered contractile cardiomyocyte sheets to repair damaged myocardium: their integration with recipient myocardium. Transplantation 80(11):1586–1595

Molne J, Bjorquist P, Andersson K et al (2008) Blood group ABO antigen expression in human embryonic stem cells and in differentiated hepatocyte- and cardiomyocyte-like cells. Transplantation 86(10):1407–1413. doi:10.1097/TP.0b013e31818a6805

Moon SH, Kang SW, Park SJ et al (2013) The use of aggregates of purified cardiomyocytes derived from human ESCs for functional engraftment after myocardial infarction. Biomaterials 34(16):4013–4026. doi:10.1016/j.biomaterials.2013.02.022

Moore JC, Fu J, Chan YC et al (2008) Distinct cardiogenic preferences of two human embryonic stem cell (hESC) lines are imprinted in their proteomes in the pluripotent state. Biochem Biophys Res Commun 372(4):553–558. doi:10.1016/j.bbrc. 2008.05.076

Mummery C, Ward-van Oostwaard D, Doevendans P et al (2003) Differentiation of human embryonic stem cells to cardiomyocytes: role of coculture with visceral endoderm-like cells. Circulation 107(21):2733–2740. doi:10.1161/01.CIR.0000068356.38592.68

Murry CE, Wiseman RW, Schwartz SM et al (1996) Skeletal myoblast transplantation for repair of myocardial necrosis. J Clin Invest 98(11):2512–2523. doi:10.1172/JCI119070

Murry CE, Soonpaa MH, Reinecke H et al (2004) Haematopoietic stem cells do not transdifferentiate into cardiac myocytes in myocardial infarcts. Nature 428(6983):664–668. doi:10.1038/nature02446

Murry CE, Reinecke H, Pabon LM (2006) Regeneration gaps: observations on stem cells and cardiac repair. J Am Coll Cardiol 47(9):1777–1785. doi:10.1016/j.jacc.2006.02.002

Nabauer M, Beuckelmann DJ, Uberfuhr P et al (1996) Regional differences in current density and rate-dependent properties of the transient outward current in subepicardial and subendocardial myocytes of human left ventricle. Circulation 93(1):168–177

Nagaya N, Fujii T, Iwase T et al (2004) Intravenous administration of mesenchymal stem cells improves cardiac function in rats with acute myocardial infarction through angiogenesis and myogenesis. Am J Physiol Heart Circ Physiol 287(6):H2670–H2676. doi:10.1152/ajpheart.01071.2003

Naito H, Nishizaki K, Yoshikawa M et al (2004) Xenogeneic embryonic stem cell-derived cardiomyocyte transplantation. Transplant Proc 36(8):2507–2508. doi:10.1016/j.transproceed.2004.06.031

Naito AT, Shiojima I, Akazawa H et al (2006) Developmental stage-specific biphasic roles of Wnt/beta-catenin signaling in cardiomyogenesis and hematopoiesis. Proc Natl Acad Sci U S A 103(52):19812–19817. doi:10.1073/pnas.0605768103

Nakajima F, Tokunaga K, Nakatsuji N (2007) Human leukocyte antigen matching estimations in a hypothetical bank of human embryonic stem cell lines in the Japanese population for use in cell transplantation therapy. Stem Cells 25(4):983–985. doi:10.1634/stemcells.2006-0566

Nam YJ, Song K, Luo X et al (2013) Reprogramming of human fibroblasts toward a cardiac fate. Proc Natl Acad Sci U S A 110(14):5588–5593. doi:10.1073/pnas.1301019110

Norol F, Bonnet N, Peinnequin A et al (2007) GFP-transduced CD34+ and Lin- CD34- hematopoietic stem cells did not adopt a cardiac phenotype in a nonhuman primate model of myocardial infarct. Exp Hematol 35(4):653–661. doi:10.1016/j.exphem. 2006.12.003

Nunes SS, Miklas JW, Liu J et al (2013) Biowire: a platform for maturation of human pluripotent stem cell-derived cardiomyocytes. Nat Methods 10(8):781–787. doi:10.1038/nmeth.2524

Nussbaum J, Minami E, Laflamme MA et al (2007) Transplantation of undifferentiated murine embryonic stem cells in the heart: teratoma formation and immune response. FASEB J 21(7):1345–1357. doi:10.1096/fj.06-6769com

Oberwallner B, Brodarac A, Choi YH et al (2014) Preparation of cardiac extracellular matrix scaffolds by decellularization of human myocardium. J Biomed Mater Res A 102(9):3263–3272. doi:10.1002/jbma.35000

Otsuji TG, Minami I, Kurose Y et al (2010) Progressive maturation in contracting cardiomyocytes derived from human embryonic stem cells: qualitative effects on electrophysiological responses to drugs. Stem Cell Res 4(3):201–213. doi:10.1016/j.scr.2010.01.002

Ott HC, Matthiesen TS, Goh SK et al (2008) Perfusion-decellularized matrix: using nature's platform to engineer a bioartificial heart. Nat Med 14(2):213–221. doi:10.1038/nm1684

Paige SL, Osugi T, Afanasiev OK et al (2010) Endogenous Wnt/beta-catenin signaling is required for cardiac differentiation in human embryonic stem cells. PLoS One 5(6):e11134. doi:10.1371/journal.pone.0011134

Pasumarthi KB, Field LJ (2002) Cardiomyocyte cell cycle regulation. Circ Res 90(10):1044–1054

Piacentino V III, Weber CR, Chen X et al (2003) Cellular basis of abnormal calcium transients of failing human ventricular myocytes. Circ Res 92(6):651–658. doi:10.1161/01.RES.0000062469.83985.9B

Ponten A, Walsh S, Malan D et al (2013) FACS-based isolation, propagation and characterization of mouse embryonic cardiomyocytes based on VCAM-1 surface marker expression. PLoS One 8(12):e82403. doi:10.1371/journal.pone.0082403

Porrello ER, Mahmoud AI, Simpson E et al (2011) Transient regenerative potential of the neonatal mouse heart. Science 331(6020):1078–1080. doi:10.1126/science.1200708

Poss KD, Wilson LG, Keating MT (2002) Heart regeneration in zebrafish. Science 298(5601):2188–2190. doi:10.1126/science.1077857

Prowse AB, Timmins NE, Yau TM et al (2014) Transforming the promise of pluripotent stem cell-derived cardiomyocytes to a therapy: challenges and solutions for clinical trials. Can J Cardiol 30(11):1335–1349. doi:10.1016/j.cjca.2014.08.005

Pumir A, Arutunyan A, Krinsky V et al (2005) Genesis of ectopic waves: role of coupling, automaticity, and heterogeneity. Biophys J 89(4):2332–2349. doi:10.1529/biophysj.105.061820

Qian L, Huang Y, Spencer CI et al (2012) In vivo reprogramming of murine cardiac fibroblasts into induced cardiomyocytes. Nature 485(7400):593–598. doi:10.1038/nature11044

Qian L, Berry EC, Fu JD et al (2013) Reprogramming of mouse fibroblasts into cardiomyocyte-like cells in vitro. Nat Protoc 8(6):1204–1215. doi:10.1038/nprot.2013.067

Reinecke H, Zhang M, Bartosek T et al (1999) Survival, integration, and differentiation of cardiomyocyte grafts: a study in normal and injured rat hearts. Circulation 100(2):193–202

Reinecke H, Poppa V, Murry CE (2002) Skeletal muscle stem cells do not transdifferentiate into cardiomyocytes after cardiac grafting. J Mol Cell Cardiol 34(2):241–249. doi:10.1006/jmcc.2001.1507

Reinecke H, Robey TE, Mignone JL et al (2013) Lack of thrombospondin-2 reduces fibrosis and increases vascularity around cardiac cell grafts. Cardiovasc Pathol 22(1):91–95. doi:10.1016/j.carpath.2012.03.005

Revazova ES, Turovets NA, Kochetkova OD et al (2007) Patient-specific stem cell lines derived from human parthenogenetic blastocysts. Cloning Stem Cells 9(3):432–449. doi:10.1089/clo.2007.0033

Riegler J, Tiburcy M, Ebert A et al (2015) Human engineered heart muscles engraft and survive long-term in a rodent myocardial infarction model. Circ Res 117(8):720–730. doi:10.1161/CIRCRESAHA.115.306985

Robertson C, Tran DD, George SC (2013) Concise review: maturation phases of human pluripotent stem cell-derived cardiomyocytes. Stem Cells 31(5):829–837. doi:10.1002/stem.1331

Robertson MJ, Dries-Devlin JL, Kren SM et al (2014) Optimizing recellularization of whole decellularized heart extracellular matrix. PLoS One 9(2):e90406. doi:10.1371/journal.pone.0090406
Robey TE, Saiget MK, Reinecke H et al (2008) Systems approaches to preventing transplanted cell death in cardiac repair. J Mol Cell Cardiol 45(4):567–581. doi:10.1016/j.yjmcc.2008.03.009
Rodriguez ML, Graham BT, Pabon LM et al (2014) Measuring the contractile forces of human induced pluripotent stem cell-derived cardiomyocytes with arrays of microposts. J Biomech Eng 136(5):051005. doi:10.1115/1.4027145
Roell W, Lewalter T, Sasse P et al (2007) Engraftment of connexin 43-expressing cells prevents post-infarct arrhythmia. Nature 450(7171):819–824. doi:10.1038/nature06321
Rogers NT, Hobson E, Pickering S et al (2004) Phospholipase Czeta causes Ca2+ oscillations and parthenogenetic activation of human oocytes. Reproduction 128(6):697–702. doi:10.1530/rep.1.00484
Rong Z, Wang M, Hu Z et al (2014) An effective approach to prevent immune rejection of human ESC-derived allografts. Cell Stem Cell 14(1):121–130. doi:10.1016/j.stem.2013.11.014
Ruan JL, Tulloch NL, Saiget M et al (2015) Mechanical stress promotes maturation of human myocardium from pluripotent stem cell-derived progenitors. Stem Cells 33(7):2148–2157. doi:10.1002/stem.2036
Rubart M, Pasumarthi KB, Nakajima H et al (2003) Physiological coupling of donor and host cardiomyocytes after cellular transplantation. Circ Res 92(11):1217–1224. doi:10.1161/01.RES.0000075089.39335.8C
Sadat K, Ather S, Aljaroudi W et al (2014) The effect of bone marrow mononuclear stem cell therapy on left ventricular function and myocardial perfusion. J Nucl Cardiol 21(2):351–367. doi:10.1007/s12350-013-9846-4
Sartiani L, Bettiol E, Stillitano F et al (2007) Developmental changes in cardiomyocytes differentiated from human embryonic stem cells: a molecular and electrophysiological approach. Stem Cells 25(5):1136–1144. doi:10.1634/stemcells.2006-0466
Satin J, Kehat I, Caspi O et al (2004) Mechanism of spontaneous excitability in human embryonic stem cell derived cardiomyocytes. J Physiol 559(Pt 2):479–496. doi:10.1113/jphysiol.2004.068213
Schaper J, Meiser E, Stammler G (1985) Ultrastructural morphometric analysis of myocardium from dogs, rats, hamsters, mice, and from human hearts. Circ Res 56(3):377–391
Scorsin M, Hagege AA, Marotte F et al (1997) Does transplantation of cardiomyocytes improve function of infarcted myocardium? Circulation 96(9 Suppl):II-188–II-193
Scorsin M, Hagege A, Vilquin JT et al (2000) Comparison of the effects of fetal cardiomyocyte and skeletal myoblast transplantation on postinfarction left ventricular function. J Thorac Cardiovasc Surg 119(6):1169–1175. doi:10.1067/mtc.2000.104865
Sekine H, Shimizu T, Yang J et al (2006) Pulsatile myocardial tubes fabricated with cell sheet engineering. Circulation 114(1 Suppl):I87–I93. doi:10.1161/CIRCULATIONAHA.105.000273
Sekine H, Shimizu T, Hobo K et al (2008) Endothelial cell coculture within tissue-engineered cardiomyocyte sheets enhances neovascularization and improves cardiac function of ischemic hearts. Circulation 118(14 Suppl):S145–S152. doi:10.1161/CIRCULATIONAHA.107.757286
Sekiya S, Shimizu T, Yamato M et al (2006) Bioengineered cardiac cell sheet grafts have intrinsic angiogenic potential. Biochem Biophys Res Commun 341(2):573–582. doi:10.1016/j.bbrc.2005.12.217
Sharma S, Raman S, Sun B et al (2011) Anterior wall viability and low ejection fraction predict functional improvement after CABG. J Surg Res 171(2):416–421. doi:10.1016/j.jss.2010.03.068
Sheng X, Reppel M, Nguemo F et al (2012) Human pluripotent stem cell-derived cardiomyocytes: response to TTX and lidocain reveals strong cell to cell variability. PLoS One 7(9):e45963. doi:10.1371/journal.pone.0045963
Shevach M, Soffer-Tsur N, Fleischer S et al (2014) Fabrication of omentum-based matrix for engineering vascularized cardiac tissues. Biofabrication 6(2). doi:10.1088/1758-5082/6/2/024101

Shiba Y, Fernandes S, Zhu WZ et al (2012) Human ES-cell-derived cardiomyocytes electrically couple and suppress arrhythmias in injured hearts. Nature 489(7415):322–325. doi:10.1038/nature11317

Shiba Y, Filice D, Fernandes S et al (2014) Electrical integration of human embryonic stem cell-derived cardiomyocytes in a guinea pig chronic infarct model. J Cardiovasc Pharmacol Ther 19(4):368–381. doi:10.1177/1074248413520344

Shimizu T, Yamato M, Isoi Y et al (2002) Fabrication of pulsatile cardiac tissue grafts using a novel 3-dimensional cell sheet manipulation technique and temperature-responsive cell culture surfaces. Circ Res 90(3):e40

Shiotani M, Harada T, Abe J et al (2007) Methodological validation of an existing telemetry system for QT evaluation in conscious guinea pigs. J Pharmacol Toxicol Methods 55(1):27–34. doi:10.1016/j.vascn.2006.04.008

Simari RD, Pepine CJ, Traverse JH et al (2014) Bone marrow mononuclear cell therapy for acute myocardial infarction: a perspective from the cardiovascular cell therapy research network. Circ Res 114(10):1564–1568. doi:10.1161/CIRCRESAHA.114.303720

Skobel E, Schuh A, Schwarz ER et al (2004) Transplantation of fetal cardiomyocytes into infarcted rat hearts results in long-term functional improvement. Tissue Eng 10(5–6):849–864. doi:10.1089/1076327041348491

Snir M, Kehat I, Gepstein A et al (2003) Assessment of the ultrastructural and proliferative properties of human embryonic stem cell-derived cardiomyocytes. Am J Physiol Heart Circ Physiol 285(6):H2355–H2363. doi:10.1152/ajpheart.00020.2003

Solomon S, Pitossi F, Rao MS (2015) Banking on iPSC--is it doable and is it worthwhile. Stem Cell Rev 11(1):1-10. doi:10.1007/s12015-014-9574-4

Stelzer Y, Yanuka O, Benvenisty N (2011) Global analysis of parental imprinting in human parthenogenetic induced pluripotent stem cells. Nat Struct Mol Biol 18(6):735–741. doi:10.1038/nsmb.2050

Stoppel WL, Hu D, Domian IJ et al (2015) Anisotropic silk biomaterials containing cardiac extracellular matrix for cardiac tissue engineering. Biomed Mater 10(3):034105-6041/10/3/034105. doi:10.1088/1748-6041/10/3/034105

Sun N, Yazawa M, Liu J et al (2012) Patient-specific induced pluripotent stem cells as a model for familial dilated cardiomyopathy. Sci Transl Med 4(130):130ra47. doi:10.1126/scitranslmed.3003552

Swoap SJ, Overton JM, Garber G (2004) Effect of ambient temperature on cardiovascular parameters in rats and mice: a comparative approach. Am J Phys Regul Integr Comp Phys 287(2):R391–R396. doi:10.1152/ajpregu.00731.2003

Takahashi K, Tanabe K, Ohnuki M et al (2007) Induction of pluripotent stem cells from adult human fibroblasts by defined factors. Cell 131(5):861–872. doi:10.1016/j.cell.2007.11.019

Tang YL, Zhao Q, Zhang YC et al (2004) Autologous mesenchymal stem cell transplantation induce VEGF and neovascularization in ischemic myocardium. Regul Pept 117(1):3–10

Tannenbaum SE, Turetsky TT, Singer O et al (2012) Derivation of xeno-free and GMP-grade human embryonic stem cells--platforms for future clinical applications. PLoS One 7(6):e35325. doi:10.1371/journal.pone.0035325

Taylor DA, Atkins BZ, Hungspreugs P et al (1998) Regenerating functional myocardium: improved performance after skeletal myoblast transplantation. Nat Med 4(8):929–933

Taylor CJ, Bolton EM, Pocock S et al (2005) Banking on human embryonic stem cells: estimating the number of donor cell lines needed for HLA matching. Lancet 366(9502):2019–2025. doi:10.1016/S0140-6736(05)67813-0

Taylor RE, Kim K, Sun N et al (2013) Sacrificial layer technique for axial force post assay of immature cardiomyocytes. Biomed Microdevices 15(1):171–181. doi:10.1007/s10544-012-9710-3

Terrenoire C, Wang K, Tung KW et al (2013) Induced pluripotent stem cells used to reveal drug actions in a long QT syndrome family with complex genetics. J Gen Physiol 141(1):61–72. doi:10.1085/jgp.201210899

Thavandiran N, Dubois N, Mikryukov A et al (2013) Design and formulation of functional pluripotent stem cell-derived cardiac microtissues. Proc Natl Acad Sci U S A 110(49):E4698–E4707. doi:10.1073/pnas.1311120110

Thomson JA, Itskovitz-Eldor J, Shapiro SS et al (1998) Embryonic stem cell lines derived from human blastocysts. Science 282(5391):1145–1147

Tomita S, Li RK, Weisel RD et al (1999) Autologous transplantation of bone marrow cells improves damaged heart function. Circulation 100(19 Suppl):II247–II256

Tseliou E, de Couto G, Terrovitis J et al (2014) Angiogenesis, cardiomyocyte proliferation and anti-fibrotic effects underlie structural preservation post-infarction by intramyocardially-injected cardiospheres. PLoS One 9(2):e88590. doi:10.1371/journal.pone.0088590

Tulloch NL, Muskheli V, Razumova MV et al (2011) Growth of engineered human myocardium with mechanical loading and vascular coculture. Circ Res 109(1):47–59. doi:10.1161/CIRCRESAHA.110.237206

Turner WS, Wang X, Johnson S et al (2012) Cardiac tissue development for delivery of embryonic stem cell-derived endothelial and cardiac cells in natural matrices. J Biomed Mater Res B Appl Biomater 100(8):2060–2072. doi:10.1002/jbm.b.32770

Ueno S, Weidinger G, Osugi T et al (2007) Biphasic role for Wnt/beta-catenin signaling in cardiac specification in zebrafish and embryonic stem cells. Proc Natl Acad Sci U S A 104(23):9685–9690. doi:10.1073/pnas.0702859104

Valderrábano M (2007) Influence of anisotropic conduction properties in the propagation of the cardiac action potential. Prog Biophys Mol Biol 94(1–2):144–168. doi:10.1016/j.pbiomolbio.2007.03.014

van der Velden J, Klein LJ, van der Bijl M et al (1999) Isometric tension development and its calcium sensitivity in skinned myocyte-sized preparations from different regions of the human heart. Cardiovasc Res 42(3):706–719

Wainwright JM, Czajka CA, Patel UB et al (2010) Preparation of cardiac extracellular matrix from an intact porcine heart. Tissue Eng Part C Methods 16(3):525–532. doi:10.1089/ten.TEC.2009.0392

Wang X, Jameel MN, Li Q et al (2009) Stem cells for myocardial repair with use of a transarterial catheter. Circulation 120(11 Suppl):S238–S246. doi:10.1161/CIRCULATIONAHA.109.885236

Watanabe E, Smith DM Jr, Delcarpio JB et al (1998) Cardiomyocyte transplantation in a porcine myocardial infarction model. Cell Transplant 7(3):239–246

Weisbrod D, Peretz A, Ziskind A et al (2013) SK4 Ca2+ activated K+ channel is a critical player in cardiac pacemaker derived from human embryonic stem cells. Proc Natl Acad Sci U S A 110(18):E1685–E1694. doi:10.1073/pnas.1221022110

Weng Z, Kong CW, Ren L et al (2014) A simple, cost-effective but highly efficient system for deriving ventricular cardiomyocytes from human pluripotent stem cells. Stem Cells Dev 23(14):1704–1716. doi:10.1089/scd.2013.0509

Weymann A, Patil NP, Sabashnikov A et al (2014) Bioartificial heart: a human-sized porcine model--the way ahead. PLoS One 9(11):e111591. doi:10.1371/journal.pone.0111591

Weymann A, Patil NP, Sabashnikov A et al (2015) Perfusion-decellularization of porcine lung and trachea for respiratory bioengineering. Artif Organs 39(12):1024–1032. doi:10.1111/aor.12481

Witjas-Paalberends ER, Piroddi N, Stam K et al (2013) Mutations in MYH7 reduce the force generating capacity of sarcomeres in human familial hypertrophic cardiomyopathy. Cardiovasc Res 99(3):432–441. doi:10.1093/cvr/cvt119

Xi J, Khalil M, Spitkovsky D et al (2011) Fibroblasts support functional integration of purified embryonic stem cell-derived cardiomyocytes into avital myocardial tissue. Stem Cells Dev 20(5):821–830. doi:10.1089/scd.2010.0398

Xu C, Police S, Rao N et al (2002) Characterization and enrichment of cardiomyocytes derived from human embryonic stem cells. Circ Res 91(6):501–508

Xu C, Police S, Hassanipour M et al (2011) Efficient generation and cryopreservation of cardiomyocytes derived from human embryonic stem cells. Regen Med 6(1):53–66. doi:10.2217/rme.10.91

Xu B, Li Y, Fang X et al (2013) Mechanically tissue-like elastomeric polymers and their potential as a vehicle to deliver functional cardiomyocytes. J Mech Behav Biomed Mater 28:354–365. doi:10.1016/j.jmbbm.2013.06.005

Xue T, Cho HC, Akar FG et al (2005) Functional integration of electrically active cardiac derivatives from genetically engineered human embryonic stem cells with quiescent recipient ventricular cardiomyocytes: insights into the development of cell-based pacemakers. Circulation 111(1):11–20. doi:10.1161/01.CIR.0000151313.18547.A2

Yang L, Soonpaa MH, Adler ED et al (2008) Human cardiovascular progenitor cells develop from a KDR+ embryonic-stem-cell-derived population. Nature 453(7194):524–528. doi:10.1038/nature06894

Yang Z, Zhang F, Ma W et al (2010) A novel approach to transplanting bone marrow stem cells to repair human myocardial infarction: delivery via a noninfarct-relative artery. Cardiovasc Ther 28(6):380–385. doi:10.1111/j.1755-5922.2009.00116.x

Yang WJ, Li SH, Weisel RD et al (2012) Cell fusion contributes to the rescue of apoptotic cardiomyocytes by bone marrow cells. J Cell Mol Med 16(12):3085–3095. doi:10.1111/j.1582-4934.2012.01600.x

Yang X, Rodriguez M, Pabon L et al (2014) Tri-iodo-l-thyronine promotes the maturation of human cardiomyocytes-derived from induced pluripotent stem cells. J Mol Cell Cardiol 72:296–304. doi:10.1016/j.yjmcc.2014.04.005

Yang T, Rubart M, Soonpaa MH et al (2015) Cardiac engraftment of genetically-selected parthenogenetic stem cell-derived cardiomyocytes. PLoS One 10(6):e0131511. doi:10.1371/journal.pone.0131511

Yasui H, Lee JK, Yoshida A et al (2014) Excitation propagation in three-dimensional engineered hearts using decellularized extracellular matrix. Biomaterials 35(27):7839–7850. doi:10.1016/j.biomaterials.2014.05.080

Ye L, Chang Y, Xiong Q et al (2014) Cardiac repair in a porcine model of acute myocardial infarction with human induced pluripotent stem cell-derived cardiovascular cells. Cell Stem Cell 15(6):750–761. doi:10.1016/j.stem.2014.11.009

Yu J, Vodyanik MA, Smuga-Otto K et al (2007) Induced pluripotent stem cell lines derived from human somatic cells. Science 318(5858):1917–1920. doi:10.1126/science.1151526

Zhang YM, Hartzell C, Narlow M et al (2002) Stem cell-derived cardiomyocytes demonstrate arrhythmic potential. Circulation 106(10):1294–1299

Zhang J, Wilson GF, Soerens AG et al (2009) Functional cardiomyocytes derived from human induced pluripotent stem cells. Circ Res 104(4):e30–e41. doi:10.1161/CIRCRESAHA.108.192237

Zhang J, Klos M, Wilson GF et al (2012) Extracellular matrix promotes highly efficient cardiac differentiation of human pluripotent stem cells: the matrix sandwich method. Circ Res 111(9):1125–1136. doi:10.1161/CIRCRESAHA.112.273144

Zhang XH, Haviland S, Wei H et al (2013a) Ca2+ signaling in human induced pluripotent stem cell-derived cardiomyocytes (iPS-CM) from normal and catecholaminergic polymorphic ventricular tachycardia (CPVT)-afflicted subjects. Cell Calcium 54(2):57–70. doi:10.1016/j.ceca.2013.04.004

Zhang D, Shadrin IY, Lam J et al (2013b) Tissue-engineered cardiac patch for advanced functional maturation of human ESC-derived cardiomyocytes. Biomaterials 34(23):5813–5820. doi:10.1016/j.biomaterials.2013.04.026

Zhang L, Pan Y, Qin G et al (2014) Inhibition of stearoyl-coA desaturase selectively eliminates tumorigenic Nanog-positive cells: improving the safety of iPS cell transplantation to myocardium. Cell Cycle 13(5):762–771. doi:10.4161/cc.27677

Zhou YY, Zeng F (2013) Integration-free methods for generating induced pluripotent stem cells. Genom Proteom Bioinfo 11(5):284–287. doi:10.1016/j.gpb.2013.09.008

Zhu WZ, Santana LF, Laflamme MA (2009) Local control of excitation-contraction coupling in human embryonic stem cell-derived cardiomyocytes. PLoS One 4(4):e5407. doi:10.1371/journal.pone.0005407

Zhu WZ, Xie Y, Moyes KW et al (2010) Neuregulin/ErbB signaling regulates cardiac subtype specification in differentiating human embryonic stem cells. Circ Res 107(6):776–786. doi:10.1161/CIRCRESAHA.110.223917

Zhu WZ, Van Biber B, Laflamme MA (2011) Methods for the derivation and use of cardiomyocytes from human pluripotent stem cells. Methods Mol Biol 767:419–431. doi:10.1007/978-1-61779-201-4_31

10 State-of-the-Art in Tissue-Engineered Heart Repair

Buntaro Fujita, Malte Tiburcy, Stephan Ensminger, and Wolfram-Hubertus Zimmermann

Abstract

Heart muscle restoration with in vitro engineered tissue constructs is an exciting and rapidly advancing field. Feasibility, safety, and efficacy data have been obtained in animal models. First clinical trials are on the way to explore the therapeutic utility of cell-free and non-contractile cell-containing grafts. Engineering of contractile patches according to current good manufacturing practice (cGMP) for bona fide myocardial re-muscularization and scalability to address clinical demands remains challenging. Proof-of-concept for solutions to address obvious technical hurdles exists, and it can be anticipated that the first generation of clinically applicable engineered heart muscle (EHM) grafts will become available in the near future. Foreseeable, but likely manageable risks include arrhythmia induction and teratoma formation. Remaining biomedical challenges pertain to the requirement of immune suppression and the strategic

B. Fujita
Institute of Pharmacology and Toxicology, University Medical Center Göttingen, Göttingen, Germany

DZHK (German Center for Cardiovascular Research), Partner Site Göttingen, Göttingen, Germany

Department of Thoracic and Cardiovascular Surgery, Heart and Diabetes Center NRW, Ruhr-University Bochum, Bad Oeynhausen, Germany

M. Tiburcy • W.-H. Zimmermann (✉)
Institute of Pharmacology and Toxicology, University Medical Center Göttingen, Georg-August University, Göttingen, Germany

DZHK (German Center for Cardiovascular Research), Partner Site Göttingen, Göttingen, Germany
e-mail: w.zimmermann@med.uni-goettingen.de

S. Ensminger
Department of Thoracic and Cardiovascular Surgery, Heart and Diabetes Center NRW, Ruhr-University Bochum, Bad Oeynhausen, Germany

M. Ieda, W.-H. Zimmermann (eds.), *Cardiac Regeneration*, Cardiac and Vascular Biology 4, DOI 10.1007/978-3-319-56106-6_10

approach to optimize immune suppression without subjecting the target patient population to an unacceptable risk. This chapter summarizes the current state of tissue-engineered heart repair with a special emphasis on knowledge gained from in vitro and in vivo studies as well as issues pertaining to transplant immunology and cGMP process development.

10.1 Introduction

Chronic heart failure is a major health burden affecting >5 million patients in the USA (Mozaffarian et al. 2015) and >6 million patients in Europe (Nichols et al. 2014). Regardless of its etiology, the hallmark feature of chronic heart failure is progressive myocardial remodeling characterized by myocardial fibrosis, cardiomyocyte death, and consecutive impairment of cardiac function (Travers et al. 2016). This myocardial remodeling process is preceded and exacerbated by a mechanistically not fully understood chain of events, best described as the vicious cycle of heart failure ((Kemp and Conte 2012); Fig. 10.1). Present guidelines focus on pharmacological inhibition of neurohumoral pathways involved in myocardial remodeling as well as interventional, device, or surgical treatment. None of these approaches result in sustained reversal of the remodeling process or re-muscularization of the failing heart. To date, heart transplantation with approximately 4000 procedures performed annually is the only curative therapeutic option

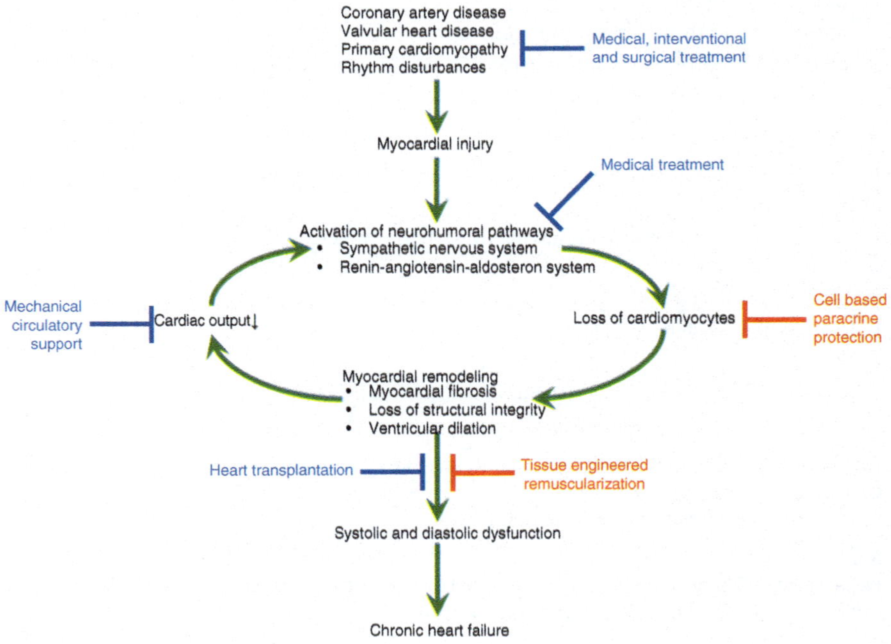

Fig. 10.1 Overview of the pathophysiologic chain of events leading to chronic heart failure. *Blue* current state-of-the-art therapy for chronic heart failure. *Red* cell therapy candidates

with excellent long-term outcome (Long et al. 2014). It will, however, remain a rare intervention because of donor organ shortage and thus does not provide a solution to the pressing medical burden imposed on our societies by the evolving heart failure epidemic (Lund et al. 2015; Roger 2010). Given the urgent need for alternative therapies, restoration of cardiac function is attempted by implantation of acellular patches and cells of different origin either directly or as engineered tissue patches (Laflamme and Murry 2011; Nguyen et al. 2016; Ye et al. 2013). In this chapter, we provide an overview of the current state of development in tissue-engineered heart repair. For a detailed discussion of the current tissue engineering principles, which include (1) cell seeding of synthetic and biological preformed matrices, (2) cell sheet engineering, and (3) hydrogel-based tissue engineering, we refer the interested reader to more specialized reviews (Ye et al. 2013; Zimmermann 2009). We touch upon what we foresee as the specific challenges for heart re-muscularization and its translation into clinical application. These include anticipated side effects, graft retention, transplant immunology, and cGMP processing.

10.2 From Paracrine Protection to Myocardial Re-muscularization

Direct delivery of cardiac stem/progenitor cells (SCIPIO (Bolli et al. 2011)), cardiosphere-derived cells (CADUSCEUS (Makkar et al. 2012)), naïve (POSEIDON—(Hare et al. 2012)) or modified (CHART-1 (Bartunek et al. 2016)) mesenchymal stem cells, and bone marrow-derived cells (REPAIR-AMI (Schachinger et al. 2006)) have been under clinical investigation for more than 10 years with mixed results (Nguyen et al. 2016). Most of these studies were designed to explore safety and efficacy of autologous cell therapy candidates. More recently this strategy has been reconsidered in favor of an allogeneic off-the-shelf approach (e.g., in ALLSTAR (Chakravarty et al. 2016)) to better address the immediate medical unmet need in patients with heart failure and render the procedure cost-effective. The anticipated modes of action of these cell therapy candidates entail (1) activation of endogenous heart repair mechanisms, (2) induction of angiogenesis for the rescue of hibernating myocardium, and (3) activation of pro-survival pathways in disease-affected cardiomyocytes. A concurrent finding of preclinical and clinical studies is the lack of long-term retention of implanted cells (Aicher et al. 2003; Blocklet et al. 2006; Hofmann et al. 2005; Hou et al. 2005). Accordingly, the interpretation of current data is that cell grafts protect rather than re-muscularize the failing heart. Consequently, applications in acute and subacute disease states rather than in terminal heart failure with substantial scarring seem mechanistically plausible (Fig. 10.1). Similar limitations as to cell retention apply to cardiomyocyte implantation (Müller-Ehmsen et al. 2002). Heat shock treatment with co-administration of survival factors (Laflamme et al. 2007; Shiba et al. 2016) or the utilization of engineered heart muscle (EHM; Riegler et al. 2015; Zimmermann et al. 2006) have been introduced successfully to address this limitation. Collectively, the available data on stem cell-based heart repair allow for the following conclusions: (1) stem cells can offer protection to the failing heart by various, so far, still not fully defined means; (2) sustainable re-muscularization requires cardiomyocyte grafts.

10.3 Tissue Engineering Strategies: Lessons Learned from In Vitro Studies

Several tissue formats have been developed for applications in tissue-engineered heart repair (Fig. 10.2): (1) contractile patches or pouches, (2) non-contractile cell-containing patches and pouches, and (3) cell-free patches (reviewed in (Ye et al. 2013; Zimmermann 2009)). Contractile patches contain cardiomyocytes (Riegler et al. 2015; Weinberger et al. 2016; Zimmermann et al. 2006). The intention is to achieve *contractile support* by bona fide re-muscularization of the chronically failing heart. An outcome measure is increased myocardial muscle mass, leading to enhanced contractile function. Alternatively, non-contractile patches and pouches are engineered to target (stem) cell therapy candidates to a defined epicardial region for spatially restricted *paracrine protection* (Bellamy et al. 2015; Chachques et al. 2008; Christman et al. 2004). Cell-free patches or hydrogels are applied topically for *mechanical stabilization* and reduction of wall stress according to the Law of Laplace (Mann et al. 2016). Mechanistically, cell-containing patches or pouches are capable of conferring mixed activities to the failing heart, including contractile support, paracrine protection, and mechanical stabilization. Cell-free patches, primarily applied for mechanical support, can be loaded with therapeutic agents for targeted delivery of, for example, defined growth factors to the epicardial surface of the heart (Wei et al. 2015; Zhang et al. 2007). Tuning of the viscoelastic properties of synthetic and biological matrix material can be attempted to optimize mechanical support of the failing heart (Sanchez et al.

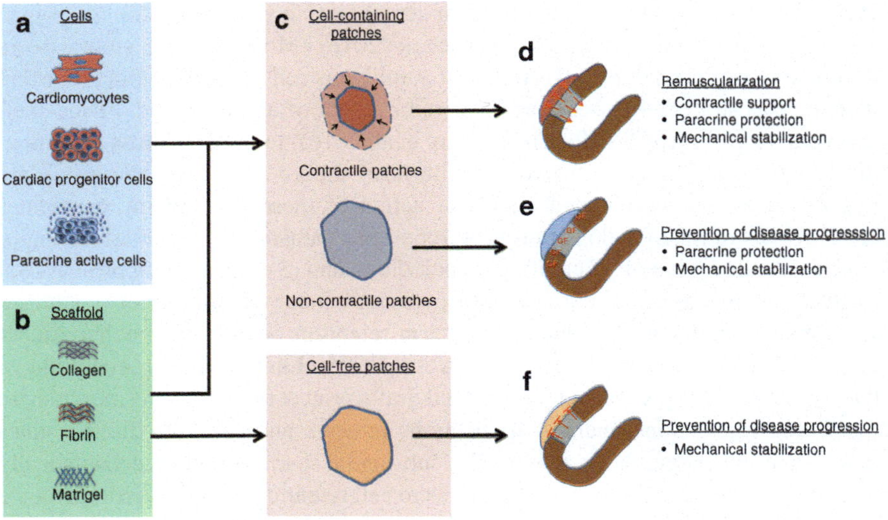

Fig. 10.2 Schematic overview of the different strategies in tissue-engineered heart repair. Different cell types (**a**) can be mixed with scaffold materials (**b**) to engineer contractile or non-contractile cell-containing patches or cell-free patches (**c**). Depending on the cell types applied, the mode of action of engineered tissue patches differs: Cell-containing contractile patches aim at the contractile support of the failing heart as well as providing paracrine protection and mechanical stabilization (**d**). Cell-containing non-contractile patches aim at paracrine support and mechanical stabilization (**e**) while cell-free patches solely provide mechanical stabilization (**f**)

2015). These various activities and means to control them by design principles make tissue-engineered heart repair a multimodal therapeutic intervention.

Most current tissue engineering approaches utilize collagen and fibrin with or without Matrigel™ (a mixture of basal membrane proteins and growth factors) to support self-assembly of cardiomyocytes into a synchronously contracting three-dimensional functional syncytium (Ye et al. 2013). The viscoelastic properties of these hydrogel-cell mixtures change dramatically in the presence of non-myocytes to reach similar values as observed in native myocardium (E-modulus in human-engineered heart muscle (EHM): ~20 kPa; own unpublished data). How cardiomyocytes and non-myocytes interact during the self-organization process and whether tuning of this process will enable enhanced cellularity and maturation is currently under investigation. The observation that non-myocytes, and in particular fibroblasts or cells with fibroblast activity (i.e., extracellular matrix production), are important for in vitro tissue self-organization came initially as a surprise (Naito et al. 2006; Zimmermann and Eschenhagen 2003), but follows the logic that the normal heart is assembled from an ensemble of multiple cell types with specific "core" functions: (1) cardiomyocytes—contractility; (2) fibroblasts—extracellular matrix production; (3) endothelial cells and smooth muscle cells—vascularization; (4) neurons—autonomic control of contractility; (5) immune cells—tissue homeostasis. All cells are in constant paracrine and biomechanical cross-talk. Cardiomyocytes, fibroblasts, and endothelial cells comprise the largest cell fraction in the heart (Naito et al. 2006; Pinto et al. 2016). All current myocardial tissue engineering approaches apply multiple defined or undefined cell types to achieve optimal contractility (Kensah et al. 2013; Radisic et al. 2008; Riegler et al. 2015; Tiburcy et al. 2017; Tulloch et al. 2011; Weinberger et al. 2016; Zhang et al. 2013; Zimmermann et al. 2006); whether heart muscle with in vivo cellularity will have to be engineered for optimal results remains to be elucidated. For clinical translation and regulatory approval, it will be best to keep the tissue-engineered therapeutic candidate also with respect to its cellular components as simple as possible. Interestingly, non-myocytes appear to vanish upon implantation of engineered grafts originally constructed from mixed cell populations, leaving essentially pure cardiomyocyte grafts (Riegler et al. 2015; Weinberger et al. 2016).

Mechanical loading and electrical stimulation have been identified as essential components for structural, functional, metabolic, and molecular maturation of tissue-engineered myocardium (Fink et al. 2000; Liaw and Zimmermann 2015; Radisic et al. 2004, 2007; Shimko and Claycomb 2008; Vandenburgh et al. 1996; Zimmermann et al. 2000, 2002b, 2006). Mechanical stimulation is one of the main driving factors for proper heart development. Hence, different mechanical loading approaches have been investigated to enhance tissue formation and maturation. These experiments revealed that mechanical loading improves force generation and structural organization, increases the number of gap junctions, and leads to enhanced paracrine factor release (Fink et al. 2000; Lammerding et al. 2004; Vandenburgh et al. 1996; Zimmermann et al. 2000, 2002b, 2006). In terms of the mode of loading, support of auxotonic contractions by mounting EHM on flexible poles resulted in optimal functional results (Naito et al. 2006; Zimmermann et al. 2006). Electrical pacing of auxotonically contracting EHM at near-physiological frequency led to further enhanced structural and functional maturation (Godier-Furnemont et al. 2015). On the cellular level, electrical stimulation resulted in different models of tissue-engineered myocardium in improved alignment of myofibrils, assembly of contractile proteins, and calcium handling (Godier-Furnemont et al. 2015; Hirt et al. 2014; Radisic et al. 2004; Tandon et al. 2009). Collectively, these data suggest that electromechanical stimulation contributes substantially to all so far investigated aspects of heart maturation. For heart repair purposes, it is unclear whether a state of optimal maturity exists or would even be desirable, because of the anticipated limited capacity of terminally differentiated and metabolically highly demanding matured heart muscle to adapt to the recipient heart environment upon implantation.

10.4 Tissue-Engineered Heart Repair: Lessons Learned from In Vivo Studies

In early proof-of-concept and more recent follow-up studies, we could demonstrate that the epicardial implantation of EHM allografts is feasible, safe, and effective (Didié et al. 2013; Zimmermann et al. 2006). Structural and functional support by EHM grafts was confirmed by echocardiography, magnetic resonance imaging, and left heart catheterization, demonstrating enhanced left ventricular free wall thickness (+0.5 mm in systole) and enhanced systolic thickening (+20%) of a targeted left ventricular free wall injury (Zimmermann et al. 2006). Similarly as observed in a clinical trial utilizing intracoronary bone marrow-derived mononuclear cell infusion (Schachinger et al. 2009), there was an improvement in parameters of global heart function such as fractional area shortening (+40%) and ejection fraction (+15%) in recipients (rat model) with severely compromised heart function, i.e., a reduction in fractional area shortening (FAS) to below 30% 2 weeks after myocardial infarction from a FAS of ~60% in healthy rats (Zimmermann et al. 2006). Efficacy of EHM implantation was subsequently confirmed in a mouse model of acute myocardial infarction (Didié et al. 2013). Electrical integration of allografts was demonstrated by epicardial mapping experiments and was further supported by the observation of enhanced systolic thickening of the EHM-treated heart wall (Zimmermann et al. 2006). However, it cannot be ruled out that additional mechanisms contributed to the observed functional benefits. These include mechanical stabilization to prevent deterioration of left ventricular geometry and paracrine protection. A direct comparison to non-contractile grafts provided however solid evidence as to the superiority of bona fide re-muscularization versus mechanical stabilization and paracrine protection alone (Didié et al. 2013; Zimmermann et al. 2006).

With the availability of human pluripotent stem cells, robust directed differentiation protocols and controlled bioreactor systems, large-scale production of human cardiomyocytes and tissue-engineered constructs is now possible (Burridge et al. 2012; Chen et al. 2015; Soong et al. 2012; Streckfuss-Bomeke et al. 2013; Tiburcy et al. 2014; Tiburcy and Zimmermann 2014; Tiburcy et al. 2017). Early studies suggested that human embryonic stem cell (hESC)- and induced pluripotent stem cell (hiPSC)-derived cardiomyocytes can improve the function of injured hearts in rodent models (Caspi et al. 2007a; Funakoshi et al. 2016; Laflamme et al. 2007; Nelson et al. 2009; van Laake et al. 2007). Whether these effects were related to electromechanical integration or indirect effects (milieu modulation, mechanical support) remains unclear. Myocardial re-muscularization could also be demonstrated in a large animal xenograft study (human cardiomyocytes in pigtail macaque—*Macaca nemestrina*), while confirming the anticipated arrhythmogenicity of cardiomyocyte grafts (Chong et al. 2014). This study did not detect beneficial or detrimental functional consequences of cardiomyocyte engraftment, very likely because of only mild impairment of left ventricular function after ischemia/reperfusion injury (2.4–10.4% of the left ventricle was affected with an ejection fraction of ~70%).

Human EHM made of hESC derivatives enriched for cardiomyocytes showed sustained survival with progressive vascularization and cardiomyocyte maturation in a chronic ischemia/reperfusion rodent model over a study duration of >200 days (Riegler et al. 2015). Interestingly, enhanced ejection fraction (+5%) was observed independently of the presence or absence of cardiomyocytes in this model of mild myocardial damage with an ejection fraction of ~55% at the time of EHM implantation (Riegler et al. 2015). These observations support the notion that cardiomyocyte-independent effects can mediate functional improvements and that more severe models of end-stage heart failure would be required to evaluate in detail the therapeutic effectiveness of re-muscularization. Similarly, hiPSC-derived cardiomyocyte sheets and patches improved cardiac function in a porcine ischemic cardiomyopathy model (Kawamura et al. 2012) and a cryo-injury guinea pig model (Weinberger et al. 2016). Despite these encouraging data, xenogeneic models provide only limited information as to safety concerns pertaining primarily to tumor formation and arrhythmia induction because of physiological (e.g., heart rate, heart size), molecular (e.g., differences in the expression of sarcomeric proteins and ion channels), and immunological (e.g., differences in AB0, MHC proteins) mismatches. We posit that the use of homologous (especially as to organ physiology and immunogenicity) large animal models will be instrumental to translate existing preclinical models. Here, not only the close resemblance to the anatomy, physiology, and pathophysiology to the human system is important, but also the availability of allogeneic pluripotent stem cells for the production of tissue-engineered allografts. To date, the only model that fulfills these criteria is the macaque model. The usefulness of this model has been clearly demonstrated in several late preclinical studies (Bel et al. 2010; Shiba et al. 2016) and was found to be instrumental for the approval of the first-in-patient ESCORT trial in Europe, which tests non-contractile patches comprising hESC-derived cardiac progenitor cells identified by the expression of stage-specific embryonic antigen 1 (SSEA-1 *aka* CD15) and the second heart field marker Isl1 in patients with severe heart failure (Menasche et al. 2015). In an earlier study, feasibility and safety of collagen meshes seeded with mesenchymal stem cells were demonstrated (Chachques et al. 2007). Interestingly, an autologous engineered myoblast cell sheet product (TCD-51073; Terumo Corporation) was conditionally approved in 2015 by the Japanese Council of the Ministry of Health, Labour and Welfare for applications in heart failure. A direct contribution to myocardial contractility via electromechanical integration of myoblasts or myotubes is however unlikely (Menasche 2008). Mechanical stabilization and paracrine mechanisms may have contributed to the observed therapeutic effects in animal models and patients with heart failure (Sawa et al. 2015; Shirasaka et al. 2016).

Cell-free therapeutics are considered biologically less complex and thus from a regulatory point of view less complicated alternatives for myocardial preservation. The multicenter phase II AUGMENT-HF trial investigates the treatment of advanced heart failure patients with an alginate-hydrogel formulation (Algisyl®—LoneStar Heart, Inc.) in a non-blinded randomized fashion (Anker et al. 2015). After 1-year follow-up, improved exercise capacity, quality of life, and NYHA class, but also a trend to higher mortality (9 of 35 vs. 4 of 38 patients in the control group), were observed in the Algisyl®-treated group (Mann et al. 2016). Further studies also with alterative materials, for example, with natural extracellular matrix (Wassenaar et al. 2016), are warranted to determine the therapeutic value of this approach.

10.5 Metabolic Engineering for Enhanced Survival Upon Transplantation

For implantation, metabolic immaturity appears advantageous and may finally be key for enhanced cell survival if immediate blood perfusion cannot be established. After implantation and graft retention with concomitant vascularization, metabolic maturation will likely be advantageous for optimal in situ performance. Despite fast and extensive vascularization of engineered heart tissue grafts shortly after implantation (Leor et al. 2000; Riegler et al. 2015; Tulloch et al. 2011; Zimmermann et al. 2002a, 2006), there is clearly limited oxygen and nutrient supply in the absence of immediate tissue perfusion upon epicardial implantation. To shorten the time to optimal blood perfusion, several groups including our own have employed pre-vascularization strategies either by taking advantage of endothelial cell components in naïve heart cell mixtures (Naito et al. 2006) or by the specific addition of endothelial and smooth muscle cells (Caspi et al. 2007b; Lesman et al. 2010; Tulloch et al. 2011; Weinberger et al. 2016). Despite the observation that vascularization of the heart muscle implants is supported by co-engrafted endothelial cells (Tulloch et al. 2011; Zimmermann et al. 2006), we are not aware of compelling evidence as to an impact of pre-endothelialization or pre-vascularization on performance of engineered heart tissue grafts. Interestingly, thin multilayered engineered cell sheet implants (>80 μm in thickness) demonstrated core necrosis after implantation and required a multistep implant strategy to enable augmentation of epicardial cardiomyocyte grafts (Shimizu et al. 2006). The reason for this discrepancy between hydrogel and cell sheet approaches is not resolved, but may be related to differences in metabolic demand in high-density multilayered cardiomyocyte constructs versus thick hydrogel models comprising loosely organized muscle networks with the capacity for hypertrophic growth in vivo (Zimmermann et al. 2006; Riegler et al. 2015). Whether this proposed metabolic advantage will hold up in centimeter scale clinical tissue grafts remains to be determined. Large animal studies will be instrumental to scrutinize whether means to prevascularize in vitro or enhance vascularization in vivo will be necessary to match oxygen/nutrient supply and demand for optimal therapeutic outcome.

10.6 Immunological Considerations

Implantation of cells and tissue bears the risk of triggering an immune response, leading to rejection of the graft. Accurate characterization of the immunogenicity of stem cell-based tissue-engineered constructs is therefore of utmost importance for safe and effective therapy. The principal mechanism underlying graft rejection is recognition of the donor graft (or parts of it) by the immune system of the recipient as "non-self." For rejection of solid organ transplants, three important alloantigen systems have been identified: (1) the AB0 system, (2) the major histocompatibility complex (MHC), and (3) minor histocompatibility antigens (miHA). Experience from solid organ transplantation indicated that single mismatches in any of these systems may trigger an immune response. The current heart allocation system exclusively considers the AB0 system, whereas HLA matching is not performed, mainly for logistic reasons, i.e., to not increase ischemia time beyond 4 h. In kidney transplantation HLA matching is feasible and was found to improve outcome (Montgomery et al. 2012).

Immune responses appear to be organ specific, which is most likely related to differences in organ cell composition, variable antigen expression, and a variable degree of vascularization and thus accessibility by circulating immune cells. AB0 antigens are highly expressed in cardiac vascular endothelial cells and typically not found on cardiomyocytes (Thorpe et al. 1991). Similarly, HLA antigens are strongly expressed on cardiac endothelial cells and represent the first point of contact between immune cells of the recipient in AB0-matched grafts (Wijngaard et al. 1991). The presence of foreign HLA antigens on endothelial cells is therefore considered a major factor in the immune response after heart transplantation.

To prevent rejection of highly vascularized AB0-matched cardiac allografts, pharmacological immune suppression is applied. Common immunosuppressive regimens include a combination of steroids and drugs that inhibit T-cell function (Table 10.1). Steroids are typically reduced within the first weeks after transplantation and, if possible, tapered off completely in the subsequent clinical course. Maintenance immunosuppression typically consists of a calcineurin inhibitor (cyclosporine A or tacrolimus) in combination with an antiproliferative agent (nowadays preferably mycophenolate mofetil, formerly azathioprine) (Lund et al. 2015). In addition, in about 50% of heart transplant recipients, induction therapy is performed with the interleukin-2-receptor antagonists (basiliximab, daclizumab) or polyclonal antilymphocyte globulin/antithymocyte globulin (ATG) (Lund et al. 2015). For specific patient subgroups with certain comorbidities and drug intolerance, additional immunosuppressants are available, including mTOR-inhibitors (sirolimus, everolimus) and co-stimulatory signal blocking agents (abatacept, belatacept). Under these immune suppression therapies, acute rejection episodes can be prevented effectively. Chronic rejection remains however a serious clinical concern (Lund et al. 2015). Acute and chronic heart rejection episodes are probably triggered mainly by unfavorable HLA mismatches and could be further aggravated by miHA mismatches.

Table 10.1 Immune suppressants in clinical application in heart transplant recipients. Typical doses and target trough levels assessed by therapeutic drug monitoring (TDM) are indicated

	Dose	TDM: target trough levels
Steroids		
Methylprednisolone	*Intra-/perioperatively* • Pulse therapy: 3 g	Not applied
	Maintenance • Low dose (5–10 mg/day) tapered off if possible	
Calcineurin inhibitors		
Cyclosporine A	*Early postoperative* • 4–8 mg/kg/day PO or 1–2 mg/kg/day IV	200–250 ng/mL
	Maintenance • Dosing to target trough levels	140–180 ng/mL
Tacrolimus	*Early postoperative* • 0.05–0.1 mg/kg/day PO or • 0.01–0.02 mg/kg/day IV	10–15 ng/mL
	Maintenance • Dosing to target trough levels	5–8 ng/mL
Antiproliferative agents		
Mycophenolate mofetil	500–1500 mg BID PO or IV	2.5–5.0 μg/mL
Azathioprine	1–2 mg/kg/day PO or IV	Not applied

In tissue-engineered heart repair, similar precautions will have to be taken as in solid organ transplantation. In an ideal world, EHM with a perfect AB0 and HLA match would be implanted. This could be achieved by employing autologous iPS-cells if all cell components would be generated from the patient's own (autologous) iPS-cells. Immunosuppression should in theory be dispensable under these conditions. However, a large body of literature suggests that even autologous iPSC-derived cells trigger an immune response due to residual expression of embryonic self-antigens, reprogramming induced DNA mutations, or incomplete epigenetic resetting (Gore et al. 2011; Lister et al. 2011). In addition, immunogenic profiles would likely be variable in different iPSC clones and derivatives thereof (Koyanagi-Aoi et al. 2013; Zhao et al. 2015). In the clinical scenario of autografting, the individual immunological signatures of the cell/tissue grafts would have to be characterized prior to implantation and after a time-consuming period of primary cell isolation, reprogramming, differentiation, and tissue construction. This would in our view preclude a timely allocation of an autologous cell therapy to patients with advanced heart failure and an accordingly high 1-year mortality.

A more realistic approach would be the establishment of an immunologically well-characterized cell or even tissue repository to facilitate AB0 and HLA matching in an allogeneic off-the-shelf setting. It has been estimated that up to 160,000 potential donors would have to be screened to develop a repository, allowing immunologically relevant HLA matching in 90% of the Japanese population (Okita et al. 2011). A more recent report noted that this number can be significantly reduced if

HLA-haploidentical donor cells would be used (Taylor et al. 2012). Finding these HLA-haploidentical donors appears to be another major task, given the high allelic variability of the HLA locus.

A unique twist of developmental biology, i.e., the possibility to derive parthenogenetic stem cells (pSCs) from unfertilized oocytes, could be instrumental in establishing a manageable cell bank, because pSCs derived from metaphase II (MII) oocytes contain a largely haploidentical chromosome set (Kim et al. 2007). In fact, experimental data from mouse and human models confirmed haploidentity of their respective H2 and HLA loci in ~75% of the derived pSC-lines (Didié et al. 2013; Revazova et al. 2008). The advantage of MHC-haploidentity was recently demonstrated by the retention of EHM allografts constructed from H2-haploidentical pSC-derived cardiomyocytes without the need for comprehensive immune suppression under MHC-matching (Didié et al. 2013).

Despite this exciting data, safe and effective treatment with human stem cell-derived tissue-engineered constructs will in our view first be attempted under immune suppression. Several immune suppression strategies have been investigated in rodent and large animal models (Table 10.2). In rodent models, rejection of allogeneic EHM was successfully prevented by a clinically relevant immunosuppressive regimen consisting of cyclosporine A, azathioprine, and methylprednisolone (Zimmermann et al. 2002a, 2006). Human cardiomyocytes containing tissue constructs implanted in Sprague Dawley rats receiving tacrolimus (Riegler et al. 2015) and guinea pigs receiving cyclosporine A and methylprednisolone (Weinberger et al. 2016) showed good survival. Furthermore, human cardiomyocytes injected into *Macaca nemestrina* receiving methylprednisolone, cyclosporine A, and abatacept showed no obvious signs of rejection at the time of tissue harvest (Chong et al. 2014). In two similarly designed studies, cardiac progenitor cells from rhesus embryonic stem cells were implanted in rhesus monkeys receiving tacrolimus or cyclosporine A at clinically relevant doses (Bel et al. 2010; Blin et al. 2010). In both studies, no obvious signs of graft rejection were observed. Interestingly, rejection of MHC-mismatched rhesus macaque iPSC-derived cardiomyocytes was noted even under "clinically appropriate" immune suppression (Shiba et al. 2016), suggesting an alteration of antigen presentation by the reprogramming procedure or procedure related bystander effects leading to immunological priming.

Apart from antigen mismatches, further factors may play a role in eliciting an immune response after implantation of EHM constructs. A particular risk is associated with the presence of circulating donor-specific antibodies (DSA) (Lister et al. 2011; Zhang et al. 2011). DSA are considered the primary cause of hyperacute rejection, and it appears imperative to screen for such antibodies also in tissue-engineered heart repair recipients.

Collectively, we conclude that (1) both the cell source and the types of cells employed for generation of tissue constructs will determine the immunogenic profile of a tissue-engineered heart muscle graft—cardiomyocytes appear to be the least immunogenic cellular component; (2) rejection of engineered tissue constructs should be manageable under current immune suppression regimens, but an increased risk of infection will have to be considered; and (3) preformed circulating DSA should be screened for to avoid hyperacute rejection. Autologous transplantation of iPSC-derived cells has been commonly suggested since their first introduction (Takahashi et al. 2007) as a solution to the

Table 10.2 Overview of immune suppression regimens employed in animal studies (selection)

Cell source	Cell types	Recipient	Immune suppression	Rejection	Reference
Neonatal rat heart	Mixed heart cells - CM ~63% [ACTN2^{+}] - Fib ~33% [P4HB^{+}]	Fisher rats (syngenic model)	None	Yes	(Zimmermann et al., 2002a)
			• CSA: 5 mg/kg/d s.c. • AZA: 2 mg/kg/d s.c. • MP: 2 mg/kg/d s.c.	No	
Neonatal rat heart	Mixed heart cells - CM ~47% [ACTN2^{+}] - Fib ~49% [P4HB^{+}]	Wistar rats (allograft)	• CSA: 5 mg/kg/d s.c. • AZA: 2 mg/kg/d s.c. • MP: 5 mg/kg/d s.c.	No	(Zimmermann et al., 2006)
Embryonic stem cells (human and Macaca mulatta)	Cardiac progenitor cells - purity: 20-60% [FUT4^{+}]	Macaca mulatta (xenograft/allograft)	• TAC: 1 mg/kg/d • or CSA: adjusted to trough level of 100 ng/mL from day -5 (before cell implantation)	No	(Blin et al., 2010)
Embryonic stem cells (human and Macaca mulatta)	Cardiac progenitor cells - purity: ~95% [FUT4^{+}]	Macaca mulatta (xenograft/allograft)	• CSA: 10 mg/kg/d for 3 days i.m., thereafter 150 mg/kg 2x/week p.o. (trough level 100 ng/mL)	No	(Bel et al., 2010)
Embryonic stem cells (human)	CM - purity: ~63% [ACTN2^{+}]	Guinea pigs (xenograft)	• CSA: 15 mg/kg/d s.c. for 7 days from day -2, thereafter 7.5 mg/kg/d • MP: 2 mg/kg/d IP starting day -2	No	(Shiba et al., 2012)
Parthenogenetic stem cells (mouse)	Cell mixture - 75% CM: purity >95% [ACTN2^{+}] - 25% Fib	Mice (MHC-matched allograft)	• MP: 5 mg/kg/d s.c.	No	(Didié et al., 2013)
Induced pluripotent stem cells (human)	Cell mixture 33% CMs, 33% ECs, 33% SMCs	Pigs (xenograft)	• CSA; 15 mg/kg/d p.o. from day -3	Yes	(Ye et al., 2014)

Embryonic stem cells (human)	CM - purity: ~73% [$TNNT2^+$]	Macaca nemestrina (xenograft)	• MP: 500 mg i.v. bolus on day -1 followed by 0.1-1.5 mg/kg/d • CSA: 200-250 μg/L trough levels from day -5 • ABA: 12.5 mg/kg/14 days from day -1	No	(Chong et al., 2014)
Embryonic stem cells (human)	CM - purity: 71-95% [$ACTN2^+$]	Nude rats (xenograft)	• None	No	(Riegler et al., 2015)
		Sprague-Dawley rats (xenograft)	• None	Yes	
			• TAC: 8 mg/kg/d	No	
Induced pluripotent stem cells (Macaca fascicularis	CM - purity ~84% [$TNNT2^+$]	Macaca fascicularis MHC-matched allograft	• MP: 10 mg/kg/d i.m. from day -1 for 3 days followed by 1 mg/kg/d i.m. • TAC: 0.1 mg/kg/d i.m. from day -2	No	(Shiba et al., 2016)
		MHC-mismatched allograft		Yes	
Induced pluripotent stem cells (human)	Cell mixture - ~70% CM: purity 65-95% [$ACTN2^+$ and $TNNT2^+$] - ~30% endothelial cells: purity >90% [$PECAM1^+$]	Guinea pigs (xenograft)	• MP: 2 mg/kg/d • CSA: 5 mg/kg/d	No	(Weinberger et al., 2016)

CSA cyclosporine A, *AZA* azathioprine, *MP* methylprednisolone, *TAC* tacrolimus (FK506), *ABA* abatacept, *CM* cardiomyocytes, *Fib* fibroblasts obvious immunological challenge. Recent data on immune rejection of iPSC autografts (Zhao et al. 2015) and the apparent need for immune suppression after application of MHC-mismatched iPSC allografts (Shiba et al. 2016) challenged this concept. Moreover, we consider the need for a timely allocation of a heart muscle graft, cost-effectiveness, and difficulties to ensure reproducibility of therapeutic efficiency in autografts as additional and potentially even more important practical roadblocks. Currently, an off-the-shelf allograft approach under established immune suppression protocols seems in light of the medical unmet need the most likely scenario for the translation of the first-generation tissue-engineered heart muscle patch into a clinical application.

10.7 Good Manufacturing Practice and Procedural Safety

Clinical translation will not be possible unless tissue-engineered grafts can be produced according to current good manufacturing practice (cGMP). A meticulously defined production process will finally ensure safety and reproducibility in clinical applications. Several studies have confirmed that re-muscularization can be achieved by human cardiomyocyte implantation, either by direct injection (Chong et al. 2014; Shiba et al. 2016) or application as tissue-engineered myocardium (Riegler et al. 2015; Weinberger et al. 2016). There is also considerable evidence for electromechanical integration of human cardiomyocyte grafts even in non-human recipient myocardium (Chong et al. 2014; Kehat et al. 2004; Shiba et al. 2012, 2016; Weinberger et al. 2016). Risks of arrhythmia induction and teratoma formation are clearly described. Collectively, the available data provides a solid underpinning for the clinical translation of cardiomyocyte injection and tissue-engineered heart repair. As to tissue-engineered heart repair and the construction of tissue-engineered heart muscle under cGMP conditions, several steps will have to be taken:

1. The scaffold material must be defined and of clinical grade—as to collagen and fibrin, there will be no issues because of their use in clinical practice. Conversely, Matrigel™, secreted protein from the murine Engelbreth-Holm-Swarm sarcoma cells, will most likely not be acceptable.
2. Cardiomyocytes and support cells will have to be produced by specialized laboratories to meet cGMP requirements.
3. Culture medium will have to be defined and ideally devoid of serum components for optimal reproducibility.

Classical product testing including sterility, identity, and potency assays need to be in place. Safety assays of regulatory relevance should be performed with the cGMP tissue-engineered product primarily in immune-compromised mice or rats. Respective models are well established to investigate growth of human tumor cells and pluripotent stem cells. Arrhythmogenicity will remain a concern, and according to recent studies, its occurrence must be anticipated in all treated patients at least transiently (Chong et al. 2014; Shiba et al. 2016). Highly synchronized grafts, electrical pacing, and concomitant implantation of defibrillation devices may address this concern. Whether therapeutic efficacy will outweigh the risk of cardiomyocyte-based heart repair will finally only be predictable after the completion of clinical studies.

Conclusion

Tissue engineering technologies have made tremendous advances throughout recent years and are rapidly approaching the clinical stage. Cell-free (Mann et al. 2016) and noncontractile patches (Chachques et al. 2007; Menasche et al. 2015) are already under clinical investigation (Table 10.3); contractile patches and pouches will follow. We do not consider scale to meet clinical demands and costs to be limiting factors. Process development to meet cGMP quality has overcome important obstacles with early proof of concept for the cGMP-compatible manufacturing of contractile EHM patches available today (Riegler et al. 2015; Tiburcy et al. 2017). Further assessments of safety and efficacy in immune-compromised rodents and homologous large animal models, preferably the macaque as the most commonly used and predictive model for solid organ transplantation, will inform the design of

Table 10.3 Overview of preclinical and studies on (tissue) engineered heart repair

Cell-free alginate	*Clinical trial:*	
	– Algisyl™ tested in AUGMENT-HF study	Mann et al. (2016) NCT01311791
	– VentriGel™ Study	NCT02305602
	Preclinical animal study:	
	– Alginate in pig with myocardial infarction	Leor et al. (2009)
	– Decellularized ECM from pig heart (VentriGel™)	Wassenaar et al. (2016)
Non-contractile patches	*Clinical trials:*	
	– ESCORT (hESC-derived progenitors in fibrin patch)	Menasche et al. (2015) NCT02057900
	– Autologous myoblast cell sheet (TCD-51073) Study	Sawa et al. (2015) UMIN000008013
	– MAGNUM (mesenchymal stem cells in collagen patch)	Chachques et al. (2008)
	Preclinical animal studies:	
	– ESC-derived progenitor sheets in macaque model	Bel et al. (2010)
	– Autologous skeletal cell sheets in pig	Miyagawa et al. (2010)
	– Fibrin patch with ESC-derived endothelial cells and smooth muscle cells in pig	Xiong et al. (2011)
Contractile patches	*Clinical trials:*	
	None	
	Preclinical animal studies:	
	– Collagen patch with ESC-derived cardiomyocytes in RNU rats	Riegler et al. (2015)
	– Fibrin/Matrigel™ with iPSC-derived cardiomyocytes/endothelial cells in guinea pigs	Weinberger et al. (2016)

AUGMENT-HF A randomized, controlled study to evaluate Algisyl-LVR™ as a method of left ventricular augmentation for heart failure, *ESCORT* Transplantation of human embryonic stem cell-derived progenitors in severe heart failure, *MAGNUM* Myocardial assistance by grafting a new bioartificial upgraded myocardium, *EF* ejection fraction, *NYHA* New York Heart Association, *ECM* extracellular matrix early clinical trials. The optimal route of administration and dosing remain to be tested in large animal models, the latter under allogeneic conditions with clinically acceptable immune suppression if re-muscularization is the proposed mode of action. Therapeutic effects by mechanical stabilization and paracrine protection, if its target is known, do in our view not necessarily require preclinical allograft models. After 20 years of work in the field of myocardial tissue engineering, there is in our view a clear rationale for the clinical translation of tissue-engineered heart repair. Close interactions with regulatory authorities are needed to design pivotal (i.e., IND-enabling) preclinical trials and ensure that patients will not be subjected to inacceptable risks in the anticipated first-in-patient studies.

Acknowledgments The authors are supported by the ADUMED foundation (B.F.), the German Research Foundation (DFG ZI 708/10-1, SFB 1002 TP C04/S, SFB 937 A18, IRTG 1816; M.T., W.H.Z.), the Foundation Leducq (W.H.Z.), and the German Federal Ministry for Science and Education (BMBF FKZ 13GW0007A [BMBF/CIRM ETIII Award] and DZHK; W.H.Z.).

Compliance with Ethical Standards

Conflict of Interest W.H.Z is cofounder and advisor of Repairon GmbH.

Ethical Approval This article does not contain any studies with human participants performed by any of the authors.

References

Aicher A, Brenner W, Zuhayra M, Badorff C, Massoudi S, Assmus B, Eckey T, Henze E, Zeiher AM, Dimmeler S (2003) Assessment of the tissue distribution of transplanted human endothelial progenitor cells by radioactive labeling. Circulation 107:2134–2139

Anker SD, Coats AJ, Cristian G, Dragomir D, Pusineri E, Piredda M, Bettari L, Dowling R, Volterrani M, Kirwan BA et al (2015) A prospective comparison of alginate-hydrogel with standard medical therapy to determine impact on functional capacity and clinical outcomes in patients with advanced heart failure (AUGMENT-HF trial). Eur Heart J 36:2297–2309

Bartunek J, Davison B, Sherman W, Povsic T, Henry TD, Gersh B, Metra M, Filippatos G, Hajjar R, Behfar A et al (2016) Congestive heart failure cardiopoietic regenerative therapy (CHART-1) trial design. Eur J Heart Fail 18:160–168

Bel A, Planat-Bernard V, Saito A, Bonnevie L, Bellamy V, Sabbah L, Bellabas L, Brinon B, Vanneaux V, Pradeau P et al (2010) Composite cell sheets: a further step toward safe and effective myocardial regeneration by cardiac progenitors derived from embryonic stem cells. Circulation 122:S118–S123

Bellamy V, Vanneaux V, Bel A, Nemetalla H, Emmanuelle Boitard S, Farouz Y, Joanne P, Perier MC, Robidel E, Mandet C et al (2015) Long-term functional benefits of human embryonic stem cell-derived cardiac progenitors embedded into a fibrin scaffold. J Heart Lung Transplant 34:1198–1207

Blin G, Nury D, Stefanovic S, Neri T, Guillevic O, Brinon B, Bellamy V, Rucker-Martin C, Barbry P, Bel A et al (2010) A purified population of multipotent cardiovascular progenitors derived from primate pluripotent stem cells engrafts in postmyocardial infarcted nonhuman primates. J Clin Invest 120:1125–1139

Blocklet D, Toungouz M, Berkenboom G, Lambermont M, Unger P, Preumont N, Stoupel E, Egrise D, Degaute JP, Goldman M et al (2006) Myocardial homing of nonmobilized peripheral-blood CD34+ cells after intracoronary injection. Stem Cells 24:333–336

Bolli R, Chugh AR, D'Amario D, Loughran JH, Stoddard MF, Ikram S, Beache GM, Wagner SG, Leri A, Hosoda T et al (2011) Cardiac stem cells in patients with ischaemic cardiomyopathy (SCIPIO): initial results of a randomised phase 1 trial. Lancet 378:1847–1857

Burridge PW, Keller G, Gold JD, Wu JC (2012) Production of de novo cardiomyocytes: human pluripotent stem cell differentiation and direct reprogramming. Cell Stem Cell 10:16–28

Caspi O, Huber I, Kehat I, Habib M, Arbel G, Gepstein A, Yankelson L, Aronson D, Beyar R, Gepstein L (2007a) Transplantation of human embryonic stem cell-derived cardiomyocytes improves myocardial performance in infarcted rat hearts. J Am Coll Cardiol 50:1884–1893

Caspi O, Lesman A, Basevitch Y, Gepstein A, Arbel G, Habib IH, Gepstein L, Levenberg S (2007b) Tissue engineering of vascularized cardiac muscle from human embryonic stem cells. Circ Res 100:263–272

Chachques JC, Trainini JC, Lago N, Cortes-Morichetti M, Schussler O, Carpentier A (2008) Myocardial assistance by grafting a new bioartificial upgraded myocardium (MAGNUM trial): clinical feasibility study. Ann Thorac Surg 85:901–908

Chachques JC, Trainini JC, Lago N, Masoli OH, Barisani JL, Cortes-Morichetti M, Schussler O, Carpentier A (2007) Myocardial assistance by grafting a new bioartificial upgraded myocardium (MAGNUM clinical trial): one year follow-up. Cell Transplant 16:927–934

Chakravarty T, Makkar RR, Ascheim D, Traverse JH, Schatz R, DeMaria A, Francis GS, Povsic TJ, Smith R, Lima JA et al (2016) ALLogeneic heart stem cells to achieve myocardial regeneration (ALLSTAR) trial: rationale & design. Cell Transplant 26(2):205–214

Chen VC, Ye J, Shukla P, Hua G, Chen D, Lin Z, Liu JC, Chai J, Gold J, Wu J et al (2015) Development of a scalable suspension culture for cardiac differentiation from human pluripotent stem cells. Stem Cell Res 15:365–375

Chong JJ, Yang X, Don CW, Minami E, Liu YW, Weyers JJ, Mahoney WM, Van Biber B, Cook SM, Palpant NJ et al (2014) Human embryonic-stem-cell-derived cardiomyocytes regenerate non-human primate hearts. Nature 510:273–277

Christman KL, Vardanian AJ, Fang Q, Sievers RE, Fok HH, Lee RJ (2004) Injectable fibrin scaffold improves cell transplant survival, reduces infarct expansion, and induces neovasculature formation in ischemic myocardium. J Am Coll Cardiol 44:654–660

Didié M, Christalla P, Rubart M, Muppala V, Döker S, Unsöld B, El-Armouche A, Rau T, Eschenhagen T, Schwoerer AP et al (2013) Parthenogenetic stem cells for tissue-engineered heart repair. J Clin Invest 123:1285–1298

Fink C, Ergun S, Kralisch D, Remmers U, Weil J, Eschenhagen T (2000) Chronic stretch of engineered heart tissue induces hypertrophy and functional improvement. FASEB J 14:669–679

Funakoshi S, Miki K, Takaki T, Okubo C, Hatani T, Chonabayashi K, Nishikawa M, Takei I, Oishi A, Narita M et al (2016) Enhanced engraftment, proliferation, and therapeutic potential in heart using optimized human iPSC-derived cardiomyocytes. Sci Rep 6:19111

Godier-Furnemont AF, Tiburcy M, Wagner E, Dewenter M, Lammle S, El-Armouche A, Lehnart SE, Vunjak-Novakovic G, Zimmermann WH (2015) Physiologic force-frequency response in engineered heart muscle by electromechanical stimulation. Biomaterials 60:82–91

Gore A, Li Z, Fung HL, Young JE, Agarwal S, Antosiewicz-Bourget J, Canto I, Giorgetti A, Israel MA, Kiskinis E et al (2011) Somatic coding mutations in human induced pluripotent stem cells. Nature 471:63–67

Hare JM, Fishman JE, Gerstenblith G, DiFede Velazquez DL, Zambrano JP, Suncion VY, Tracy M, Ghersin E, Johnston PV, Brinker JA et al (2012) Comparison of allogeneic vs autologous bone marrow-derived mesenchymal stem cells delivered by transendocardial injection in patients with ischemic cardiomyopathy: the POSEIDON randomized trial. JAMA 308:2369–2379

Hirt MN, Boeddinghaus J, Mitchell A, Schaaf S, Bornchen C, Muller C, Schulz H, Hubner N, Stenzig J, Stoehr A et al (2014) Functional improvement and maturation of rat and human engineered heart tissue by chronic electrical stimulation. J Mol Cell Cardiol 74:151–161

Hofmann M, Wollert KC, Meyer GP, Menke A, Arseniev L, Hertenstein B, Ganser A, Knapp WH, Drexler H (2005) Monitoring of bone marrow cell homing into the infarcted human myocardium. Circulation 111:2198–2202

Hou D, Youssef EA, Brinton TJ, Zhang P, Rogers P, Price ET, Yeung AC, Johnstone BH, Yock PG, March KL (2005) Radiolabeled cell distribution after intramyocardial, intracoronary, and interstitial retrograde coronary venous delivery: implications for current clinical trials. Circulation 112:I150–I156

Kawamura M, Miyagawa S, Miki K, Saito A, Fukushima S, Higuchi T, Kawamura T, Kuratani T, Daimon T, Shimizu T et al (2012) Feasibility, safety, and therapeutic efficacy of human induced pluripotent stem cell-derived cardiomyocyte sheets in a porcine ischemic cardiomyopathy model. Circulation 126:S29–S37

Kehat I, Khimovich L, Caspi O, Gepstein A, Shofti R, Arbel G, Huber I, Satin J, Itskovitz-Eldor J, Gepstein L (2004) Electromechanical integration of cardiomyocytes derived from human embryonic stem cells. Nat Biotechnol 22:1282–1289

Kemp CD, Conte JV (2012) The pathophysiology of heart failure. Cardiovasc Pathol 21:365–371

Kensah G, Roa Lara A, Dahlmann J, Zweigerdt R, Schwanke K, Hegermann J, Skvorc D, Gawol A, Azizian A, Wagner S et al (2013) Murine and human pluripotent stem cell-derived cardiac bodies form contractile myocardial tissue in vitro. Eur Heart J 34:1134–1146

Kim K, Lerou P, Yabuuchi A, Lengerke C, Ng K, West J, Kirby A, Daly MJ, Daley GQ (2007) Histocompatible embryonic stem cells by parthenogenesis. Science 315:482–486

Koyanagi-Aoi M, Ohnuki M, Takahashi K, Okita K, Noma H, Sawamura Y, Teramoto I, Narita M, Sato Y, Ichisaka T et al (2013) Differentiation-defective phenotypes revealed by large-scale analyses of human pluripotent stem cells. Proc Natl Acad Sci U S A 110:20569–20574

Laflamme MA, Chen KY, Naumova AV, Muskheli V, Fugate JA, Dupras SK, Reinecke H, Xu C, Hassanipour M, Police S et al (2007) Cardiomyocytes derived from human embryonic stem cells in pro-survival factors enhance function of infarcted rat hearts. Nat Biotechnol 25:1015–1024

Laflamme MA, Murry CE (2011) Heart regeneration. Nature 473:326–335

Lammerding J, Kamm RD, Lee RT (2004) Mechanotransduction in cardiac myocytes. Ann N Y Acad Sci 1015:53–70

Leor J, Aboulafia-Etzion S, Dar A, Shapiro L, Barbash IM, Battler A, Granot Y, Cohen S (2000) Bioengineered cardiac grafts: a new approach to repair the infarcted myocardium? Circulation 102:III56–III61

Leor J, Tuvia S, Guetta V, Manczur F, Castel D, Willenz U, Petnehazy O, Landa N, Feinberg MS, Konen E et al (2009) Intracoronary injection of in situ forming alginate hydrogel reverses left ventricular remodeling after myocardial infarction in Swine. J Am Coll Cardiol 54:1014–1023

Lesman A, Habib M, Caspi O, Gepstein A, Arbel G, Levenberg S, Gepstein L (2010) Transplantation of a tissue-engineered human vascularized cardiac muscle. Tissue Eng Part A 16:115–125

Liaw NY, Zimmermann WH (2015) Mechanical stimulation in the engineering of heart muscle. Adv Drug Deliv Rev 96:156–160

Lister R, Pelizzola M, Kida YS, Hawkins RD, Nery JR, Hon G, Antosiewicz-Bourget J, O'Malley R, Castanon R, Klugman S et al (2011) Hotspots of aberrant epigenomic reprogramming in human induced pluripotent stem cells. Nature 471:68–73

Long EF, Swain GW, Mangi AA (2014) Comparative survival and cost-effectiveness of advanced therapies for end-stage heart failure. Circ Heart Fail 7:470–478

Lund LH, Edwards LB, Kucheryavaya AY, Benden C, Dipchand AI, Goldfarb S, Levvey BJ, Meiser B, Rossano JW, Yusen RD et al (2015) The Registry of the International Society for Heart and Lung Transplantation: thirty-second official adult heart transplantation report--2015; focus theme: early graft failure. J Heart Lung Transplant 34:1244–1254

Makkar RR, Smith RR, Cheng K, Malliaras K, Thomson LE, Berman D, Czer LS, Marban L, Mendizabal A, Johnston PV et al (2012) Intracoronary cardiosphere-derived cells for heart regeneration after myocardial infarction (CADUCEUS): a prospective, randomised phase 1 trial. Lancet 379:895–904

Mann DL, Lee RJ, Coats AJ, Neagoe G, Dragomir D, Pusineri E, Piredda M, Bettari L, Kirwan BA, Dowling R et al (2016) One-year follow-up results from AUGMENT-HF: a multicentre

randomized controlled clinical trial of the efficacy of left ventricular augmentation with Algisyl in the treatment of heart failure. Eur J Heart Fail 18:314–325
Menasche P (2008) Skeletal myoblasts and cardiac repair. J Mol Cell Cardiol 45:545–553
Menasche P, Vanneaux V, Hagege A, Bel A, Cholley B, Cacciapuoti I, Parouchev A, Benhamouda N, Tachdjian G, Tosca L et al (2015) Human embryonic stem cell-derived cardiac progenitors for severe heart failure treatment: first clinical case report. Eur Heart J 36:2011–2017
Miyagawa S, Saito A, Sakaguchi T, Yoshikawa Y, Yamauchi T, Imanishi Y, Kawaguchi N, Teramoto N, Matsuura N, Iida H et al (2010) Impaired myocardium regeneration with skeletal cell sheets--a preclinical trial for tissue-engineered regeneration therapy. Transplantation 90:364–372
Montgomery JR, Berger JC, Warren DS, James NT, Montgomery RA, Segev DL (2012) Outcomes of ABO-incompatible kidney transplantation in the United States. Transplantation 93:603–609
Mozaffarian D, Benjamin EJ, Go AS, Arnett DK, Blaha MJ, Cushman M, de Ferranti S, Despres JP, Fullerton HJ, Howard VJ et al (2015) Heart disease and stroke statistics--2015 update: a report from the American Heart Association. Circulation 131:e29–322
Müller-Ehmsen J, Whittaker P, Kloner RA, Dow JS, Sakoda T, Long TI, Laird PW, Kedes L (2002) Survival and development of neonatal rat cardiomyocytes transplanted into adult myocardium. J Mol Cell Cardiol 34:107–116
Naito H, Melnychenko I, Didie M, Schneiderbanger K, Schubert P, Rosenkranz S, Eschenhagen T, Zimmermann WH (2006) Optimizing engineered heart tissue for therapeutic applications as surrogate heart muscle. Circulation 114:I72–I78
Nelson TJ, Martinez-Fernandez A, Yamada S, Perez-Terzic C, Ikeda Y, Terzic A (2009) Repair of acute myocardial infarction by human stemness factors induced pluripotent stem cells. Circulation 120:408–416
Nguyen PK, Rhee JW, Wu JC (2016) Adult stem cell therapy and heart failure, 2000 to 2016: a systematic review. JAMA Cardiol 1:831–841
Nichols M, Townsend N, Scarborough P, Rayner M (2014) Cardiovascular disease in Europe 2014: epidemiological update. Eur Heart J 35:2950–2959
Okita K, Matsumura Y, Sato Y, Okada A, Morizane A, Okamoto S, Hong H, Nakagawa M, Tanabe K, Tezuka K et al (2011) A more efficient method to generate integration-free human iPS cells. Nat Methods 8:409–412
Pinto AR, Ilinykh A, Ivey MJ, Kuwabara JT, D'Antoni ML, Debuque R, Chandran A, Wang L, Arora K, Rosenthal NA et al (2016) Revisiting cardiac cellular composition. Circ Res 118:400–409
Radisic M, Marsano A, Maidhof R, Wang Y, Vunjak-Novakovic G (2008) Cardiac tissue engineering using perfusion bioreactor systems. Nat Protoc 3:719–738
Radisic M, Park H, Gerecht S, Cannizzaro C, Langer R, Vunjak-Novakovic G (2007) Biomimetic approach to cardiac tissue engineering. Philos Trans R Soc Lond Ser B Biol Sci 362:1357–1368
Radisic M, Park H, Shing H, Consi T, Schoen FJ, Langer R, Freed LE, Vunjak-Novakovic G (2004) Functional assembly of engineered myocardium by electrical stimulation of cardiac myocytes cultured on scaffolds. Proc Natl Acad Sci U S A 101:18129–18134
Revazova ES, Turovets NA, Kochetkova OD, Agapova LS, Sebastian JL, Pryzhkova MV, Smolnikova VI, Kuzmichev LN, Janus JD (2008) HLA homozygous stem cell lines derived from human parthenogenetic blastocysts. Cloning Stem Cells 10:11–24
Riegler J, Tiburcy M, Ebert A, Tzatzalos E, Raaz U, Abilez OJ, Shen Q, Kooreman NG, Neofytou E, Chen VC et al (2015) Human engineered heart muscles engraft and survive long term in a rodent myocardial infarction model. Circ Res 117:720–730
Roger VL (2010) The heart failure epidemic. Int J Environ Res Public Health 7:1807–1830
Sanchez PL, Fernandez-Santos ME, Costanza S, Climent AM, Moscoso I, Gonzalez-Nicolas MA, Sanz-Ruiz R, Rodriguez H, Kren SM, Garrido G et al (2015) Acellular human heart matrix: a critical step toward whole heart grafts. Biomaterials 61:279–289
Sawa Y, Yoshikawa Y, Toda K, Fukushima S, Yamazaki K, Ono M, Sakata Y, Hagiwara N, Kinugawa K, Miyagawa S (2015) Safety and efficacy of autologous skeletal myoblast sheets

(TCD-51073) for the treatment of severe chronic heart failure due to ischemic heart disease. Circ J 79:991–999

Schachinger V, Assmus B, Erbs S, Elsasser A, Haberbosch W, Hambrecht R, Yu J, Corti R, Mathey DG, Hamm CW et al (2009) Intracoronary infusion of bone marrow-derived mononuclear cells abrogates adverse left ventricular remodelling post-acute myocardial infarction: insights from the reinfusion of enriched progenitor cells and infarct remodelling in acute myocardial infarction (REPAIR-AMI) trial. Eur J Heart Fail 11:973–979

Schachinger V, Erbs S, Elsasser A, Haberbosch W, Hambrecht R, Holschermann H, Yu J, Corti R, Mathey DG, Hamm CW et al (2006) Intracoronary bone marrow-derived progenitor cells in acute myocardial infarction. N Engl J Med 355:1210–1221

Shiba Y, Fernandes S, Zhu WZ, Filice D, Muskheli V, Kim J, Palpant NJ, Gantz J, Moyes KW, Reinecke H et al (2012) Human ES-cell-derived cardiomyocytes electrically couple and suppress arrhythmias in injured hearts. Nature 489:322–325

Shiba Y, Gomibuchi T, Seto T, Wada Y, Ichimura H, Tanaka Y, Ogasawara T, Okada K, Shiba N, Sakamoto K et al (2016) Allogeneic transplantation of iPS cell-derived cardiomyocytes regenerates primate hearts. Nature 538:388–391

Shimizu T, Sekine H, Yang J, Isoi Y, Yamato M, Kikuchi A, Kobayashi E, Okano T (2006) Polysurgery of cell sheet grafts overcomes diffusion limits to produce thick, vascularized myocardial tissues. FASEB J 20:708–710

Shimko VF, Claycomb WC (2008) Effect of mechanical loading on three-dimensional cultures of embryonic stem cell-derived cardiomyocytes. Tissue Eng Part A 14:49–58

Shirasaka T, Miyagawa S, Fukushima S, Kawaguchi N, Nakatani S, Daimon T, Okita Y, Sawa Y (2016) Skeletal myoblast cell sheet implantation ameliorates both systolic and diastolic cardiac performance in canine dilated cardiomyopathy model. Transplantation 100:295–302

Soong PL, Tiburcy M, Zimmermann WH (2012) Cardiac differentiation of human embryonic stem cells and their assembly into engineered heart muscle. Curr Protoc Cell Biol. doi:10.1002/0471143030.cb2308s55

Streckfuss-Bomeke K, Wolf F, Azizian A, Stauske M, Tiburcy M, Wagner S, Hubscher D, Dressel R, Chen S, Jende J et al (2013) Comparative study of human-induced pluripotent stem cells derived from bone marrow cells, hair keratinocytes, and skin fibroblasts. Eur Heart J 34:2618–2629

Takahashi K, Tanabe K, Ohnuki M, Narita M, Ichisaka T, Tomoda K, Yamanaka S (2007) Induction of pluripotent stem cells from adult human fibroblasts by defined factors. Cell 131:861–872

Tandon N, Cannizzaro C, Chao PH, Maidhof R, Marsano A, Au HT, Radisic M, Vunjak-Novakovic G (2009) Electrical stimulation systems for cardiac tissue engineering. Nat Protoc 4:155–173

Taylor CJ, Peacock S, Chaudhry AN, Bradley JA, Bolton EM (2012) Generating an iPSC bank for HLA-matched tissue transplantation based on known donor and recipient HLA types. Cell Stem Cell 11:147–152

Thorpe SJ, Hunt B, Yacoub M (1991) Expression of ABH blood group antigens in human heart tissue and its relevance to cardiac transplantation. Transplantation 51:1290–1295

Tiburcy M, Meyer T, Soong PL, Zimmermann WH (2014) Collagen-based engineered heart muscle. Methods Mol Biol 1181:167–176

Tiburcy M, Zimmermann WH (2014) Modeling myocardial growth and hypertrophy in engineered heart muscle. Trends Cardiovasc Med 24:7–13

Tiburcy M, Hudson JE, Balfanz P, Schlick S, Meyer T et al (2017) Defined engineered human myocardium with advanced maturation for applications in heart failure modeling and repair. Circulation 135:1832–1847

Travers JG, Kamal FA, Robbins J, Yutzey KE, Blaxall BC (2016) Cardiac fibrosis: the fibroblast awakens. Circ Res 118:1021–1040

Tulloch NL, Muskheli V, Razumova MV, Korte FS, Regnier M, Hauch KD, Pabon L, Reinecke H, Murry CE (2011) Growth of engineered human myocardium with mechanical loading and vascular coculture. Circ Res 109:47–59

van Laake LW, Passier R, Monshouwer-Kloots J, Verkleij AJ, Lips DJ, Freund C, den Ouden K, Ward-van Oostwaard D, Korving J, Tertoolen LG et al (2007) Human embryonic stem cell-derived cardiomyocytes survive and mature in the mouse heart and transiently improve function after myocardial infarction. Stem Cell Res 1:9–24

Vandenburgh HH, Solerssi R, Shansky J, Adams JW, Henderson SA (1996) Mechanical stimulation of organogenic cardiomyocyte growth in vitro. Am J Phys 270:C1284–C1292

Wassenaar JW, Gaetani R, Garcia JJ, Braden RL, Luo CG, Huang D, DeMaria AN, Omens JH, Christman KL (2016) Evidence for mechanisms underlying the functional benefits of a myocardial matrix hydrogel for post-MI treatment. J Am Coll Cardiol 67:1074–1086

Wei K, Serpooshan V, Hurtado C, Diez-Cunado M, Zhao M, Maruyama S, Zhu W, Fajardo G, Noseda M, Nakamura K et al (2015) Epicardial FSTL1 reconstitution regenerates the adult mammalian heart. Nature 525:479–485

Weinberger F, Breckwoldt K, Pecha S, Kelly A, Geertz B, Starbatty J, Yorgan T, Cheng KH, Lessmann K, Stolen T et al (2016) Cardiac repair in guinea pigs with human engineered heart tissue from induced pluripotent stem cells. Sci Transl Med 8(363):363ra148–363ra148

Wijngaard PL, de Bresser JM, de Groot PG, Gmelig-Meyling FH, Schuurman HJ, Jambroes G, Borleffs JC (1991) Endothelial and smooth muscle cells in the heart allograft response: isolation procedure and immunocytochemical features. J Heart Lung Transplant 10:416–423

Xiong Q, Hill KL, Li Q, Suntharalingam P, Mansoor A, Wang X, Jameel MN, Zhang P, Swingen C, Kaufman DS et al (2011) A fibrin patch-based enhanced delivery of human embryonic stem cell-derived vascular cell transplantation in a porcine model of postinfarction left ventricular remodeling. Stem Cells 29:367–375

Ye L, Chang YH, Xiong Q, Zhang P, Zhang L, Somasundaram P, Lepley M, Swingen C, Su L, Wendel JS et al (2014) Cardiac repair in a porcine model of acute myocardial infarction with human induced pluripotent stem cell-derived cardiovascular cells. Cell Stem Cell 15:750–761

Ye L, Zimmermann WH, Garry DJ, Zhang J (2013) Patching the heart: cardiac repair from within and outside. Circ Res 113:922–932

Zhang Q, Cecka JM, Gjertson DW, Ge P, Rose ML, Patel JK, Ardehali A, Kobashigawa JA, Fishbein MC, Reed EF (2011) HLA and MICA: targets of antibody-mediated rejection in heart transplantation. Transplantation 91:1153–1158

Zhang G, Nakamura Y, Wang X, Hu Q, Suggs LJ, Zhang J (2007) Controlled release of stromal cell-derived factor-1 alpha in situ increases c-kit+ cell homing to the infarcted heart. Tissue Eng 13:2063–2071

Zhang D, Shadrin IY, Lam J, Xian HQ, Snodgrass HR, Bursac N (2013) Tissue-engineered cardiac patch for advanced functional maturation of human ESC-derived cardiomyocytes. Biomaterials 34:5813–5820

Zhao T, Zhang ZN, Westenskow PD, Todorova D, Hu Z, Lin T, Rong Z, Kim J, He J, Wang M et al (2015) Humanized mice reveal differential immunogenicity of cells derived from autologous induced pluripotent stem cells. Cell Stem Cell 17:353–359

Zimmermann WH (2009) Remuscularizing failing hearts with tissue engineered myocardium. Antioxid Redox Signal 11:2011–2023

Zimmermann WH, Didie M, Wasmeier GH, Nixdorff U, Hess A, Melnychenko I, Boy O, Neuhuber WL, Weyand M, Eschenhagen T (2002a) Cardiac grafting of engineered heart tissue in syngenic rats. Circulation 106:I151–I157

Zimmermann WH, Eschenhagen T (2003) Cardiac tissue engineering for replacement therapy. Heart Fail Rev 8:259–269

Zimmermann WH, Fink C, Kralisch D, Remmers U, Weil J, Eschenhagen T (2000) Three-dimensional engineered heart tissue from neonatal rat cardiac myocytes. Biotechnol Bioeng 68:106–114

Zimmermann WH, Melnychenko I, Wasmeier G, Didie M, Naito H, Nixdorff U, Hess A, Budinsky L, Brune K, Michaelis B et al (2006) Engineered heart tissue grafts improve systolic and diastolic function in infarcted rat hearts. Nat Med 12:452–458

Zimmermann WH, Schneiderbanger K, Schubert P, Didie M, Munzel F, Heubach JF, Kostin S, Neuhuber WL, Eschenhagen T (2002b) Tissue engineering of a differentiated cardiac muscle construct. Circ Res 90:223–230

11 Imaging Cardiac Stem Cell Therapy

Xulei Qin, Ian Y. Chen, and Joseph C. Wu

Abstract

Cardiac stem cell therapy is a promising approach to repair an injured heart. Noninvasive imaging can be tremendously useful for characterizing its therapeutic mechanisms and efficacy. In this chapter, we summarize the roles of various imaging modalities in assessing cell fate and cardiac function following stem cell therapy. The advancement of these imaging technologies is crucial for the full clinical translation of cardiac stem cell therapy.

11.1 Introduction

Ischemic heart disease is a condition characterized by compromised coronary blood flow relative to myocardial demand. When severe, the interruption of blood supply to the heart itself can damage heart muscle, causing a myocardial infarction (MI). This condition is accompanied by both contractile dysfunction and arrhythmia, which contribute significantly to its high morbidity and mortality (Zipes and Jalife 2009; Chen et al. 2013). Unfortunately, the adult heart has a limited capacity to repair itself in the setting of an acute MI. Therefore, approaches to regenerate the injured myocardium via exogenous delivery of stem cells have been proposed and

X. Qin • I.Y. Chen • J.C. Wu (✉)
Stanford Cardiovascular Institute, Stanford University School of Medicine, Stanford, CA 94305, USA

Departments of Medicine and Radiology, Stanford University School of Medicine, Stanford, CA 94305, USA

Institute of Stem Cell Biology and Regenerative Medicine, Stanford University School of Medicine, Stanford, CA 94305, USA
e-mail: joewu@stanford.edu

M. Ieda, W.-H. Zimmermann (eds.), *Cardiac Regeneration*, Cardiac and Vascular Biology 4, DOI 10.1007/978-3-319-56106-6_11

tested in preclinical animal models, with some also undergoing further evaluation in clinical trials.

To date, three major types of stem cells have been used for myocardial repair, including adult stem cells, embryonic stem cells (ESCs), and induced pluripotent stem cells (iPSCs). In clinical trials, adult stem cells such as bone marrow stem cells (BMSCs), mesenchymal stem cells (MSCs), endothelial progenitor cells (EPCs), and cardiac progenitor cells (CPCs) have been the primary choices. Although not yet extensively tested in humans partly due to ethical concerns, human ESCs have characteristics that are ideal for cardiac cell transplantation. For instance, compared to adult stem cells, human ESCs are pluripotent and capable of unlimited self-renewal. Both human ESC-derived cardiomyocytes (ESC-CMs) and ESC-derived endothelial cells (ESC-ECs) have been used for heart repair in MI animal models (Li et al. 2007; Chong et al. 2014). A recent clinical study further demonstrated the feasibility of transplanting human ESC-derived cardiac progenitor cells in a tissue-engineered graft to the failing heart of a patient (Menasche et al. 2015). Compared to human ESCs, human iPSCs have similar pluripotency and self-renewal capacities but raise fewer ethical concerns. They can be derived from the patient's somatic cells (e.g., skin, blood, urinary epithelial, or adipose stromal cells). Transplantation of human iPSC-derived cardiomyocytes (iPSC-CMs) has been shown to improve cardiac function in a mouse MI model (Ong et al. 2015). However, for both human ESCs and iPSCs, their clinical transplantation is hampered by issues related to cost, immunogenicity, and tumorigenicity (Lee et al. 2013; Neofytou et al. 2015).

Despite the exciting progresses in stem cell biology, several issues in cardiac stem cell therapy still need to be addressed before its full clinical translation: (1) How can we optimally deliver these cells to the target site and track their distribution? (2) How can we monitor the fate of the transplanted cells, including their survival, integration, proliferation, and differentiation? (3) How can we best evaluate their therapeutic efficacy? (4) What is the best way to evaluate their long-term safety? To answer these important questions, researchers have developed noninvasive imaging techniques using ultrasound, magnetic resonance imaging (MRI), single photon emission computed tomography (SPECT), and positron emission tomography (PET) imaging to investigate cardiac stem cell therapy in both preclinical and clinical studies. Molecular imaging techniques optimized for cell tracking have been particularly useful in elucidating the fate, efficacy, and safety of these transplanted cells and will continue to help accelerate the translation of cardiac stem cell therapy.

11.2 Imaging Modalities for Cardiac Stem Cell Therapy

Different imaging modalities have been used in various capacities to evaluate cardiac stem cell therapy in both preclinical and clinical studies (Chen and Wu 2011; Nguyen et al. 2014). Molecular imaging techniques have been primarily used for monitoring the fate of transplanted cells following implantation (e.g., viability, proliferation, and distribution), whereas conventional imaging modalities have been used mostly for evaluating the therapeutic efficacy of cell therapy. The strengths and weaknesses of each modality need to be specifically matched to its spatiotemporal resolution, target sensitivity, and penetration depth in order to maximize its diagnostic performance (Table 11.1).

Table 11.1 Comparisons among different imaging modalities for imaging cardiac stem cell therapy

Imaging modality	Imaging principal	Advantage	Limitation	Imaging cardiac stem cell therapy
Radionuclide imaging	Imaging organs or transplanted cells following the injection of radioactive tracer	High sensitivity; quantification ability	Radioactivity; expensive; poor spatial resolution; short imaging time	• Biodistribution • Cell viability, proliferation, and differentiation • Myocardial viability and perfusion
Magnetic resonance imaging (MRI)	Using magnetic fields and radio frequency pulses to detect the anatomic and tissue structures	High spatial resolution; excellent soft tissue contrast; non-radiation	Expensive; complicated operations	• Cell delivery • Cell biodistribution • Therapeutic efficacies • Teratoma risk
Ultrasound	Using the reflected sound waves to image organ anatomy and blood flow	Real-time; inexpensive; easy operation	Low tissue contrast; operator dependency	• Cell delivery • Cardiac functions
Computed tomography (CT)	Imaging organ anatomy	Detailed structures with high spatial resolution	Ionizing radiation, limited soft tissue discrimination	• Cell delivery
Optical imaging	Using an optical detector to capture the emission light from target cells	High sensitivity, convenient, inexpensive	Poor penetration depth; poor spatial resolution, preclinical only	• Biodistribution • Cell viability, proliferation, and differentiation
Magnetic particle imaging (MPI)	Using low-frequency magnetic fields to detect iron oxide nanoparticles	High sensitivity, high tissue contrast, fast, non-radiational	Inability to discriminate live cells from dead cells	• Biodistribution

11.2.1 Imaging Modalities

Clinical imaging modalities including radionuclide imaging, ultrasound, MRI, and computed tomography (CT) have different roles in the management of various cardiac conditions due to their distinct imaging capabilities. In terms of evaluating cardiac stem cell therapy, one strength of imaging techniques is in providing detailed anatomical and functional information of the heart as illustrated in Fig. 11.1a.

Radionuclide imaging methods such as SPECT and PET using different radioisotopes (e.g., ^{99m}Tc for SPECT; ^{18}F for PET) have been used to detect myocardial perfusion, scar size, and viability as endpoints of cardiac stem cell therapy trials. Cardiac MRI, due to its superior spatial resolution, offers a more precise assessment of cardiac function and myocardial scar than PET (La Gerche et al. 2013; Friedrich and Marcotte 2013). Cardiac ultrasound (e.g., echocardiography) is routinely used in the clinic to image structure, function, and blood flow of the heart in real time and has been widely applied to evaluate cardiac stem cell therapy (Huikuri et al. 2008; Wendel et al. 2014; Chong et al. 2014). Although CT has received increased attention for its role in the diagnosis of coronary artery disease, it has been primarily used as an adjunct modality to SPECT or PET (i.e., SPECT/CT or PET/CT) by providing the anatomical information useful for determining the distribution of transplanted cells.

For preclinical studies involving small animals, many of these imaging modalities have miniature versions with higher sensitivities and spatial resolutions than their clinical counterparts. Imaging modalities that are exclusively used in small animals include bioluminescence imaging (BLI) and fluorescence imaging (FI), both of which use visible lights as signal and are therefore limited by the depth of light penetration. BLI has been used mainly to track stem cells that are genetically marked with the firefly luciferase reporter gene. Imaging of stem cell viability or

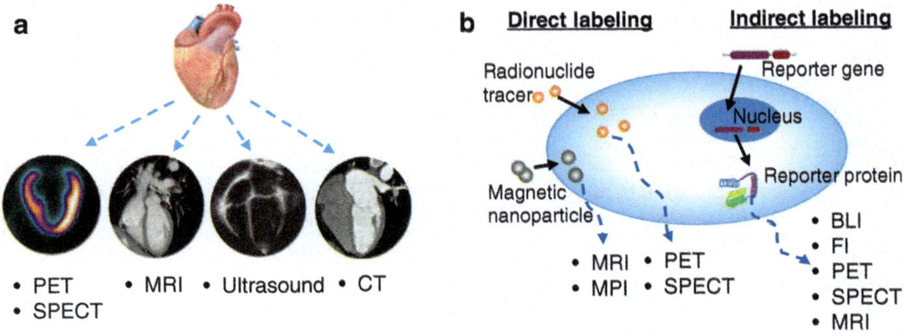

Fig. 11.1 Imaging for cardiac stem cell therapy. (**a**) Clinical imaging plays a fundamental role in the evaluation of cardiac anatomy and function. (**b**) Both direct and indirect cell labeling strategies have been developed to monitor the fate of transplanted cells in vivo. Direct cell labeling enables the imaging of labeled cells using an exogenous signal source (e.g., nanoparticle or nuclear tracer). Indirect cell labeling allows the imaging of cells overexpressing a reporter gene, which encodes for either a cell surface receptor or an intracellular enzyme to produce an endogenous signal source

distribution is performed by exogenously delivering the substrate D-luciferin, which upon encountering the firefly luciferase enzyme in these cells will cause emission of visible light as signal. In contrast, FI of stem cells is performed by first marking the cells with either a fluorescent reporter gene or a fluorescent dye and then imaging the cells using a dedicated FI system with the appropriate laser and excitation/emission filters. These optical imaging techniques are highly sensitive and have been broadly used in preclinical studies. Magnetic particle imaging (MPI) that relies on the use of superparamagnetic iron oxide (SPIO) contrast agents as cell markers have also been used extensively to sensitively image transplanted cells in vivo (Zheng et al. 2015).

11.2.2 Cell Labeling Strategies

Current clinical imaging methods are unable to directly track the transplanted stem cells in vivo unless the cells were specifically labeled a priori. An ideal cell label is one that has high specificity for the target, generates ample signal to enable sensitive detection, but does not perturb cell integrity or function. Figure 11.1b illustrates two major cell labeling strategies. The first strategy involves direct labeling of cells with an exogenous signal source (e.g., nanoparticle or nuclear tracer). The second strategy involves labeling the cells with a reporter gene, whose product as a protein, a receptor, or an intracellular enzyme will then partake in signal generation.

Direct labeling is typically achieved by incubating cells with an imaging probe that either binds to the cell surface, diffuses freely through the cell membrane, or is taken up into the cell via endocytosis. Superparamagnetic nanoparticles (e.g., SPIO), paramagnetic ion chelates, and radionuclide tracers are the most frequently used probes for direct labeling of cardiac stem cells. SPIO nanoparticles cause a highly sensitive hypointense signal on T2- or T2*-weighted images, yielding high-resolution tracking of the migration and integration of injected cells. However, because SPIO nanoparticles can be released after cell death and engulfed by macrophages, SPIO-based cell labeling and imaging lead to overestimation of cell viability (Li et al. 2008; Winter et al. 2010). In contrast to SPIO, gadolinium (Gd) chelates shorten longitudinal (T1) relaxation times to generate hyperintense signal on T1-weighted images. Labeling of cells with Gd chelates in general provides lower imaging sensitivity than labeling cells with SPIO. Moreover, the potential dechelation of free Gd^{3+} ions raises potential toxicity concerns for Gd-based cell labeling (Budde and Frank 2009). ^{19}F-labeled perfluorocarbon nanoparticles have also been used to label stem cells so that MR spectroscopy can be performed to track their locations (Partlow et al. 2007). Cell tracking using this approach requires a prolonged imaging time due to its lower sensitivity compared to conventional ^{1}H-based MRI.

Direct cell labeling with radionuclides is necessary when using SPECT or PET for cell tracking. Several radionuclides have been used to monitor transplanted cells for myocardial restoration, including ^{111}In-oxine (^{111}In; half-life ~2.8 days) and

^{99m}Tc-hexamethylprophylene amine oxine (^{99m}Tc-HMPAO; half-life ~6 h) for SPECT imaging and ^{18}F-fluorodeoxyglucose (^{18}F-FDG; half-life ~110 min) for PET imaging. Although these tracers provide highly sensitive cell detection, their relatively short half-lives preclude long-term cell monitoring. Their other drawbacks include limited spatial resolution and radioactivity.

Indirect (genetic) labeling involves transfecting cells with a reporter gene whose expression as either intracellular proteins, enzymes, or surface receptors can then be feasibly imaged. This cell labeling method has been widely used for optical and radionuclide imaging of stem cells in preclinical studies. To date, BLI is the most widely used optical imaging method for tracking stem cells in small living animals. It requires the creation of genetically modified stem cells a priori to express a luciferase enzyme (e.g., firefly luciferase or *Renilla* luciferase). During imaging, the enzyme substrate (e.g., D-luciferin for firefly luciferase or coelenterazine for *Renilla* luciferase) is used as an imaging probe, which on encountering the luciferase enzyme will emit visible light as signal to indicate cell location. This imaging approach is highly sensitive, semiquantitative, and easy to perform, making it suitable for tracking the location and survival of transplanted stem cells in preclinical studies (Nguyen et al. 2014). However, BLI's poor depth-dependent spatial resolution and the inability of visible light to penetrate thick tissues make it unsuitable for human studies.

Reporter gene-based labeling is also frequently utilized in association with radionuclide imaging, which commonly uses herpes simplex virus type 1 thymidine kinase (HSV1-*tk*) and its mutant (HSV1-*sr39tk*) (Cao et al. 2006). The reporter enzymes expressed from these genes can phosphorylate and trap intracellularly various PET imaging probes, including 9-[4-[^{18}F]fluoro-3-(hydroxymethyl)butyl]guanine (^{18}F-FHBG) or similar agents (Sun et al. 2009; Yaghoubi et al. 2009), thus providing an intrinsic mechanism for signal amplification to ensure high imaging sensitivity. Other reporter genes have also been validated for radionuclide imaging of stem cells, including dopamine type 2 receptor (D2R) for PET imaging and sodium iodide symporter (NIS) for both PET and SPECT imaging (Penheiter et al. 2012).

The main advantage of reporter gene-based labeling is that the detected imaging signal only comes from the viable cells expressing the reporter gene. Thus, it can be used for long-term monitoring of viability, proliferation, and differentiation of the transplanted cells in vivo. However, its clinical application is hampered by potential insertional mutagenesis from transgene integration and a possible significant immune response against the genetically modified cells.

11.3 Noninvasive Imaging for Cardiac Stem Cell Therapy

To date, noninvasive imaging has been used to investigate various parameters of cardiac stem cell therapy, including cell delivery, cell fate after transplantation, therapeutic efficacy, and safety (Fig. 11.2).

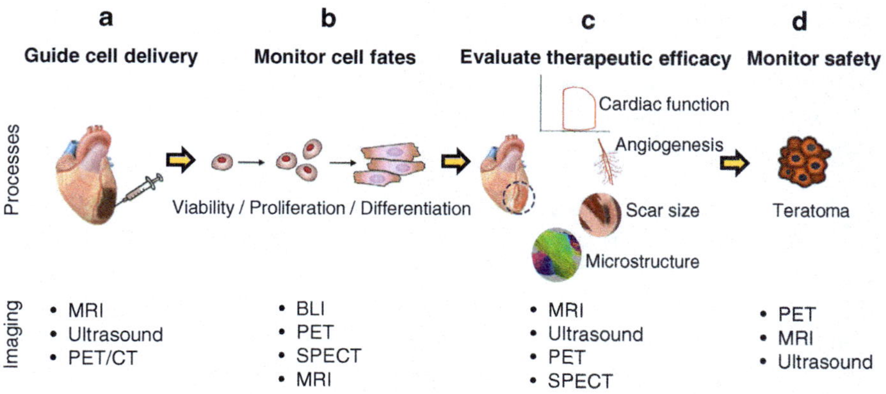

Fig. 11.2 Monitoring different processes of cardiac stem cell therapy in vivo. Cardiac imaging provides noninvasive assessments of different processes associated with cardiac stem cell therapy. (**a**) Clinical imaging modalities (MRI, ultrasound, and PET/CT) are used to guide cardiac stem cell delivery. (**b**) Cell labeling-based imaging methods (BLI, PET, SPECT, and MRI) are used to monitor cell fate after transplantation. (**c**) Clinical imaging modalities (MRI, ultrasound, PET, and SPECT) are used to evaluate the therapeutic efficacy. (**d**) PET, MRI, or ultrasound is used to monitor teratoma risks

11.3.1 Image-Guided Cell Delivery

Efficient cell delivery is a prerequisite for successful cardiac stem cell therapy. The process of delivering cells involves finding the target site(s) and delivering therapeutic cells in order to maximize cell retention. Different delivery methods have been used to deliver cells to the infarcted heart, including intravascular injection (Kang et al. 2012; Choudry et al. 2015), intramyocardial injection, and matrix-based delivery (Ye et al. 2014). Cardiac imaging plays an important role in guiding cell delivery and evaluating cell retention and long-term engraftment.

A clinical stem cell trial using ^{18}F-FDG as cell labels found under 5% of injected cells in the infarcted myocardium following intracoronary injection (Hofmann et al. 2005). Compared to intracoronary injection, intramyocardial and transendocardial injections have been shown to maintain greater cell retention in the myocardium (Hou et al. 2005; Vrtovec et al. 2013). Several multimodality guidance methods have been developed to enhance the accuracy of intramyocardial delivery. For instance, an integrated imaging platform was developed to combine MRI angiography and electroanatomical mapping unipolar voltage to accurately identify infarct border zone for cell injections (Williams et al. 2013). Alternatively, ultrasound has also shown effectiveness in guiding the delivery of stem cells labeled with nanoparticles (Jokerst et al. 2013). Although these methods have helped improve the precision of cell delivery, a recent study involving tracking of stem cells using ^{18}F-FDG PET found that injected cells were rapidly lost within 2 h of injection (Lang et al. 2013). Riegler et al. offered a solution to this problem by using tissue-engineered matrix-based delivery of cells in the form of engineered heart muscles (EHMs), which significantly improved cell retention long-term in post-MI rats (Riegler et al. 2015).

11.3.2 Imaging Transplanted Cell Fate In Vivo

To improve the efficacy of stem cell therapy, noninvasive imaging techniques have been used to assess various parameters such as the distribution, viability, proliferation, and differentiation of stem cells.

11.3.2.1 Imaging Cell Distribution and Viability

The distribution of transplanted cells can be imaged by direct cell labeling. For example, stem cells can be labeled with iron oxide nanoparticles to obtain highly sensitive and high-resolution imaging of cell location following delivery (Emmert et al. 2013a, b; Vallee et al. 2012; Qiao et al. 2009). Using SPIO nanoparticles for cell labeling, autologous bone marrow stem cells were detected in the infarcted myocardium 3 weeks after transplantation (Peng et al. 2013). Besides iron oxide nanoparticles, nanoparticles incorporating Gd as a contrast agent were used in MRI to track the location of transplanted cells up to 2 weeks (Jokerst et al. 2013). Cells can also be labeled with radioactive compounds for short-term radionuclide imaging of cell distribution. For instance, a PET-based study showed that 5% of the injected ESC-CMs or fibroblasts labeled with ^{18}F-FDG remained at the site of injection, with most of the cells having leaked to the lungs (Lang et al. 2014).

Despite its usefulness in short-term tracking cell distribution, direct labeling does not lend itself to long-term monitoring of cell viability because the cell labels are diluted with cell division and do not accurately reflect the number of viable cells. Furthermore, these labels can be released from cells following cell death and be taken up by scavenging macrophages, thus rendering the signal nonspecific to the transplanted cells. Radionuclide-based labels are limited by half-lives, which make their use unsuitable for long-term cell tracking. Thus, for long-term cell viability monitoring, indirect genetic labeling is more ideal.

As mentioned earlier, BLI is currently the most popular method for imaging the long-term survival of transplanted cells in preclinical studies (Hu et al. 2011; Ong et al. 2015; Freeman et al. 2015). Using BLI, Riegler et al. found that almost 75% of the transplanted cells in EHMs died within the first 2 weeks, though a fraction of the cells retained viability for up to 85 days (Riegler et al. 2015). Besides BLI, cell viability can be also tracked using PET, SPECT, or MRI in conjunction with appropriate reporter genes. For instance, ^{18}F-FHBG PET imaging was successfully used to track the reduction of cell viability of transplanted human cardiac progenitor cells labeled with the HSV1-*tk* reporter gene over a 4-week period (Liu et al. 2012). ^{99m}Tc-methoxyisobutylisonitrile (^{99m}Tc-MIBI) and ^{99m}Tc-pertechnetate ($^{99m}TcO_4^-$) SPECT imaging of stem cells transduced with the NIS reporter gene was achieved for up to 9 days (Lee et al. 2015). MRI in conjunction with intravenous injection of SPIO-conjugated monoclonal antibody targeting was used to serially image the viability of transplanted ESCs overexpressing the HA-fluc-*myc* antigen as a reporter (Chung et al. 2011). Lastly, multimodality imaging of BMSCs implanted in the infarcted rat myocardium has been successfully performed by indirectly labeling the cell with a triple-fusion reporter gene composed of HSV1-*tk*, enhanced green fluorescence protein, and firefly luciferase (Pei et al. 2014).

11.3.2.2 Imaging Cell Proliferation

Besides viability, cell proliferation also determines the total number of therapeutic cells in the infarcted myocardium. Generally speaking, indirect labeling of cells with a reporter gene is necessary for imaging evaluation of cell proliferation. In a pioneering study, Cao et al. showed the feasibility of labeling undifferentiated ESCs with a triple-fusion reporter gene that expresses firefly luciferase, monomeric red fluorescent protein, and truncated thymidine kinase, and they were able to image the proliferation of ESCs in the rat myocardium over a 4-week period using BLI and PET (Cao et al. 2006). Chan et al. further showed that cardiosphere-derived cells labeled with the hNIS reporter gene can be imaged for up to 3 days using $^{99m}TcO_4$ with SPECT, during which the cell signal increased, indicating possible short-term cell proliferation following implantation (Chan et al. 2015). Lastly, MRI has also been used to assess the proliferation of transplanted cells that overexpress the ferritin reporter gene following their implantation into infarcted mouse hearts (Naumova et al. 2012).

11.3.2.3 Imaging Cell Differentiation

Imaging stem cell differentiation in an adult heart remains challenging because there is not a robust protocol for directly differentiating implanted pluripotent stem cells into functional cardiomyocytes in vivo. Nevertheless, several imaging methods have been attempted. For instance, Kammili et al. successfully used BLI to monitor the differentiation of implanted ESCs into cardiomyocytes by genetically modifying the stem cells with firefly luciferase reporter gene to express firefly luciferase under the regulation of the cardiac sodium calcium exchanger-1 promoter (Kammili et al. 2010). This approach is feasible because firefly luciferase is expressed only when the cardiac-specific promoter is turned on during cardiac differentiation. Using this method, Kammili et al. found that the transplanted ESCs continued to differentiate for 2–4 weeks after transplantation into neonatal mouse hearts. Later, Wang et al. applied a similar approach to image the in vivo endothelial differentiation of hMSCs following injection into infarcted mouse hearts. Specifically, hMSCs were genetically modified to express firefly luciferase under the regulation of endothelial-specific promoter Tie-2 (Wang et al. 2012). hMSCs were capable of differentiating into endothelial cells within 48 h of injection into MI hearts. Although imaging of stem cell differentiation remains challenging, these pioneering studies suggest that molecular imaging with further optimization of reporter gene technology may hold the key to pinpointing the evasive mechanisms of stem cell differentiation in vivo.

11.3.3 Evaluating Therapeutic Efficacy

The major goal of cardiac stem cell therapy is to improve cardiac function via enhanced regeneration of the infarcted myocardium. Various clinical imaging modalities have been used to specifically assess its efficacy in terms of cardiac function, myocardial perfusion, myocardial viability, myocardial scar, and myocardial microvasculature (Chen and Wu 2011; Nguyen et al. 2014).

11.3.3.1 Evaluating Cardiac Functions

Systolic and diastolic cardiac functions measure the capacity of the heart to pump and fill blood. These parameters can be assessed using both ultrasound and MRI and have been the cornerstone of standard evaluation for any cardiac therapy.

Systolic function is a surrogate measure of myocardial contractility. Most stem cell trials to date have relied on ultrasound (echocardiography) or MRI-based estimates of global systolic function, which can be inferred from both left ventricular ejection fraction (EF) and volumes (Huikuri et al. 2008; Wendel et al. 2014; Chong et al. 2014; Laflamme et al. 2007; Surder et al. 2013; Macarthur et al. 2014; Riegler et al. 2015). Although these global measures are convenient for evaluating systolic function, they are less sensitive in detecting regional changes in myocardial function that can result from direct cell implantation. Therefore, a more accurate assessment of regional contractile function in terms of regional wall thickening or strain should be an integral component of routine evaluation for cardiac stem cell therapy. Indeed, several groups have shown that regional wall thickening in the scar or border zone can be a sensitive measure of recovered systolic function following cell-based treatments (Zimmermann et al. 2006; Kang et al. 2012). Strain imaging using either cardiac ultrasound or MRI has also been successfully used to assess the direct effect of cell implantation on regional contractile function (Yamada et al. 2013; Macarthur et al. 2014; Chen et al. 2015).

Diastolic dysfunction is a condition that describes the abnormal relaxation of the ventricular myocardium following cardiac contraction and twist. It often occurs in patients with ischemic heart failure and can be a useful measure for assessing the efficacy of cardiac stem cell therapy. So far, several methods have been used to noninvasively evaluate the recovery of diastolic dysfunction following cardiac stem cell therapy. For instance, Schneider et al. have applied ultrasound-based strain rate imaging to document the improved diastolic function following BMSC transplantation (Schneider et al. 2009). Similarly, Riegler et al. applied tissue Doppler from cardiac ultrasound to measure possible changes in diastolic function following EHM transplantation in a chronic rat MI model (Riegler et al. 2015).

11.3.3.2 Evaluating Myocardial Perfusion

The potential effects of stem cell therapy on vasculogenesis can be assessed in terms of myocardial perfusion, which can be imaged using SPECT, PET, or MRI. A randomized clinical trial used ^{99m}Tc-MIBI rest and stress SPECT imaging to assess left ventricle perfusion found a better microcirculation in the BMSC treatment group compared to control group (Grajek et al. 2010). Using SPECT imaging with ^{99m}Tc-MIBI, investigators found improved myocardial perfusion of the left ventricle at 8 weeks after MSC transplantation (Peng et al. 2013). Similarly, ^{13}N-ammonia PET was used to show improved hibernating myocardium following intracoronary BMSC infusion in 15 patients (Castellani et al. 2010). In addition, Gowdak et al. showed that Gd-based contrast agents can be used to sensitively detect the increased myocardial perfusion by MRI in 21 patients after injection of autologous BMSCs (Gowdak et al. 2011).

11.3.3.3 Evaluating Myocardial Viability and Scar Changes

Increased myocardial viability is a main goal for applying cardiac stem cell therapy to treat MI. Various imaging methods have been used to assess myocardial viability. ^{18}F-FDG PET, which measures the glucose metabolism, is considered the gold standard for measuring myocardial viability. It has been widely used to measure myocardial viability following stem cell transplantation (Hofmann et al. 2005; Castellani et al. 2010; Sheikh et al. 2012). Additionally, SPECT/CT imaging with ^{99m}Tc-HMPAO was applied to evaluate cardiomyocyte apoptosis in a stem cell-engineered delivery platform (Godier-Furnemont et al. 2013). Several MRI technologies have also been used to evaluate myocardial viability. For instance, an MR spectroscopic saturation transfer technique has been developed to detect the myocardial ATP turnover rate after human iPSC transplantation. Using this technique, Xiong et al. showed markedly attenuated ATP utilization in the border zone after treatment of human iPSC-derived vascular cells (Xiong et al. 2013). Manganese-enhanced MRI has also been applied to detect myocardial viability in MI mice treated with human placenta-derived amniotic MSCs (Kim et al. 2015).

Besides increased myocardial viability, shrinkage of scar size is also considered a desirable outcome of stem cell therapy. Late gadolinium enhancement (LGE) MRI is considered the gold standard for quantifying myocardial scar in the clinical setting (Kwong and Farzaneh-Far 2011). This technique has been successfully used to assess changes in scar size after stem cell therapies in both preclinical and clinical studies (Golpanian et al. 2015; Kraehenbuehl et al. 2011; Riegler et al. 2015).

11.3.3.4 Evaluating Myocardial Microstructure

The myocardial microstructure (i.e., cardiac fiber orientation) is closely related to cardiac contractile and conduction functions. Therefore, successful cardiac repair is expected to improve myocardial microstructure, and a comprehensive evaluation of cardiac stem cell therapy should encompass an assessment of myocardial microstructure. However, the evaluation of myocardial microstructures has been primarily limited to ex vivo studies because of the requirements for high resolution and limited motion (Chen et al. 2015). Fortunately, recent advances in MR diffusion tensor imaging (DTI) have made it possible to directly assess the effect of stem cell therapy on the microstructure of infarcted myocardium in a beating heart. DTI tractography method has been used to directly measure the reconstructed microstructures after BMMC therapy in an in vivo mouse MI model (Sosnovik et al. 2014).

11.3.4 Evaluating the Safety of Stem Cell Therapy

Depending on the cell type, stem cell therapy can cause untoward side effects. For example, undifferentiated pluripotent stem cells, even at low numbers following implantation, can cause teratoma (Lee et al. 2013; Neofytou et al. 2015). Genetic modification of stem cells, especially with the use of integrating vectors, can also cause insertional mutagenesis and thus potentially undesirable cellular phenotypes. Because of these concerns, it is imperative to include in the assessment of cardiac stem cell therapy an evaluation for potential side effects.

Several imaging methods have been utilized to detect the potential for teratoma formation after pluripotent stem cell transplantation. A simple approach is to directly identify the abnormally shaped mass using cardiac ultrasound (Oommen et al. 2015). This approach, however, suffers from low accuracy and sensitivity, especially when used for very small tumors. BLI provides an alternative approach to assessing teratoma by detecting the rapid increases in signal associated with enhanced cellular proliferation (Cao et al. 2006; Wyles et al. 2014). Another approach to detect teratoma is by assessing the enhanced angiogenesis that often accompanies teratoma formation, which can be detected using ^{64}Cu-DOTA-RGD4 PET (Cao et al. 2009). MRI using serial intravenous administration of SPIO-conjugated monoclonal antibody targeting the HA-fluc-*myc* antigen can also be employed to characterize teratoma formation (Chung et al. 2011). A recent study using T2-weighted, T2*-weighted, and LGE MRI showed the feasibility of detecting iPSC-derived teratoma of >8 mm^3 (Riegler et al. 2016). This study also demonstrated the potential utility of combining MRI with serum biomarkers (i.e., carcinoembryonic antigen, a-fetoprotein, and human chorionic gonadotropin) for more sensitive detection of teratoma.

Conclusion

Although cardiac stem cell therapy for myocardial repair has been well established in preclinical studies, it has yet to confer consistent benefits in humans due the lack of a systemic approach to optimize this therapeutic modality in the clinical setting (Nguyen et al. 2016a, b). In this regard, noninvasive imaging can play an important role by providing a valuable tool to optimize cell delivery, investigate unknown biological mechanisms, evaluate therapeutic efficacy, and detect potential stem cell misbehavior. Although it is now feasible to perform image-guided cell delivery and evaluate therapeutic efficacy, challenges remain for long-term monitoring of transplanted cells because of limitations associated with direct cell labeling, as well as safety concerns associated with indirect cell labeling using reporter genes. Therefore, the need to develop a safe and sensitive approach to monitor long-term stem cell fates in humans is of the utmost clinical importance. The rapid progress in developing powerful noninvasive and molecular imaging techniques will make them indispensable in the clinical translation of promising cardiac stem cell therapy.

Acknowledgment We thank funding support from American Heart Association 13EIA14420025, National Institutes of Health (NIH) R01 HL123968-03S1, NIH R01 HL133272, NIH R01 HL132875, NIH R01 HL130020, California Institute of Regenerative Medicine (CIRM) DR2A-05394, TR3-05556, and RT3-07798 (JCW), American Heart Association 17SDG33460212 (XQ), and NIH T32 EB009035 (IYC).

Compliance with Ethical Standards

Conflict of Interest The authors declare that they have no conflict of interest.

Ethical Approval This article does not contain any studies with human participants or animals performed by any of the authors.

References

Budde MD, Frank JA (2009) Magnetic tagging of therapeutic cells for MRI. J Nucl Med 50(2):171–174. doi:10.2967/jnumed.108.053546

Cao F, Li Z, Lee A, Liu Z, Chen K, Wang H, Cai W, Chen X, Wu JC (2009) Noninvasive de novo imaging of human embryonic stem cell-derived teratoma formation. Cancer Res 69(7):2709–2713. doi:10.1158/0008-5472.CAN-08-4122

Cao F, Lin S, Xie X, Ray P, Patel M, Zhang X, Drukker M, Dylla SJ, Connolly AJ, Chen X, Weissman IL, Gambhir SS, Wu JC (2006) In vivo visualization of embryonic stem cell survival, proliferation, and migration after cardiac delivery. Circulation 113(7):1005–1014. doi:10.1161/circulationaha.105.588954

Castellani M, Colombo A, Giordano R, Pusineri E, Canzi C, Longari V, Piccaluga E, Palatresi S, Dellavedova L, Soligo D, Rebulla P, Gerundini P (2010) The role of PET with 13N-ammonia and 18F-FDG in the assessment of myocardial perfusion and metabolism in patients with recent AMI and intracoronary stem cell injection. J Nucl Med 51(12):1908–1916. doi:10.2967/jnumed.110.078469

Chan AT, Karakas MF, Vakrou S, Afzal J, Rittenbach A, Lin X, Wahl RL, Pomper MG, Steenbergen CJ, Tsui BM, Elisseeff JH, Abraham MR (2015) Hyaluronic acid-serum hydrogels rapidly restore metabolism of encapsulated stem cells and promote engraftment. Biomaterials 73:1–11. doi:10.1016/j.biomaterials.2015.09.001

Chen IY, Wu JC (2011) Cardiovascular molecular imaging: focus on clinical translation. Circulation 123(4):425–443. doi:10.1161/CIRCULATIONAHA.109.916338

Chen J, Hsieh AF, Dharmarajan K, Masoudi FA, Krumholz HM (2013) National trends in heart failure hospitalization after acute myocardial infarction for Medicare beneficiaries: 1998–2010. Circulation 128(24):2577–2584. doi:10.1161/CIRCULATIONAHA.113.003668

Chen Y, Ye L, Zhong J, Li X, Yan C, Chandler MP, Calvin S, Xiao F, Negia M, Low WC, Zhang J, Yu X (2015) The structural basis of functional improvement in response to human umbilical cord blood stem cell transplantation in hearts with postinfarct LV remodeling. Cell Transplant 24(6):971–983. doi:10.3727/096368913x675746

Chong JJ, Yang X, Don CW, Minami E, Liu YW, Weyers JJ, Mahoney WM, Van Biber B, Cook SM, Palpant NJ, Gantz JA, Fugate JA, Muskheli V, Gough GM, Vogel KW, Astley CA, Hotchkiss CE, Baldessari A, Pabon L, Reinecke H, Gill EA, Nelson V, Kiem HP, Laflamme MA, Murry CE (2014) Human embryonic-stem-cell-derived cardiomyocytes regenerate non-human primate hearts. Nature 510(7504):273–277. doi:10.1038/nature13233

Choudry F, Hamshere S, Saunders N, Veerapen J, Bavnbek K, Knight C, Pellerin D, Locca D, Westwood M, Rakhit R, Crake T, Kastrup J, Parmar M, Agrawal S, Jones D, Martin J, Mathur A (2015) A randomized double-blind control study of early intra-coronary autologous bone marrow cell infusion in acute myocardial infarction: the REGENERATE-AMI clinical trialdagger. Eur Heart J. doi:10.1093/eurheartj/ehv493

Chung J, Kee K, Barral JK, Dash R, Kosuge H, Wang X, Weissman I, Robbins RC, Nishimura D, Quertermous T, Reijo-Pera RA, Yang PC (2011) In vivo molecular MRI of cell survival and teratoma formation following embryonic stem cell transplantation into the injured murine myocardium. Magn Reson Med 66(5):1374–1381. doi:10.1002/mrm.22929

Emmert MY, Weber B, Wolint P, Frauenfelder T, Zeisberger SM, Behr L, Sammut S, Scherman J, Brokopp CE, Schwartlander R, Vogel V, Vogt P, Grunenfelder J, Alkadhi H, Falk V, Boss A, Hoerstrup SP (2013a) Intramyocardial transplantation and tracking of human mesenchymal stem cells in a novel intra-uterine pre-immune fetal sheep myocardial infarction model: a proof of concept study. PLoS One 8(3):e57759. doi:10.1371/journal.pone.0057759

Emmert MY, Wolint P, Winklhofer S, Stolzmann P, Cesarovic N, Fleischmann T, Nguyen TD, Frauenfelder T, Boni R, Scherman J, Bettex D, Grunenfelder J, Schwartlander R, Vogel V, Gyongyosi M, Alkadhi H, Falk V, Hoerstrup SP (2013b) Transcatheter based electromechanical mapping guided intramyocardial transplantation and in vivo tracking of human stem cell based three dimensional microtissues in the porcine heart. Biomaterials 34(10):2428–2441. doi:10.1016/j.biomaterials.2012.12.021

Freeman BT, Kouris NA, Ogle BM (2015) Tracking fusion of human mesenchymal stem cells after transplantation to the heart. Stem Cells Transl Med 4(6):685–694. doi:10.5966/sctm.2014-0198

Friedrich MG, Marcotte F (2013) Cardiac magnetic resonance assessment of myocarditis. Circ Cardiovasc Imaging 6(5):833–839. doi:10.1161/CIRCIMAGING.113.000416

Godier-Furnemont AF, Tekabe Y, Kollaros M, Eng G, Morales A, Vunjak-Novakovic G, Johnson LL (2013) Noninvasive imaging of myocyte apoptosis following application of a stem cell-engineered delivery platform to acutely infarcted myocardium. J Nucl Med 54(6):977–983. doi:10.2967/jnumed.112.112979

Golpanian S, El-Khorazaty J, Mendizabal A, DiFede DL, Suncion VY, Karantalis V, Fishman JE, Ghersin E, Balkan W, Hare JM (2015) Effect of aging on human mesenchymal stem cell therapy in ischemic cardiomyopathy patients. J Am Coll Cardiol 65(2):125–132. doi:10.1016/j.jacc.2014.10.040

Gowdak LH, Schettert IT, Rochitte CE, Lisboa LA, Dallan LA, Cesar LA, de Oliveira SA, Krieger JE (2011) Early increase in myocardial perfusion after stem cell therapy in patients undergoing incomplete coronary artery bypass surgery. J Cardiovasc Transl Res 4(1):106–113. doi:10.1007/s12265-010-9234-2

Grajek S, Popiel M, Gil L, Breborowicz P, Lesiak M, Czepczynski R, Sawinski K, Straburzynska-Migaj E, Araszkiewicz A, Czyz A, Kozlowska-Skrzypczak M, Komarnicki M (2010) Influence of bone marrow stem cells on left ventricle perfusion and ejection fraction in patients with acute myocardial infarction of anterior wall: randomized clinical trial: impact of bone marrow stem cell intracoronary infusion on improvement of microcirculation. Eur Heart J 31(6):691–702. doi:10.1093/eurheartj/ehp536

Hofmann M, Wollert KC, Meyer GP, Menke A, Arseniev L, Hertenstein B, Ganser A, Knapp WH, Drexler H (2005) Monitoring of bone marrow cell homing into the infarcted human myocardium. Circulation 111(17):2198–2202. doi:10.1161/01.Cir.0000163546.27639.Aa

Hou D, Youssef EA, Brinton TJ, Zhang P, Rogers P, Price ET, Yeung AC, Johnstone BH, Yock PG, March KL (2005) Radiolabeled cell distribution after intramyocardial, intracoronary, and interstitial retrograde coronary venous delivery: implications for current clinical trials. Circulation 112(9 Suppl):I150–I156. doi:10.1161/CIRCULATIONAHA.104.526749

Hu S, Huang M, Nguyen PK, Gong Y, Li Z, Jia F, Lan F, Liu J, Nag D, Robbins RC, Wu JC (2011) Novel microRNA prosurvival cocktail for improving engraftment and function of cardiac progenitor cell transplantation. Circulation 124(11 Suppl):S27–S34. doi:10.1161/circulationaha.111.017954

Huikuri HV, Kervinen K, Niemela M, Ylitalo K, Saily M, Koistinen P, Savolainen ER, Ukkonen H, Pietila M, Airaksinen JK, Knuuti J, Makikallio TH, Investigators F (2008) Effects of intracoronary injection of mononuclear bone marrow cells on left ventricular function, arrhythmia risk profile, and restenosis after thrombolytic therapy of acute myocardial infarction. Eur Heart J 29(22):2723–2732. doi:10.1093/eurheartj/ehn436

Jokerst JV, Khademi C, Gambhir SS (2013) Intracellular aggregation of multimodal silica nanoparticles for ultrasound-guided stem cell implantation. Sci Transl Med 5(177):177ra135–177ra135. doi:10.1126/scitranslmed.3005228

Kammili RK, Taylor DG, Xia JX, Osuala K, Thompson K, Menick DR, Ebert SN (2010) Generation of novel reporter stem cells and their application for molecular imaging of cardiac-differentiated stem cells in vivo. Stem Cells Dev 19(9):1437–1448. doi:10.1089/scd.2009.0308

Kang HJ, Kim MK, Lee HY, Park KW, Lee W, Cho YS, Koo BK, Choi DJ, Park YB, Kim HS (2012) Five-year results of intracoronary infusion of the mobilized peripheral blood stem cells by granulocyte colony-stimulating factor in patients with myocardial infarction. Eur Heart J 33(24):3062–3069. doi:10.1093/eurheartj/ehs231

Kim PJ, Mahmoudi M, Ge X, Matsuura Y, Toma I, Metzler S, Kooreman NG, Ramunas J, Holbrook C, McConnell MV, Blau H, Harnish P, Rulifson E, Yang PC (2015) Direct evaluation of myocardial viability and stem cell engraftment demonstrates salvage of the injured myocardium. Circ Res 116(7):e40–e50. doi:10.1161/circresaha.116.304668

Kraehenbuehl TP, Ferreira LS, Hayward AM, Nahrendorf M, van der Vlies AJ, Vasile E, Weissleder R, Langer R, Hubbell JA (2011) Human embryonic stem cell-derived microvascular grafts for cardiac tissue preservation after myocardial infarction. Biomaterials 32(4):1102–1109. doi:10.1016/j.biomaterials.2010.10.005

Kwong RY, Farzaneh-Far A (2011) Measuring myocardial scar by CMR. JACC Cardiovasc Imaging 4(2):157–160. doi:10.1016/j.jcmg.2010.12.004

La Gerche A, Claessen G, Van de Bruaene A, Pattyn N, Van Cleemput J, Gewillig M, Bogaert J, Dymarkowski S, Claus P, Heidbuchel H (2013) Cardiac MRI: a new gold standard for ventricular volume quantification during high-intensity exercise. Circ Cardiovasc Imaging 6(2):329–338. doi:10.1161/CIRCIMAGING.112.980037

Laflamme MA, Chen KY, Naumova AV, Muskheli V, Fugate JA, Dupras SK, Reinecke H, Xu C, Hassanipour M, Police S, O'Sullivan C, Collins L, Chen Y, Minami E, Gill EA, Ueno S, Yuan C, Gold J, Murry CE (2007) Cardiomyocytes derived from human embryonic stem cells in pro-survival factors enhance function of infarcted rat hearts. Nat Biotechnol 25(9):1015–1024. doi:10.1038/nbt1327

Lang C, Lehner S, Todica A, Boening G, Franz WM, Bartenstein P, Hacker M, David R (2013) Positron emission tomography based in-vivo imaging of early phase stem cell retention after intramyocardial delivery in the mouse model. Eur J Nucl Med Mol Imaging 40(11):1730–1738. doi:10.1007/s00259-013-2480-1

Lang C, Lehner S, Todica A, Boening G, Zacherl M, Franz WM, Krause BJ, Bartenstein P, Hacker M, David R (2014) In-vivo comparison of the acute retention of stem cell derivatives and fibroblasts after intramyocardial transplantation in the mouse model. Eur J Nucl Med Mol Imaging 41(12):2325–2336. doi:10.1007/s00259-014-2858-8

Lee AR, Woo SK, Kang SK, Lee SY, Lee MY, Park NW, Song SH, Lee SY, Nahm SS, JE Y, Kim MH, Yoo RJ, Kang JH, Lee YJ, Eom KD (2015) Adenovirus-mediated expression of human sodium-iodide symporter gene permits in vivo tracking of adipose tissue-derived stem cells in a canine myocardial infarction model. Nucl Med Biol 42(7):621–629. doi:10.1016/j.nucmedbio.2015.03.006

Lee AS, Tang C, Rao MS, Weissman IL, JC W (2013) Tumorigenicity as a clinical hurdle for pluripotent stem cell therapies. Nat Med 19(8):998–1004. doi:10.1038/nm.3267

Li Z, Wu JC, Sheikh AY, Kraft D, Cao F, Xie X, Patel M, Gambhir SS, Robbins RC, Cooke JP, Wu JC (2007) Differentiation, survival, and function of embryonic stem cell derived endothelial cells for ischemic heart disease. Circulation 116(11 Suppl):I46–I54. doi:10.1161/circulationaha.106.680561

Li ZJ, Suzuki Y, Huang M, Cao F, Xie XY, Connolly AJ, Yang PC, JC W (2008) Comparison of reporter gene and iron particle labeling for tracking fate of human embryonic stem cells and differentiated endothelial cells in living subjects. Stem Cells 26(4):864–873. doi:10.1634/stemcells.2007-0843

Liu J, Narsinh KH, Lan F, Wang L, Nguyen PK, Hu S, Lee A, Han L, Gong Y, Huang M, Nag D, Rosenberg J, Chouldechova A, Robbins RC, Wu JC (2012) Early stem cell engraftment predicts late cardiac functional recovery: preclinical insights from molecular imaging. Circ Cardiovasc Imaging 5(4):481–490. doi:10.1161/circimaging.111.969329

Macarthur JW Jr, Cohen JE, McGarvey JR, Shudo Y, Patel JB, Trubelja A, Fairman AS, Edwards BB, Hung G, Hiesinger W, Goldstone AB, Atluri P, Wilensky RL, Pilla JJ, Gorman JH 3rd, Gorman RC, Woo YJ (2014) Preclinical evaluation of the engineered stem cell chemokine stromal cell-derived factor 1alpha analog in a translational ovine myocardial infarction model. Circ Res 114(4):650–659. doi:10.1161/circresaha.114.302884

Menasche P, Vanneaux V, Hagege A, Bel A, Cholley B, Cacciapuoti I, Parouchev A, Benhamouda N, Tachdjian G, Tosca L, Trouvin JH, Fabreguettes JR, Bellamy V, Guillemain R, Suberbielle Boissel C, Tartour E, Desnos M, Larghero J (2015) Human embryonic stem cell-derived cardiac progenitors for severe heart failure treatment: first clinical case report. Eur Heart J 36(30):2011–2017. doi:10.1093/eurheartj/ehv189

Naumova AV, Yarnykh VL, Balu N, Reinecke H, Murry CE, Yuan C (2012) Quantification of MRI signal of transgenic grafts overexpressing ferritin in murine myocardial infarcts. NMR Biomed 25(10):1187–1195. doi:10.1002/nbm.2788

Neofytou E, O'Brien CG, Couture LA, Wu JC (2015) Hurdles to clinical translation of human induced pluripotent stem cells. J Clin Invest 125(7):2551–2557. doi:10.1172/JCI80575

Nguyen PK, Neofytou E, Rhee JW, Wu JC (2016a) Potential strategies to address the major clinical hurdles facing stem cell regenerative therapy for cardiovascular disease: a review. JAMA Cardiol 1(8):953–962. doi:10.1001/jamacardio.2016.2750

Nguyen PK, Rhee JW, Wu JC (2016b) Adult stem cell therapy and heart failure, 2000 to 2016: a systematic review. JAMA Cardiol 1(7):831–841. doi:10.1001/jamacardio.2016.2225

Nguyen PK, Riegler J,Wu JC (2014) Stem cell imaging: from bench to bedside. Cell Stem Cell 14(4):431–444. doi:10.1016/j.stem.2014.03.009

Ong SG, Huber BC, Hee Lee W, Kodo K, Ebert AD, Ma Y, Nguyen PK, Diecke S, Chen WY, Wu JC (2015) Microfluidic single-cell analysis of transplanted human induced pluripotent stem cell-derived cardiomyocytes after acute myocardial infarction. Circulation 132(8):762–771. doi:10.1161/circulationaha.114.015231

Oommen S, Yamada S, Cantero Peral S, Campbell KA, Bruinsma ES, Terzic A, Nelson TJ (2015) Human umbilical cord blood-derived mononuclear cells improve murine ventricular function upon intramyocardial delivery in right ventricular chronic pressure overload. Stem Cell Res Ther 6:50. doi:10.1186/s13287-015-0044-y

Partlow KC, Chen JJ, Brant JA, Neubauer AM, Meyerrose TE, Creer MH, Nolta JA, Caruthers SD, Lanza GM, Wickline SA (2007) F-19 magnetic resonance imaging for stem/progenitor cell tracking with multiple unique perfluorocarbon nanobeacons. FASEB J 21(8):1647–1654. doi:10.1096/fj.06-6505com

Pei Z, Lan X, Cheng Z, Qin C, Xia X, Yuan H, Ding Z, Zhang Y (2014) Multimodality molecular imaging to monitor transplanted stem cells for the treatment of ischemic heart disease. PLoS One 9(3):e90543. doi:10.1371/journal.pone.0090543

Peng C, Yang K, Xiang P, Zhang C, Zou L, Wu X, Gao Y, Kang Z, He K, Liu J, Cheng M, Wang J, Chen L (2013) Effect of transplantation with autologous bone marrow stem cells on acute myocardial infarction. Int J Cardiol 162(3):158–165. doi:10.1016/j.ijcard.2011.05.077

Penheiter AR, Russell SJ, Carlson SK (2012) The sodium iodide symporter (NIS) as an imaging reporter for gene, viral, and cell-based therapies. Curr Gene Ther 12(1):33–47. doi:10.2174/156652312799789235

Qiao H, Zhang H, Zheng Y, Ponde DE, Shen D, Gao F, Bakken AB, Schmitz A, Kung HF, Ferrari VA, Zhou R (2009) Embryonic stem cell grafting in normal and infarcted myocardium: serial assessment with MR imaging and PET dual detection. Radiology 250(3):821–829. doi:10.1148/radiol.2503080205

Riegler J, Ebert A, Qin X, Shen Q, Wang M, Ameen M, Kodo K, Ong SG, Lee WH, Lee G, Neofytou E, Gold JD, Connolly AJ, Wu JC (2016) Comparison of magnetic resonance imaging and serum biomarkers for detection of human pluripotent stem cell-derived teratomas. Stem Cell Rep. doi:10.1016/j.stemcr.2015.12.008

Riegler J, Tiburcy M, Ebert A, Tzatzalos E, Raaz U, Abilez OJ, Shen Q, Kooreman NG, Neofytou E, Chen VC, Wang M, Meyer T, Tsao PS, Connolly AJ, Couture LA, Gold JD, Zimmermann WH, Wu JC (2015) Human engineered heart muscles engraft and survive long term in a rodent myocardial infarction model. Circ Res 117(8):720–730. doi:10.1161/circresaha.115.306985

Schneider C, Jaquet K, Geidel S, Rau T, Malisius R, Boczor S, Zienkiewicz T, Kuck KH, Krause K (2009) Transplantation of bone marrow-derived stem cells improves myocardial diastolic function: strain rate imaging in a model of hibernating myocardium. J Am Soc Echocardiogr 22(10):1180–1189. doi:10.1016/j.echo.2009.06.011

Sheikh AY, Huber BC, Narsinh KH, Spin JM, van der Bogt K, de Almeida PE, Ransohoff KJ, Kraft DL, Fajardo G, Ardigo D, Ransohoff J, Bernstein D, Fischbein MP, Robbins RC, Wu JC (2012) In vivo functional and transcriptional profiling of bone marrow stem cells after transplantation into ischemic myocardium. Arterioscler Thromb Vasc Biol 32(1):92–102. doi:10.1161/atvbaha.111.238618

Sosnovik DE, Mekkaoui C, Huang S, Chen HH, Dai G, Stoeck CT, Ngoy S, Guan J, Wang R, Kostis WJ, Jackowski MP, Wedeen VJ, Kozerke S, Liao R (2014) Microstructural impact of ischemia and bone marrow-derived cell therapy revealed with diffusion tensor magnetic resonance imaging tractography of the heart in vivo. Circulation 129(17):1731–1741. doi:10.1161/CIRCULATIONAHA.113.005841

Sun N, Lee A, Wu JC (2009) Long term non-invasive imaging of embryonic stem cells using reporter genes. Nat Protoc 4(8):1192–1201. doi:10.1038/nprot.2009.100

Surder D, Manka R, Lo Cicero V, Moccetti T, Rufibach K, Soncin S, Turchetto L, Radrizzani M, Astori G, Schwitter J, Erne P, Zuber M, Auf der Maur C, Jamshidi P, Gaemperli O, Windecker S, Moschovitis A, Wahl A, Buhler I, Wyss C, Kozerke S, Landmesser U, Luscher TF, Corti R (2013) Intracoronary injection of bone marrow-derived mononuclear cells early or late after acute myocardial infarction: effects on global left ventricular function. Circulation 127(19):1968–1979. doi:10.1161/CIRCULATIONAHA.112.001035

Vallee JP, Hauwel M, Lepetit-Coiffe M, Bei W, Montet-Abou K, Meda P, Gardier S, Zammaretti P, Kraehenbuehl TP, Herrmann F, Hubbell JA, Jaconi ME (2012) Embryonic stem cell-based cardiopatches improve cardiac function in infarcted rats. Stem Cells Transl Med 1(3):248–260. doi:10.5966/sctm.2011-0028

Vrtovec B, Poglajen G, Lezaic L, Sever M, Socan A, Domanovic D, Cernelc P, Torre-Amione G, Haddad F, Wu JC (2013) Comparison of transendocardial and intracoronary CD34+ cell transplantation in patients with nonischemic dilated cardiomyopathy. Circulation 128(11 Suppl 1):S42–S49. doi:10.1161/CIRCULATIONAHA.112.000230

Wang J, Najjar A, Zhang S, Rabinovich B, Willerson JT, Gelovani JG, Yeh ET (2012) Molecular imaging of mesenchymal stem cell: mechanistic insight into cardiac repair after experimental myocardial infarction. Circ Cardiovasc Imaging 5(1):94–101. doi:10.1161/circimaging.111.966424

Wendel JS, Ye L, Zhang P, Tranquillo RT, Zhang JJ (2014) Functional consequences of a tissue-engineered myocardial patch for cardiac repair in a rat infarct model. Tissue Eng Part A 20(7–8):1325–1335. doi:10.1089/ten.TEA.2013.0312

Williams AR, Suncion VY, McCall F, Guerra D, Mather J, Zambrano JP, Heldman AW, Hare JM (2013) Durable scar size reduction due to allogeneic mesenchymal stem cell therapy regulates whole-chamber remodeling. J Am Heart Assoc 2(3):e000140. doi:10.1161/jaha.113.000140

Winter EM, Hogers B, van der Graaf LM, Gittenberger-de Groot AC, Poelmann RE, van der Weerd L (2010) Cell tracking using iron oxide fails to distinguish dead from living transplanted cells in the infarcted heart. Magn Reson Med 63(3):817–821. doi:10.1002/mrm.22094

Wyles SP, Yamada S, Oommen S, Maleszewski JJ, Beraldi R, Martinez-Fernandez A, Terzic A, Nelson TJ (2014) Inhibition of DNA topoisomerase II selectively reduces the threat of tumorigenicity following induced pluripotent stem cell-based myocardial therapy. Stem Cells Dev 23(19):2274–2282. doi:10.1089/scd.2014.0259

Xiong Q, Ye L, Zhang P, Lepley M, Tian J, Li J, Zhang L, Swingen C, Vaughan JT, Kaufman DS, Zhang J (2013) Functional consequences of human induced pluripotent stem cell therapy:

myocardial ATP turnover rate in the in vivo swine heart with postinfarction remodeling. Circulation 127(9):997–1008. doi:10.1161/CIRCULATIONAHA.112.000641

Yaghoubi SS, Jensen MC, Satyamurthy N, Budhiraja S, Paik D, Czernin J, Gambhir SS (2009) Noninvasive detection of therapeutic cytolytic T cells with 18F-FHBG PET in a patient with glioma. Nat Clin Pract Oncol 6(1):53–58. doi:10.1038/ncponc1278

Yamada S, Nelson TJ, Kane GC, Martinez-Fernandez A, Crespo-Diaz RJ, Ikeda Y, Perez-Terzic C, Terzic A (2013) Induced pluripotent stem cell intervention rescues ventricular wall motion disparity, achieving biological cardiac resynchronization post-infarction. J Physiol 591(Pt 17):4335–4349. doi:10.1113/jphysiol.2013.252288

Ye L, Chang YH, Xiong Q, Zhang P, Zhang L, Somasundaram P, Lepley M, Swingen C, Su L, Wendel JS, Guo J, Jang A, Rosenbush D, Greder L, Dutton JR, Zhang J, Kamp TJ, Kaufman DS, Ge Y, Zhang J (2014) Cardiac repair in a porcine model of acute myocardial infarction with human induced pluripotent stem cell-derived cardiovascular cells. Cell Stem Cell 15(6):750–761. doi:10.1016/j.stem.2014.11.009

Zheng B, Vazin T, Goodwill PW, Conway A, Verma A, Saritas EU, Schaffer D, Conolly SM (2015) Magnetic Particle Imaging tracks the long-term fate of in vivo neural cell implants with high image contrast. Sci Rep 5:14055. doi:10.1038/srep14055

Zimmermann WH, Melnychenko I, Wasmeier G, Didie M, Naito H, Nixdorff U, Hess A, Budinsky L, Brune K, Michaelis B, Dhein S, Schwoerer A, Ehmke H, Eschenhagen T (2006) Engineered heart tissue grafts improve systolic and diastolic function in infarcted rat hearts. Nat Med 12(4):452–458. doi:10.1038/nm1394

Zipes DP, Jalife J (2009) Cardiac electrophysiology: from cell to bedside. Saunders/Elsevier, Philadelphia

Stem Cell Transplant Immunology 12

Katharine K. Miller and Sonja Schrepfer

Abstract

Stem cell transplantation is quickly developing as an attractive therapeutic option for regenerating tissues injured by cardiovascular disease. From embryonic to induced pluripotent stem cells, from injection of stem cells to differentiation of cardiac cell lineages, researchers continue to push the boundaries of how stem

K.K. Miller
University Heart Center Hamburg, Transplant and Stem Cell Immunobiology (TSI)-Lab, Martinistrasse 52, D-20246 Hamburg, Germany

Department of Surgery, Transplant and Stem Cell Immunobiology (TSI)-Lab, University California San Francisco (UCSF), 513 Parnassus Avenue, Medical Sciences S1207, San Francisco, CA 94143, USA

Cardiovascular Research Center (CVRC), University Medical Center Hamburg-Eppendorf, Martinistrasse 52, D-20246 Hamburg, Germany

German Centre for Cardiovascular Research (DZHK) e.V., University Medical Center Hamburg-Eppendorf, Martinistrasse 52, D-20246 Hamburg, Germany
e-mail: kmiller@uke.de

S. Schrepfer (✉)
University Heart Center Hamburg, Transplant and Stem Cell Immunobiology (TSI)-Lab, Martinistrasse 52, D-20246 Hamburg, Germany

Department of Surgery, Transplant and Stem Cell Immunobiology (TSI)-Lab, University California San Francisco (UCSF), 513 Parnassus Avenue, Medical Sciences S1207, San Francisco, CA 94143, USA

Cardiovascular Research Center (CVRC), University Medical Center Hamburg-Eppendorf, Martinistrasse 52, D-20246 Hamburg, Germany

German Centre for Cardiovascular Research (DZHK) e.V., University Medical Center Hamburg-Eppendorf, Martinistrasse 52, D-20246 Hamburg, Germany

Department of Cardiovascular Surgery, University Heart Center Hamburg, Martinistrasse 52, D-20246 Hamburg, Germany
e-mail: sonja.schrepfer@ucsf.edu

M. Ieda, W.-H. Zimmermann (eds.), *Cardiac Regeneration*, Cardiac and Vascular Biology 4, DOI 10.1007/978-3-319-56106-6_12

cells can be used in treatments. The major hurdle in the way of creating effective methods for tissue regeneration is immune rejection of transplanted materials; even undifferentiated stem cells can be recognized by the transplant recipients' immune system, limiting their survival and overall beneficial potential. Posttransplant rejection of cellular materials does not always follow the same immunological progression, and as such, different types of stem cells can be rejected through distinct immune pathways. Therefore, a strong understanding of the known mechanisms behind stem cell immunogenicity—including specific cases of embryonic and patient-specific stem cell rejection—is pivotal for researchers to develop more efficient therapeutics. The future of stem cell transplantation research lies in developing techniques that prevent immune recognition of transplanted cells or tissues and in generating ready-to-use stem cell lines that can be quickly and easily prepared for transplantation.

Abbreviations

ES	Embryonic stem
HLA	Human leukocyte antigen
IFN	Interferon
iPS	Induced pluripotent
MHC	Major histocompatibility complex
miHA	Minor histocompatibility antigen
NK	Natural killer
NT-ESC	Nuclear transfer embryonic stem cell
TCR	T cell receptor
SCNT	Somatic cell nuclear transfer
SNPs	Single nucleotide polymorphisms

12.1 Stem Cell Therapy: Possibilities and Drawbacks

Stem cell therapy is fast developing as one of the most intriguing prospective treatments for regenerating injured cardiovascular tissue. With the low availability of organs for transplantation and the accompanying lengthy wait, the possibility of regenerating tissue by transplanting readily available cell lines into patients is understandably appealing. Stem cell therapy has shown promising initial results for rehabilitating ischemic heart tissue after transplantation in animal models (Yang et al. 2002; Laflamme et al. 2007; Nelson et al. 2009; Carpenter et al. 2012; Zwi-Dantsis et al. 2013); however, the propensity for the transplant recipient's immune system to reject allogeneic material greatly reduces the potential efficacy of therapeutics and diminishes the possible positive effects surrounding such treatment.

Because pluripotent stem cells can be differentiated into numerous cell types, the potential application of stem cell therapy is wide-ranging. Differentiation can be performed ex vivo, allowing researchers or clinicians to closely monitor the procedure, insuring that the proper population of cells is generated prior to transplantation. In most cases, undifferentiated stem cells are avoided in clinical transplant therapies due to their propensity to form teratomas (Blum and Benvenisty 2008); rather, such therapies tend to use differentiated stem cells (e.g., stem cell-derived cardiomyocytes transplanted into the myocardium of patients suffering from heart failure).

Pluripotent stem cells can be obtained through a range of methods, from directly using embryonic stem cells, to generating induced pluripotent stem cells, and to performing somatic cell nuclear transfer. Designing more effective treatment options requires knowledge of the pros and cons behind each of the stem cell varieties, as well as the immunological reasons behind posttransplant rejection of pluripotent stem cells and their differentiated progeny. Researchers are working to design stem cell lines and transplantation methods that will not trigger rejection from the recipient's immune system. Until then, scientists continue to develop more efficacious therapeutics by avoiding materials that strongly induce immunological rejection pathways.

12.2 Immunological Mechanisms of Stem Cell Rejection

Transplanted materials—including stem cells—are easily rejected by the recipient's immune system. While acute cellular rejection can be successfully avoided through the use of general immunosuppressants, this treatment is not an ideal solution for long-term clinical applications as it can result in negative side effects (see Sect. 12.5.1). New methods must be developed in order to generate therapies that are conducive to robust cellular regeneration and the enduring health of transplant patients. In order to generate stem cell transplantation methods that effectively evade activation of the immune system, it is pivotal to understand the molecular mechanism behind their posttransplant rejection.

12.2.1 Major Histocompatibility Complexes

The immune system is designed to protect the individual from invading materials; the properties of the immune system that create effective protection are also the reason why allogeneic transplanted material is so effectively rejected. T lymphocytes continuously search for invading material and can recognize cells presenting antigens bound to major histocompatibility complexes (MHCs) at the cell surface (For reviews, see Horton et al. 2004; Neefjes et al. 2011). MHCs—human leukocyte antigens (HLAs) in humans—are cell surface molecules organized into two classes. Class I MHCs consist of three subunits and interact with a β2-microglobulin subunit, while class II MHCs are made up of four subunits and have no β2-microglobulin interaction (Fig. 12.1). MHC classes are also expressed in different cell types: class I MHCs are nearly ubiquitously expressed on cells with nuclei, and class II MHCs

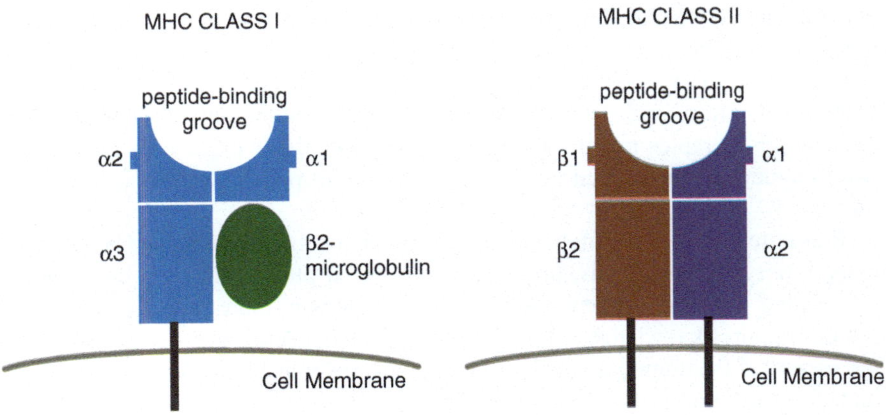

Fig. 12.1 MHC classes. Both MHC classes are membrane-bound complexes. MHC class I consists of three subunits which associate with a β2-microglobulin. MHC class II consists of four subunits. Both MHC classes contain a peptide-binding groove where antigen peptides can be presented

are found on so-called "antigen-presenting" cells, such as endothelial cells, macrophages, B cells, and dendritic cells. This separation in localization helps to ensure that the appropriate receptor can recognize the correct MHC class. When an antigen is taken into the cell, it is broken down into peptides. MHCs bind to certain frequently occurring peptide sequences and present them on the cell surface, at which point T cell receptors (TCRs) can interact directly with the MHC and peptide complex. The presence of the peptide bound to the MHC is necessary for this interaction and ensures that self-MHCs alone do not trigger a T cell reaction. TCRs are selected to recognize and avoid reacting to self-MHC variants without peptide or with self-peptide (Starr et al. 2003).

MHCs are polygenic: every individual carries multiple genes that influence the molecular makeup of their MHCs (Fig. 12.2). Each human individual carries six MHC class I alleles and 6–8 MHC class II alleles. Through this diversity, multiple different class I and II MHCs can be generated (Horton et al. 2004). Each polygene-specific MHC variant targets distinct subsets of peptide sequences, which allows immune detection of diverse antigens. Increasing the number of MHC-associated polygenes would expand the number of total peptides recognized by the varied final MHCs, which begs the question why MHC variations are relatively limited in individuals. One possible explanation relies on the fact that TCRs that can bind to and recognize self-MHCs without peptide should not be expressed. By increasing the number of MHC variants, there would have to be a corollary reduction in T cell diversity in order to prevent T cells attacking self-cells. The immune system seems to have struck a balance between the diversity of MHCs that can bind to various peptide sequences and the variety of T cells that can target antigen-bound MHCs.

Although MHC polygenes found in individuals are limited in number, the chance of infectious disease spreading throughout a population is low due to the polymorphic nature of MHCs (Fig. 12.2). Different MHC genes are expressed in individuals

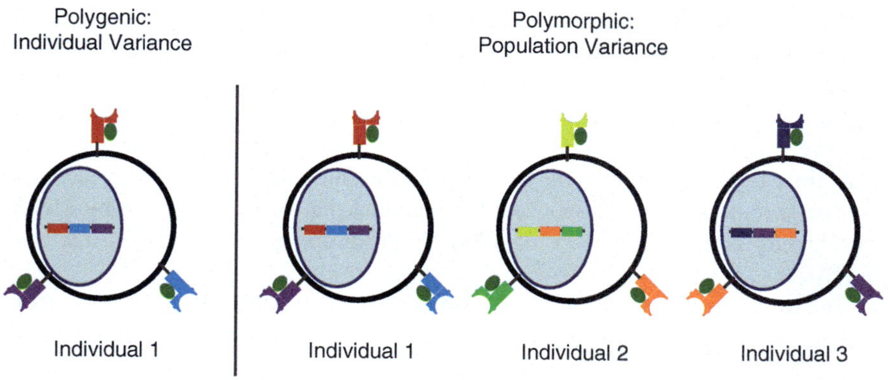

Fig. 12.2 MHCs are polygenic and polymorphic. Diversity of MHCs is generated through two methods. Diversity within the individual is generated through the polygenic nature of MHCs, with multiple genes coming together to form variants of MHCs. A broader diversity is generated population-wide through the polymorphic nature of MHCs. Different individuals often express different MHC genes, thereby increasing the overall number of peptides that can be bound and decreasing the chances for a population-wide epidemic

across the population, which allows the balance of MHC diversity and T cell diversity in the individual to be maintained while still generating an extremely diverse overall variety of peptide recognition in the population.

While the population-wide polymorphic nature of MHCs is very successful in preventing the human population from being wiped out by disease, it vastly complicates the matter of cellular, tissue, and organ transplantation. Allogeneic transplanted materials usually express different MHCs than the recipient. This causes transplanted materials to be identified as foreign invaders by the recipient's T cells. TCRs recognize specific self-MHCs bound to peptides, and they have been "trained" not to respond to self-MHCs without foreign antigen peptides. However, T cells can respond to non-autologous MHCs through a cross-reactivity process. This method for recognizing and reacting to mismatched MHCs is extremely important for the immune system to prevent invading material from attacking the host (Zerrahn et al. 1997; Macedo et al. 2009) but also complicates transplantation therapeutics.

12.2.2 Minor Histocompatibility Antigens

If MHCs were the only reason for immune rejection, MHC-matched transplantable materials would easily address this problem. However, even when using MHC-matched materials, posttransplant immune responses have still been observed (Goulmy et al. 1976; Vogt et al. 2000). Rejection can be caused by expression of minor histocompatibility antigens (miHAs). Simply, miHAs are altered peptides created from a small gene variance between individuals in a population. MHCs can bind to and present miHAs for recognition by T cells.

Most miHAs are generated by single nucleotide polymorphisms (SNPs), leading in some instances to alterations of a single amino acid within the encoded polypeptide according to the RNA codon usage. This change can potentially alter the structure and function of the protein and can even create a truncated isoform if a stop codon is produced upstream. Other mutations, such as gene deletions, can also generate miHAs. The final outcome in all cases is a small difference in the protein that is expressed between the donor and the recipient that, when presented by MHCs, can directly activate T cells. This alloantigenic property is the defining factor of miHAs.

The first reported miHA-caused rejection was identified after an HLA-matched male to female sibling transplantation rejection occurred. In this case, the presence of a miHA found on the Y chromosome increased T cell production and caused immune rejection (Goulmy et al. 1976). This study emphasized that certain rejection pathways can be activated despite controlling for matched MHCs and suggested that additional non-MHC-dependent transplant rejection pathways may exist.

In addition to Y chromosomal miHAs, several autosomal miHAs have been identified. Although the number of genes that could potentially generate miHAs is quite high, it appears that only certain gene alterations trigger recognition by the immune system. Even so, more than 50 different miHAs have been identified in humans, with more that likely exist (Table 12.1; Spierings 2014).

Table 12.1 More than 50 minor H antigens have been identified. While many antigens have the possibility to be minor H antigens, to date, around 50 minor H antigens have been identified (Spierings 2014)

Name	Gene
ACC-1Y	BCL2A1
ACC-1C	BCL2A1
ACC-2	BCL2A1
ACC-4	CTSH
ACC-5	CTSH
ACC-6	HMSD
C19orf48	C19orf48
CD19	CD19
DPH1	DPH1
HA-1/A2	HMHA1
HA-1/B60	HMHA1
HA-2	MYO1G
HA-3	AKAP13
HA-8	KIAA0020
HB-1H	HMHB1
HB-1Y	HMHB1
HEATR1	HEATR1

Table 12.1 (continued)

Name	Gene
HER2	HER-2/NEU
LB-ADIR-1	TOR3A
LB-APOBEC3B-1K	APOBEC3B
LB-ARHGDIB-1R	ARHGDIB
LB-BCAT2-1R	BCAT2
LB-EBI3-1I	EBI3
LB-ECGF-1	TYMP
LB-ERAP1-1R	ERAP1
LB-GEMIN4-1V	GEMIN4
LB-LY75-1K	LY75
LB-MR1-1R	MR1
LB-MTHFD1-1Q	MTHFD1
LB-NISCH-1A	NISCH
LB-NUP133-1R	NUP133
LB-PDCD11-1F	PDCD11
LB-PI4K2B-1S	PI4K2B
LB-PRCP-1D	PRCP
LB-PTK2B-1T	PTK2B
LB-SON-1R	SON
LB-SSR1-1S	SSR1
LB-SWAP70-1Q	SWAP70
LB-TRIP10-1EPC	TRIP10
LB-WNK1-1I	WNK1
LRH-1	P2X5
P2RX7	P2RX7
PANE1	CENPM
SLC19A1	SLC19A1
SLC1A5	SLC1A5
SP110	SP110
T4A	TRIM42
TRIM22	TRIM22
UGT2B17	UGT2B17
UGT2B17	UGT2B17
UGT2B17	UGT2B17
UTA2-1	KIAA1551
UTDP4	ZDHHC12
ZAPHIR	ZNF419

Although mismatched MHCs are considered to be the clearest cause of post-transplant rejection, miHAs have also been shown to be involved. With regard to hematopoietic stem cell transplantation, mismatched miHAs increased the occurrence of graft-versus-host disease and strongly decreased the probability of overall survival (Dzierzak-Mietla et al. 2012). Taken together, it becomes clear that known miHAs should be considered when designing transplantation therapeutics.

12.3 Embryonic Stem Cell Immunogenicity

Early developments in stem cell therapy revolved around the generation of the first human embryonic stem (ES) cell line (Thomson et al. 1998). These pluripotent cells can be obtained from the blastocyst stage of embryonic development and not only have the ability to self-regenerate but can also be differentiated into various cell types. Despite ethical concerns and consequent restrictions on their availability, the unique properties of ES cells have been integral to many important research and clinical developments.

Initial reports suggested that ES cells were afforded a level of immune privilege, largely thought to exist due to their low levels of MHC expression (Li et al. 2004; Drukker et al. 2002). It was hoped that due to their immune privilege, ES cells could be transplanted into patients without triggering an immune response. However, over time it has become clear that the concept of ES cell immune privilege is more nuanced. Although some ES cells express low levels of MHCs, these MHC expression levels seem to be enough to trigger an immune response (Swijnenburg et al. 2008a; Deuse et al. 2011). Moreover, MHC levels appear to be highly variable and to change with regard to culture time, differentiation state, and culture conditions (Drukker et al. 2002).

An example of variable MHC expression levels on ES cells can be seen after the addition of interferon (IFN)-γ, a cytokine associated with transplantation and rejection (Drukker et al. 2002). Experimental addition of IFN-γ increased expression levels of MHC-I in undifferentiated ES cells, although a similar increase is not seen after addition of IFN-α or IFN-β. However, when ES cells are differentiated, all three IFNs can cause increased expression of MHC-I. This suggests that an increase in ES cell MHC expression could occur posttransplantation, therefore initiating rejection of cellular material.

Even without MHC expression level variation, ES cells may not avoid rejection. The "missing self" hypothesis suggests that cells that present low levels of MHC-I are more likely to be targeted by natural killer (NK) cells. NK cells use an inhibitory feedback loop to prevent an attack when they recognize MHCs (Karlhofer et al. 1992; Kambayashi et al. 2001). When MHC-I is not presented on a cell, the inhibitory pathway of NK cells is not activated. NK cells can then target low MHC-I-expressing cells as invading material. Indeed, during a syngeneic transplant model, it was shown that low levels of MHC-I resulted in a nearly total destruction of the graft by NK cells (Ma et al. 2011). However, when IFN-γ was added to the cells to induce MHC-I expression, the NK attack was mitigated. It appears that a delicate balance of MHC expression in stem cells during development is maintained in order to avoid triggering an immune response.

It has been suggested that an inability to easily resolve rejection during experiments over time is the culprit for varied reports on the immune privilege (or lack thereof) of ES cells. This was addressed by monitoring human ES cell survival through noninvasive bioluminescence, which further confirmed the progression of human ES cell rejection in a xenotransplant model. When posttransplant human ES cell survival was compared in immunocompetent and immunodeficient mice, it was shown that ES cell rejection was much higher in the immunocompetent mice (Swijnenburg et al. 2008b). Additionally, upon repeat injection of human ES cells, the rejection speed increased, suggesting that rejection was promoted by the adaptive immune system.

There has been particular interest in using stem cells to regenerate injured tissue, and ES cell's suggested immune privilege made them an ideal starting material for this research. ES cells can be differentiated in vitro into beating cardiomyocytes (Mummery et al. 2002; Xu et al. 2002; He et al. 2003), and initial transplantation of ES-derived cardiomyocytes into mouse cardiac tissue showed promise, with reports of integration and partially improved heart function (Laflamme et al. 2007; Ardehali et al. 2013). However, combined with the inclination for ES cells to form teratomas, there have been reports that ES-differentiated cardiomyocytes can induce an immune response in the myocardium posttransplantation (Nussbaum et al. 2007).

The first trial of human ES cell transplantation into humans has been undertaken with regard to regenerative therapy for patients with macular degeneration (Schwartz et al. 2012; Schwartz et al. 2015). The eye is a known immune privileged organ, which naturally reduces the possibility of posttransplant rejection (Streilein 2003). In this trial, retinal pigment epithelium was derived from human ES cells and transplanted into the subretinal space. Impressively, even nearly 2 years after treatment, a continued significant improvement in visual acuity was observed in the eyes that received the transplant. This was not accompanied by obvious safety issues. Future trials will surely look into methods to further improve regeneration, as well as hopefully moving ES cell therapy, toward the ability to regenerate additional tissue types.

ES cell therapy shows great promise for use in tissue regeneration. With varied reports of MHC expression level and immune responses, it is important that future studies not take ES cell immune privilege for granted. The immunogenicity of pluripotent stem cells remains one of the major hurdles in the way of developing effective clinical stem cell applications.

12.4 Patient-Specific Stem Cell Immunogenicity

Because allogeneic material is frequently and easily rejected, developing syngeneic and autologous stem cells has been a clear goal for generating better stem cell methodology. Subsequently, multiple methods for generating such cell lines have been established, including induced pluripotent stem (iPS) cell and somatic cell nuclear transfer (SCNT) derivation. Both of these powerful technologies have been important for modern stem cell research development. However, despite their ability to generate genetically identical cellular material for transplantation, neither stem cell type fully avoids the problem of posttransplant immune rejection.

12.4.1 Induced Pluripotent Stem Cell Immunogenicity

Due to the ethical questions associated with ES cell use and the resulting restricted availability of ES cell lines, the research community welcomed the advent of iPS cells in 2006 (Takahashi and Yamanaka 2006). iPS cells are pluripotent cells generated through systematic reprogramming of adult cells by sequentially adding multiple chemicals or molecules. Once generated, iPS cells can be differentiated into many different cell lineages for research or clinical purposes.

Patient-specific differentiated cells have been targeted for their possible therapeutic applications through direct transplantation or generation of tissue grafts. Since these cells would be genetically identical to the donor recipient, it was initially believed that there would be no cause for the recipient immune system to recognize and reject them posttransplantation. However, it appears that even with a genetically identical template, patient-specific iPS cells may not always successfully avoid rejection.

Through the process of reprogramming autologous iPS cells, certain gene expression levels are increased when compared with ES cells. This altered gene expression can be recognized by the transplant recipient's immune system (Zhao et al. 2011; de Almeida et al. 2014). It may be that some of the overexpressed genes are normally turned off during development of the fetus' immune system. In the case of autologous iPS cells, their expression appears to cause the immune system to identify them as nonself cells (Fig. 12.3). A separate study found that T cell intrusion and tissue necrosis accompanied teratoma formation when autologous human iPS-derived cells were transplanted into a humanized mouse model. Moreover, depending on the type of cell derived from the human iPS cells, the level of immune response and accompanying rejection was altered (Zhao et al. 2015).

Despite difficulties with posttransplant rejection, iPS cells and iPS cell-derived cardiomyocytes continue to be a focal point for myocardial regenerative therapy. This is due in part to the fact that iPS cells can be proliferated in culture and differentiated into multiple cardiac cell lineages. Initial results have been encouraging in animal models, showing partial rescue of cardiac function (Nelson et al. 2009; Carpenter et al. 2012; Zwi-Dantsis et al. 2013). However, it seems that the number of cells that survive posttransplantation reduces significantly over time—a process that could have multiple explanations, including an immune response (Templin et al. 2012). Clearly, the immune reaction to iPS cells will have to be investigated in detail before their use in clinical applications.

Autologous-generated iPS cells for human treatment may not be ideal even if immune rejection can be avoided, since generating patient-specific iPS cells is

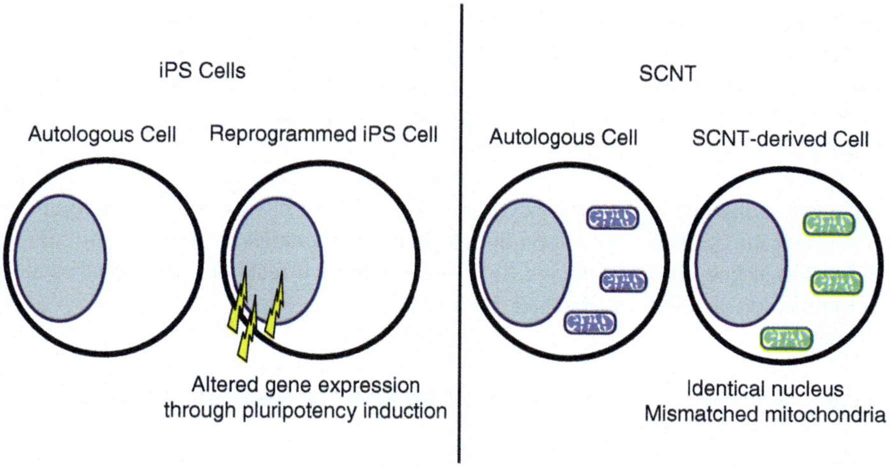

Fig. 12.3 iPS and SCNT cell immunogenicity. iPS cells can be generated to be autologous, which should prevent their rejection in theory. However, during this reprogramming, iPS cells may have altered expression of certain genes, which can cause rejection of transplanted cells. SCNT was suggested as an alternative method for generating cells with identical nuclei to the transplant recipient. However, mismatched mitochondrial DNA and consequent proteins appear to be enough to trigger rejection

extremely time, cost, and labor intensive. Speed of treatment is particularly necessary when responding to many types of cardiovascular disease. On the other hand, immune rejection of allogeneic iPS cells and iPS cell-derived cardiomyocytes undoubtedly inhibits their full regenerative potential. The ideal solution would be to create a method that could reliably reduce posttransplant rejection of allogeneic cells. If this can be achieved, an "off-the-shelf" iPS cell line or its derivatives could be kept on hand for fast response in cardiovascular disease therapy. Multiple laboratories worldwide are currently pursuing such technologies.

12.4.2 Somatic Cell Nuclear Transfer Immunogenicity

SCNT has been suggested as a method for quickly generating patient-specific stem cells (Tachibana et al. 2013). By transferring the nucleus from a patient cell into the cell body of an enucleated oocyte, the resulting pluripotent stem cell will contain an identical nuclear genome to the donor recipient. This was suggested as a potentially useful therapy for patients with mitochondrial disease, as the mitochondria are derived from the healthy oocyte donor (Tachibana et al. 2013).

While SCNT transfer has been successfully performed and used for significant stem cell research contributions, Deuse et al. found that despite generating matching nuclear genomes between the SCNT cells and the recipient, the mismatched mitochondria can stimulate an immune response due to differences in the mitochondrial DNA (Fig. 12.3; Deuse et al. 2015). Observation of embryonic stem cells generated by nuclear transfer (NT-ESC) revealed that mismatched mitochondrial proteins are able to trigger the recipient immune system, even with as few as one or two mismatched proteins. The immune response appeared to be adaptive in nature, directed against mitochondrial content, and amenable for tolerance induction (Deuse et al. 2015).

iPS cell and SCNT technology continue to be extremely important for developing new regenerative therapies; however, it is clear that posttransplant rejection is a serious issue. For future development of stem cell therapies, it will be particularly important to keep in mind the possible immunogenic effects of proteins associated with pluripotency and mismatched mitochondria.

12.5 Current and Developing Methods to Reduce Stem Cell Immunogenicity

The future of stem cell therapy relies on developing methods that can regenerate tissue without activating an immune response. Finding a method that can overcome this hurdle is one of the "holy grails" of modern cardiovascular disease research. Some methods, such as general immunosuppression, while not ideal, are presently in use to prevent rejection of transplanted materials. However, many innovate techniques are currently under development.

12.5.1 General Immunosuppression

Since there are many nuanced reasons why immune rejection of transplanted stem cells can occur, the research community will have to be innovative with regard to developing therapeutic methods. Currently, one of the main methods used to prevent posttransplant rejection is through long-term use of general immunosuppressants, including corticosteroids and calcineurin inhibitors such as cyclosporine or tacrolimus. However, use of long-term immunosuppressants is not an ideal solution, as it can lead to severe side effects including cardiovascular complications, infections, and increased risk of cancer, among others (Hsu et al. 2008; Khurana and Brennan 2011). Finding methods that prevent the rejection of transplanted material without compromising the general immune system would be better alternatives.

12.5.2 Cardiospheres

The recent generation of cardiospheres and cardiosphere-derived cells has shown promise as a method for therapeutic regeneration. Cardiospheres are cells derived from the heart that have stem capabilities, in that they can be differentiated into different cell lineages and can regenerate (Messina 2004). Importantly, autologous cardiosphere-derived cell transplantation was clinically tested through a trial named CADUCEUS (Cardiosphere-Derived Autologous Stem Cells to Reverse Ventricular

Dysfunction) (Makkar et al. 2012). The CADUCEUS trial monitored tissue regeneration and overall health in patients that had a recent myocardial infarction. Although they did not find any improvements in cardiac function, they did observe a reduction in myocardial scarring. The current ALLSTAR (Allogeneic Heart Stem Cells to Achieve Myocardial Regeneration) trial is assessing for the first time the safety and efficacy of allogeneic cardiospheres as a treatment option for patients within 12 months of a myocardial infarction (Makkar et al. 2014). Use of cardiospheres for tissue regeneration shows promise; however it will be important to identify the mechanism by which cardiosphere cells act as well as methods for increasing treatment efficacy.

12.5.3 Generation of a Molecularly Modified Stem Cell Line

One method to avoid immune rejection would be to create a molecularly modified non-immunogenic stem cell line. Such an "off-the-shelf" cell line would be very useful for clinical stem cell therapy in regenerative medicine. This cell line might be created by altering the expression levels of different molecules (e.g., MHCs) in order to modify the cell's communication and interaction with T lymphocytes, NK cells, and macrophages. However, one complication with such a method is that mismatched MHCs are targeted by T lymphocytes. Generating multiple "off-the-shelf" stem cell lines that express various MHC molecules could solve this problem. The number of MHC cell variants needed for a comprehensive cell bank for human treatment has previously been calculated, and while it varies between populations, it remains within a reasonably maintainable range (Taylor et al. 2012).

A second major complication with generating a non-immunogenic stem cell line is that certain stem cell molecular markers (e.g., OCT4) appear to strike a delicate balance between conferring stem abilities and causing cancer (Chiou et al. 2010). Because of this, most humans develop T cells against these markers in abundance, which may be necessary in adults to prevent cancer development. If generation of a non-immunogenic stem cell line relies on reducing the recognition of stem cell molecular markers, this may result in an increase in cancer development. Generation of a molecularly modified stem cell line that does not induce immune rejection or generate cancer will not be a simple undertaking, but if such a cell line could be achieved, it would be a game changer for stem cell therapy.

12.5.4 Creating a Local Hypo-immunogenic Environment

As indicated in Sect. 12.5.1, general immunosuppression can cause multiple negative long-term consequences for patients. A method that could generate a local hypo-immunogenic environment and suppresses the immune system only at the site of transplantation would be a prodigious alternative. A natural model for such a specific immune response reduction is found in fetomaternal tolerance (Guleria and Sayegh 2007). Because the fetus' genetic material is 50% paternally inherited and therefore partially allogeneic to the mother, there are multiple mechanisms in the mother's body that prevent fetal rejection. Importantly, the mother's general immune system does not seem to be significantly affected under such circumstances. It may be possible to harness the natural methods of fetomaternal tolerance to create a local hypo-immunogenic environment for transplanted materials without affecting the general immune system of the patient.

12.5.5 Final Takeaway

Potential stem cell therapies continue to quickly develop, and many show great promise in the field of regenerative medicine. The ability to continuously and consistently generate new cells to replace malfunctioning, dead, or missing tissue is an advantage to using stem cells; however, as with any transplanted material, the propensity for posttransplant rejection has constrained the possible positive results of stem cell therapy. By avoiding known transplant rejection catalysts (e.g., by matching MHCs before transplantation or by keeping in mind the possible immunogenicity of mismatched mitochondrial proteins), current transplant techniques continue to increase in efficacy. Future methods to reduce immune responses to transplantation are under development and give hope for increased success of regenerative stem cell therapy.

Compliance with Ethical Standards

Conflict of Interest The authors declare that they have no conflict of interest.

Ethical Approval This article does not contain any studies with human participants performed by any of the authors.

References

de Almeida PE et al (2014) Transplanted terminally differentiated induced pluripotent stem cells are accepted by immune mechanisms similar to self-tolerance. Nat Commun 5:3903

Ardehali R et al (2013) Prospective isolation of human embryonic stem cell-derived cardiovascular progenitors that integrate into human fetal heart tissue. Proc Natl Acad Sci U S A 110(9):3405–3410

Blum B, Benvenisty N (2008) The Tumorigenicity of human embryonic stem cells. Adv Cancer Res 100(08):133–158

Carpenter L et al (2012) Efficient differentiation of human induced pluripotent stem cells generates cardiac cells that provide protection following myocardial infarction in the rat. Stem Cells Dev 21(6):977–986

Chiou S-H et al (2010) Coexpression of Oct4 and Nanog enhances malignancy in lung adenocarcinoma by inducing cancer stem cell-like properties and epithelial-mesenchymal transdifferentiation. Cancer Res 70(24):10433–10444

Deuse T et al (2011) Human leukocyte antigen I knockdown human embryonic stem cells induce host ignorance and achieve prolonged xenogeneic survival. Circulation 124(11 Suppl):S3–S9

Deuse T et al (2015) SCNT-derived ESCs with mismatched mitochondria trigger an immune response in allogeneic hosts. Cell Stem Cell 16(1):33–38

Drukker M et al (2002) Characterization of the expression of MHC proteins in human embryonic stem cells. Proc Natl Acad Sci U S A 99(15):9864–9869

Dzierzak-Mietla M et al (2012) Occurrence and impact of minor histocompatibility antigens' disparities on outcomes of hematopoietic stem cell transplantation from HLA-matched sibling donors. Bone Marrow Res 2012:257086
Goulmy E et al (1976) Alloimmunity to human H-Y. Lancet 2(7996):1206
Guleria I, Sayegh MH (2007) Maternal acceptance of the fetus: true human tolerance. J Immunol 178(6):3345–3351
He J-Q et al (2003) Human embryonic stem cells develop into multiple types of cardiac myocytes: action potential characterization. Circ Res 93(1):32–39
Horton R et al (2004) Gene map of the extended human MHC. Nat Rev Genet 5(12):889–899
Hsu DC et al (2008) Long-term management of patients taking immunosuppressive drugs. Aust Prescr 32(3):68–71
Kambayashi T et al (2001) Purified MHC class I molecules inhibit activated NK cells in a cell-free system in vitro. Eur J Immunol 31(3):869–875
Karlhofer FM, Ribaudo RK, Yokoyama WM (1992) MHC class I alloantigen specificity of Ly-49+ IL-2-activated natural killer cells. Nature 358(6381):66–70
Khurana A, Brennan DC (2011) Pathology of solid organ transplantation. Springer-Verlag, Berlin, pp 11–31
Laflamme MA et al (2007) Cardiomyocytes derived from human embryonic stem cells in pro-survival factors enhance function of infarcted rat hearts. Nat Biotechnol 25(9):1015–1024
Li L et al (2004) Human embryonic stem cells possess immune-privileged properties. Stem Cells 22(4):448–456
Ma M et al (2011) Major histocompatibility complex-I expression on embryonic stem cell-derived vascular progenitor cells is critical for syngeneic transplant survival. Stem Cells 28(9):1465–1475
Macedo C et al (2009) Contribution of naïve and memory t-cell populations to the human alloimmune response. Am J Transpl 9(9):2057–2066
Makkar R et al (2014) Abstract 20536: allogeneic heart stem cells to achieve myocardial regeneration (ALLSTAR): the one year phase I results. Circulation 130(Suppl 2):A20536–A20536
Makkar RR et al (2012) Intracoronary cardiosphere-derived cells for heart regeneration after myocardial infarction (CADUCEUS): a prospective, randomised phase 1 trial. Lancet 379(9819):895–904
Messina E (2004) Isolation and expansion of adult cardiac stem cells from human and murine heart. Circ Res 95(9):911–921
Mummery C et al (2002) Cardiomyocyte differentiation of mouse and human embryonic stem cells*. J Anat 200(3):233–242
Neefjes J et al (2011) Towards a systems understanding of MHC class I and MHC class II antigen presentation. Nat Rev Immunol 11(12):823–836
Nelson TJ et al (2009) Repair of acute myocardial infarction with iPS induced by human stemness factors. Circulation 120(5):408
Nussbaum J et al (2007) Transplantation of undifferentiated murine embryonic stem cells in the heart: teratoma formation and immune response. FASEB J 21(7):1345–1357
Schwartz SD et al (2012) Embryonic stem cell trials for macular degeneration: a preliminary report. Lancet 379(9817):713–720
Schwartz SD et al (2015) Human embryonic stem cell-derived retinal pigment epithelium in patients with age-related macular degeneration and Stargardt's macular dystrophy: follow-up of two open-label phase 1/2 studies. Lancet 385(9967):509–516
Spierings E (2014) Minor histocompatibility antigens: past, present, and future. Tissue Antigens 84(4):374–360
Starr TK, Jameson SC, Hogquist KA (2003) Positive and negativeselection of T cells. Ann Rev Immunol 21(1):139–176
Streilein JW (2003) Ocular immune privilege: therapeutic opportunities from an experiment of nature. Nat Rev Immunol 3(11):879–889
Swijnenburg R-J, Schrepfer S, Govaert JA et al (2008a) Immunosuppressive therapy mitigates immunological rejection of human embryonic stem cell xenografts. Proc Natl Acad Sci U S A 105(35):12991–12996

Swijnenburg R-J, Schrepfer S, Cao F et al (2008b) In vivo imaging of embryonic stem cells reveals patterns of survival and immune rejection following transplantation. Stem Cells Dev 17:1023–1029
Tachibana M et al (2013) Human embryonic stem cells derived by somatic cell nuclear transfer. Cell 153(6):1228–1238
Takahashi K, Yamanaka S (2006) Induction of pluripotent stem cells from mouse embryonic and adult fibroblast cultures by defined factors. Cell 126(4):663–676
Taylor CJ et al (2012) Generating an iPSC bank for HLA-matched tissue transplantation based on known donor and recipient hla types. Cell Stem Cell 11(2):147–152
Templin C et al (2012) Transplantation and tracking of human-induced pluripotent stem cells in a pig model of myocardial infarction: assessment of cell survival, engraftment, and distribution by hybrid single photon emission computed tomography/computed tomography of sodium iod. Circulation 126(4):430–439
Thomson JA et al (1998) Embryonic stem cell lines derived from human blastocysts. Science 282(5391):1145–1147
Vogt MH et al (2000) DFFRY codes for a new human male-specific minor transplantation antigen involved in bone marrow graft rejection. Blood 95(3):1100–1105
Xu C et al (2002) Characterization and enrichment of cardiomyocytes derived from human embryonic stem cells. Circ Res 91(6):501–508
Yang Y et al (2002) VEGF enhances functional improvement of postinfarcted hearts by transplantation of ESC-differentiated cells. J Appl Physiol 93(3):1140–1151
Zerrahn J, Held W, Raulet DH (1997) The MHC reactivity of the T cell repertoire prior to positive and negative selection. Cell 88(5):627–636
Zhao T et al (2015) Humanized mice reveal differential immunogenicity of cells derived from autologous induced pluripotent stem cells. Cell Stem Cell 17(3):1–7
Zhao T et al (2011) Immunogenicity of induced pluripotent stem cells. Nature 474(7350):212–215
Zwi-Dantsis L et al (2013) Derivation and cardiomyocyte differentiation of induced pluripotent stem cells from heart failure patients. Eur Heart J 34(21):1575–1586

Correction to: Cardiac Regeneration

Masaki Ieda and Wolfram-Hubertus Zimmermann

Correction to:
M. Ieda, W.-H. Zimmermann (eds.), *Cardiac Regeneration*, Cardiac and Vascular Biology, DOI 10.1007/978-3-319-56106-6

The original version of this volume was revised as it was originally published unnumbered. The revised version now has been numbered; numbering is done following the order of appearance in the book series Cardiac and Vascular Biology.

The updated online version of the book can be found at DOI 10.1007/978-3-319-56106-6

M. Ieda, W.-H. Zimmermann (eds.), *Cardiac Regeneration*, Cardiac and Vascular Biology 4, DOI 10.1007/978-3-319-56106-6_13

The manufacturer's authorised representative in the EU is Springer Nature Customer Service Centre GmbH, Europaplatz 3, 69115 Heidelberg, Germany. If you have any concerns regarding our products, please contact ProductSafety@springernature.com

Printed and bound by CPI Group (UK) Ltd, Croydon, CR0 4YY

30/09/2026

02225454-0005